Vectors in Physics and Engineering

Vectors in Physics and Engineering

A.V. Durrant
Senior Lecturer in Physics
The Open University
UK

CRC Press is an imprint of the
Taylor & Francis Group, an **informa** business

CRC Press
Taylor & Francis Group
6000 Broken Sound Parkway NW, Suite 300
Boca Raton, FL 33487-2742

Visit the Taylor & Francis Web site at
http://www.taylorandfrancis.com

and the CRC Press Web site at
http://www.crcpress.com

Contents

Preface

This book is intended as a self-study text for students following courses in science and engineering where vectors are used. The material covered and the level of treatment should be sufficient to provide the vector algebra and vector calculus skills required for most honours courses in mechanics, electromagnetism, fluid mechanics, aerodynamics, applied mathematics and mathematical modelling. It is assumed that the student begins with minimal (school-level) skills in algebra, geometry and calculus and has no previous knowledge of vectors. There are brief reviews at appropriate points in the text on elementary mathematical topics: the definition of a function, the derivative of a function, the definite integral and partial differentiation.

The text is characterised by short two or three page sections where new concepts, terminologies and skills are introduced, followed by detailed summaries and consolidation in the form of Examples and Problems that test the objectives listed at the beginning of each chapter. Each Example is followed by a fully worked out solution, but the student is well advised to have a go at each Example before looking at the solution. Many of the Examples and Problems are set in the context of mechanics and electromagnetism but no significant previous knowledge of these subjects is assumed. Bare answers to selected Problems are given at the back of the book. Full solutions to all Problems can be found on www at http://physics.open.uk/~avdurran/vectors.html.

Although the material covered makes relatively little demand on previously acquired mathematical skills, the newcomer will find that there are many new concepts to grapple with, new notations and skills to master and a large number of technical terms to assimilate. Each new technical term is highlighted by heavy print at the point in the text where it is most fully described or defined.

Vector algebra is developed in the first two chapters as a way of describing elementary two and three dimensional spatial relationships and geometrical figures in terms of displacement vectors, and is then applied to problems involving velocities, forces and other physical vectors. Chapter 3 introduces vector functions of time and the derivatives of vector functions, with applications to circular motion, projectile motion and inertial forces in accelerating and rotating coordinate systems. Chapter 4 introduces scalar and vector fields initially in terms of contour surfaces and vector field lines and then as scalar and vector functions of position. Spatial symmetries of fields are briefly discussed, and cylindrical polar and spherical polar coordinate systems are introduced and then used where appropriate throughout the book. Chapter 5 introduces the differential calculus of scalar and vector fields. The concepts of gradient, divergence and curl are dealt with informally, and their role in the expression of physical laws is described. A section on the use of the "del" operator is included. The integral calculus of fields culminating in Stokes's and Gauss's theorems is dealt with in Chapter 6.

The approach throughout is physical and intuitive and there are no formal proofs. The emphasis is on developing calculation skills, understanding the concepts and seeing the relevance to physical processes. However, an attempt has been made to keep the mathematical structure of the subject visible so that students can make the connection with any mathematics courses they may take.

The author wishes to thank Dr S. A. Hopkins for the preparation of figures and *Beware CRC* for the production of camera-ready copy. A special word of thanks goes to Mr Cyril Drimer for reading the manuscript, making many useful suggestions and helping to eliminate errors. Any remaining errors are mine.

Finally, and most of all, I thank my wife Zosia for her support and encouragement over a long period of time.

1

Vector algebra I
Scaling and adding vectors

After you have studied this chapter you should be able to

Objectives

- Explain what a displacement vector is; use vector notation and the arrow representation of a vector (*Objective 1*).
- Scale a vector by a number and specify the magnitude and direction of the scaled vector; specify the unit vector in the direction of a given vector (*Objective 2*).
- Add and subtract two or more vectors by the vector addition rule (triangle rule) (*Objective 3*).
- Determine the magnitude and direction of the resultant of two or more vectors using Pythagoras's theorem and trigonometry (*Objective 4*).
- Describe geometrical figures and spatial relationships in terms of linear combinations of vectors (*Objective 5*).
- Carry out algebraic manipulations of linear combinations of vectors (*Objective 6*).
- Recognise a right-handed cartesian coordinate system, express a vector as a linear combination of the cartesian unit vectors and specify the cartesian components of the vector (*Objective 7*).
- Recognise and use the ordered triple (or ordered pair) representation of a vector (*Objective 8*).
- Recognise that a vector equation is equivalent to three scalar equations for the components (*Objective 9*)
- Use Pythagoras's theorem and trigonometry to determine the magnitude and the direction cosines of a cartesian vector (*Objective 10*).
- Scale and add cartesian vectors (*Objective 11*).
- Express statements and given physical laws in terms of vectors (*Objective 12*).

1.1 INTRODUCTION TO SCALARS, NUMBERS AND VECTORS

Many of the quantities of interest in physics and engineering can be classified as *scalars* or *vectors*. A scalar is specified by a single numerical value. Examples of scalar quantities are length, mass, temperature, electric charge, area, density, energy, pressure and there are many others. A vector has a magnitude and a direction in space. Some familiar vector quantities are force, displacement, velocity and acceleration. Because vector quantities have magnitude and direction in space their properties are much more complex than those of scalars. In fact we can describe most of the important properties of scalars in the following short section before getting started on vectors.

1.1.1 Scalars and numbers

The kilogram (kg) is the unit of mass in the International System of Units (SI).

The coulomb (C) is the SI unit of electric charge.

SI units are described in Appendix A.

A **scalar** is a quantity that is specified completely by a single number and a physical unit of measure. For example, the mass of an electron is a scalar specified, to four significant figures, by: 9.110×10^{-31} kg. Another example is the electric charge carried by an electron which is specified by -1.602×10^{-19}C. For most of this chapter we shall omit reference to physical units and regard scalars simply as numbers. Thus the terms scalar and number become synonymous, and the study of scalars becomes the study of numbers, that is, the study of how numbers add and multiply etc., and how we manipulate symbols representing numbers in algebra. This is something you will be fairly familiar with, but it will be useful to outline some of the technical aspects of numbers and number algebra before discussing vectors.

By numbers we mean **real numbers**, as opposed to imaginary or complex numbers. The real numbers are the familiar positive and negative numbers including zero. For example 5, -7, $1/3$, 2.44565×10^{7}, 0 and π are all real numbers. The collection of all possible real numbers is called the **set of real numbers** and is denoted in mathematics by the symbol $\Re$. We often need to refer to the positive numbers including zero; we call this subset the **non-negative** numbers. Many scalars are essentially non-negative; length, area and volume are examples of scalars that are always positive, or zero. In such cases the numerical specification of the scalar quantity is called the **magnitude** of the scalar. When a scalar quantity is negative, such as the electric charge carried by an electron, we use the term magnitude to refer to the corresponding positive quantity; for example, if you look at a table of physical constants you may see the magnitude of the charge on the electron listed as 1.602×10^{-19}C.

In algebra we use symbols to represent scalars. Roman or Greek symbols such as c, m, x, y, α, β, etc. are commonly used to denote particular scalars, such as the speed of light or the mass of an electron, or to denote variables or "unknowns" that have to be found by solving a problem.

Other operations on scalars include those found on scientific calculators such as taking square roots, exponentials and logarithms, etc.

Any two real numbers can be combined to give another real number by the familiar arithmetic operations of addition and multiplication, and the inverse operations of subtraction and division, except that division by zero is not defined. When we solve equations or simplify algebraic expressions, we manipulate numbers and symbols by sequences of additions, multiplications and other operations, according to *rules of algebra*. Some of these rules and their technical names are given below.

For any numbers x, y and z:

$$\left.\begin{array}{rl} x+y=y+x & \text{commutative rule for addition} \\ x\times y=y\times x & \text{commutative rule for multiplication} \\ (x+y)+z=x+(y+z) & \text{associative rule for addition} \\ (x\times y)\times z=x\times(y\times z) & \text{associative rule for multiplication} \\ z\times(x+y)=z\times x+z\times y & \text{distributive rule} \end{array}\right\} \quad (1.1)$$

There are also some definitions and conventions; for example

$$\begin{array}{rl} -x=-1\times x & \text{meaning of the symbol } -x \\ x-y=x+(-y) & \text{definition of subtraction} \\ |x| = x \text{ for } x\geq 0 & \text{modulus of a non-negative number} \\ |x| = -x \text{ for } x<0 & \text{modulus of a negative number} \end{array}$$

$x \geq 0$ means that x is positive or equal to zero, i.e. x is non-negative. The symbol $x < 0$ means that x is negative.

The modulus of a number x, indicated by enclosing the number between two vertical lines, as in $|x|$, is the magnitude of x and is always non-negative.

You have probably used the rules of algebra many times, although you may not have come across the technical names before. The rules of algebra were actually invented to describe the way scalar quantities combine with one another. For example, suppose you shovel coal into a sack, first 50 kg of coal followed by 20 kg. Then you will have a total of 70 kg of coal. If instead you had shovelled 20 kg in first and then 50 kg, you would again have 70 kg. Adding loads of coal into a sack illustrates the commutative rule for addition ($x + y = y + x$). You are not surprised by this because you are so familiar with scalars. It is however a remarkable fact that all scalar quantities obey the same rules of algebra. Lengths, areas, masses, electric charges, etc., all combine according to rules (1.1). We now turn to vectors.

1.1.2 Introducing vectors

Vectors have magnitude and direction in space. The magnitude of a vector is a non-negative number with a unit of measure. The magnitude of a vector is therefore a scalar. The direction of a vector is defined relative to some reference directions such as vertically upwards or the directions indicated by a compass. To specify a vector, both a magnitude and a direction must be given. For example, the acceleration of a freely falling body is a vector specified by giving the magnitude, $g = 9.81\ \mathrm{m\,s^{-2}}$, and the direction, vertically downwards. Another example of a vector is velocity. The magnitude of a velocity is called the speed and the direction is the direction of travel. For example, a car may be travelling with a velocity specified as 50 km per hour due north. When for convenience we drop reference to physical units, the magnitude of a vector is just a non-negative number. A vector is then specified by giving a non-negative number (a magnitude) and a direction in space.

Most of this chapter and the next is concerned with **vector algebra**, the study of how vectors are described mathematically and how vectors can be added, subtracted and multiplied. Unlike the rules of number algebra outlined in Section 1.1.1, the rules of vector algebra are not all obvious because we have little everyday experience of combining vectors. The rules of vector algebra have to be discovered by observing how vector quantities combine. Surprisingly, it is found that force, displacement, velocity, acceleration and many other physical quantities that have magnitude and direction obey the same rules of vector algebra, just as all scalar quantities obey the rules of number algebra, and so you can learn the rules of vector algebra by studying the behaviour of any one type of vector. The vector quantity usually chosen for this purpose is **displacement** which is the position of a point in space relative to some reference point. We choose displacement as the definitive vector quantity rather than some other vector such as force or velocity because the properties of displacements correspond directly to the properties of points and lines in space, and measurements of lengths and angles. Thus the rules of vector algebra are revealed in the context of geometry and trigonometry. This is our strategy in these first two chapters, beginning in Section 1.1.3 below.

1.1.3 Displacements and arrows

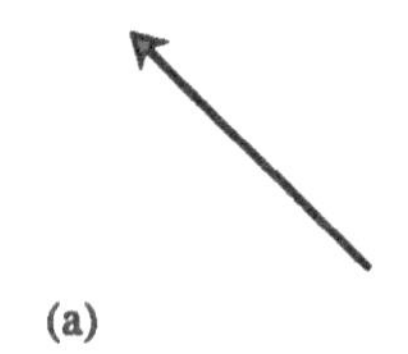

Fig 1.1a
An arrow representing a displacement.

Displacement is relative position in space. The position of the city of Birmingham, for example, is 185 km along a straight line from London in a direction 45° west of north (i.e. north-west). This specification of both *distance* and *direction* defines what is meant by the displacement of Birmingham from London. The magnitude of the displacement is just the distance, 185 km. Of course, the accuracy of this specification is limited by the finite spread of both cities over large areas and we are ignoring the curvature of the Earth's surface. Ideally we refer to the displacement of one point in space relative to some reference point. The concept of displacement is frequently used to describe the change of position of a body as a result of a movement or journey. For example, if a bus travels from London to Birmingham, by whatever route, it is said to undergo a displacement of 185 km 45° west of north.

It is important to stress that distance and direction alone completely define displacement. In particular, displacement does not depend on any particular path between the two points; nor does it depend on any particular pair of beginning and end points. Thus the displacement considered above, 185 km 45° west of north, can be made from Plymouth with an end point in the Atlantic Ocean, or from Hastings with an end point near Oxford; it is the same displacement in each case even though the beginning and end points are different.

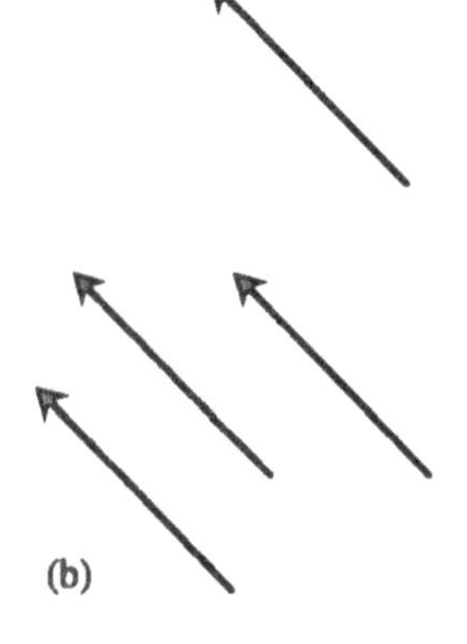

Fig 1.1b
All four arrows have the same length and direction and therefore represent the same displacement.

We can represent a displacement geometrically as a **directed line segment** which can be drawn on the page as an **arrow**. The length of the arrow represents distance and the arrowhead indicates the direction (Fig 1.1a). All arrows having the same length and direction represent the same displacement regardless of where they are drawn on the page. Different arrows representing the same displacement are shown in Fig 1.1b. Although an arrow is necessarily drawn at a particular location on the page we shall often refer to an arrow as a displacement. We shall also frequently use the term *vector* when referring to a displacement or to an arrow in a diagram. This is simply a reflection of the fact that we have chosen displacement as the definitive vector quantity; the properties of displacements define what is meant by a vector. Thus the terms *arrow*, *displacement* and *vector* are often used interchangeably.

The collection of all possible distinct displacements can be represented geometrically by arrows drawn from an arbitrary fixed reference point O to all other points in space (Fig 1.2). There is clearly an infinite number of distinct displacements corresponding to the infinite number of points in space. This infinite collection of displacements is referred to technically as the **set of vectors**. The set of vectors includes the **zero vector**, a vector of zero length and indeterminate direction; it can be thought of as the displacement of the point O from itself and is represented geometrically by the point O.

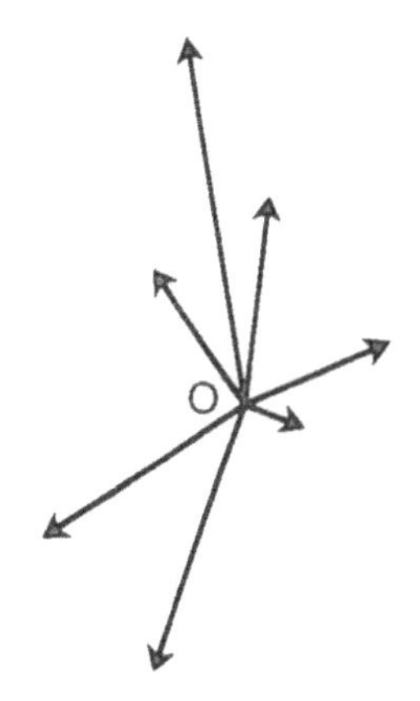

Fig 1.2
The set of vectors is represented by arrows drawn from a fixed point in space to all other points in space; of course, only a few arrows are shown.

1.1.4 Vector notation

We need algebraic symbols for denoting vectors that are distinct from those used to denote scalars or numbers. A common convention for denoting a displacement or any other vector quantity is to use a boldface symbol such as **a**, **r**, **s**, etc. The zero vector is denoted by a bold zero, **0**. The displacement of a point B from a point A is often denoted by the bold symbol **AB** or by $\overrightarrow{AB}$. Note the ordering of the letters here: first the beginning point A; then the end point B. The reverse ordering, **BA** or $\overrightarrow{BA}$, represents the displacement of point A from point B, i.e. a displacement of the same magnitude as **AB** but in the opposite direction (Fig 1.3). A non-bold symbol such as AB or BA, without an arrow symbol, does not denote a displacement; it denotes the line joining the two points A and B or sometimes the length of the line, depending on the context.

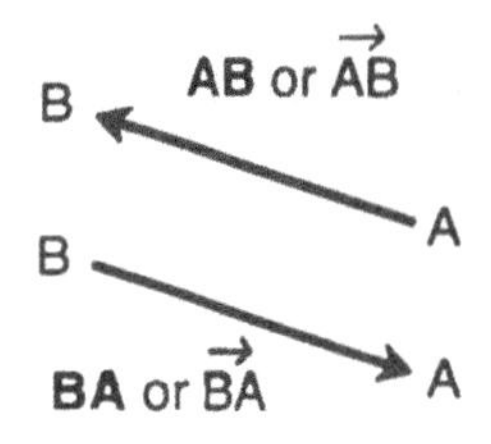

Fig 1.3
The displacement of B from A can be denoted by the bold symbol **AB** or by an arrowed symbol such as $\overrightarrow{AB}$. **BA** or $\overrightarrow{BA}$ denotes the displacement of A from B.

In hand-written manuscript where boldface is not an option, common conventions for indicating vectors are to use underlined or arrowed symbols such as $\underset{\sim}{a}$, $\underline{a}$, $\underline{AB}$, or $\overrightarrow{AB}$.

In this book we shall use only boldface symbols such as **a** and **AB**.

The **magnitude of a vector**, sometimes called the *length* of the vector, is a non-negative number denoted by enclosing the vector symbol within modulus lines:

$$|\mathbf{a}| = \text{magnitude of vector } \mathbf{a}$$

Thus if **a** denotes the displacement of Birmingham from London, 185 km north-west, then the magnitude is $|\mathbf{a}| = 185$ km. In speech we say $|\mathbf{a}|$ as "the modulus of vector **a**" or more commonly "mod **a**". An alternative convention for denoting the magnitude of a vector is to use the vector symbol printed in ordinary (italic) type instead of bold type. Thus

You met modulus lines in Section 1.1.1 in the context of numbers where the modulus of a number is its magnitude irrespective of sign, e.g. $|-5| = 5$ and $|5| = 5$.

$$a = |\mathbf{a}| = \text{magnitude of } \mathbf{a} \qquad (1.2)$$

For a displacement represented by a symbol such as **AB**, the magnitude may be denoted by $|\mathbf{AB}|$ or more simply by AB.

In hand-written work you could write $|\underset{\sim}{a}|$ or a to denote the magnitude of **a**.

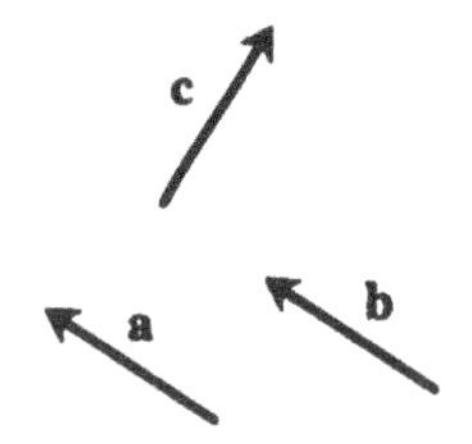

Fig 1.4
Three arrows of the same length: $|\mathbf{a}| = |\mathbf{b}| = |\mathbf{c}|$. **a** and **b** also have the same direction and so $\mathbf{a} = \mathbf{b}$. **c** is in a different direction and so **a** is not equal to **c**, i.e. $\mathbf{a} \neq \mathbf{c}$.

Two vectors are said to be equal if they have the same magnitude and direction. Thus the **equality of two vectors**, denoted by writing $\mathbf{a} = \mathbf{b}$, means that $|\mathbf{a}| = |\mathbf{b}|$ *and* the directions of **a** and **b** are the same (Fig 1.4).

Summary of section 1.1

- A **scalar** is specified by a single number. A **vector** is specified by a magnitude (a non-negative number) and a direction in space.

- **Displacement** is relative position in space. The displacement of a point B from a point A is completely specified by the distance between the two points and the direction of B from A; the distance is the magnitude of the displacement.

- A displacement can be represented geometrically by a **directed line segment** drawn on the page as an **arrow**. All arrows having the same length and direction represent the same displacement.

- Vectors are denoted by symbols such as **a**, **AB** or $\overrightarrow{AB}$. The **magnitude** or length of a vector is a scalar denoted by symbols such as $|\mathbf{a}|$, a, $|\mathbf{AB}|$ or AB.

- **The equality of two vectors a** and **b** is written as $\mathbf{a} = \mathbf{b}$ and means that $|\mathbf{a}| = |\mathbf{b}|$ and that **a** and **b** have the same direction.

Example 1.1 (*Objective 1*) Refer to Fig 1.5. Select the *false* or *meaningless* statements from the options A to G and give reasons for your choice.

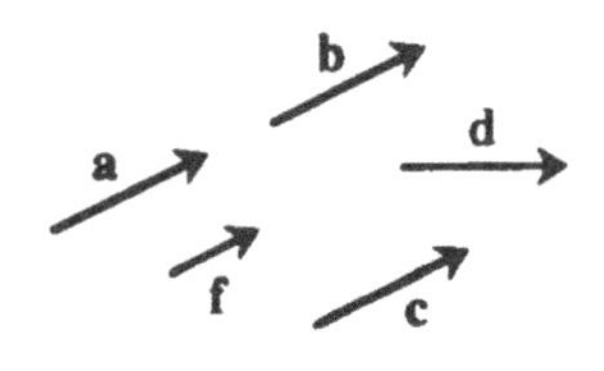

Fig 1.5
The five arrows represent vectors; four of them have the same length and four of them have the same direction.

A	$\mathbf{b} = \mathbf{c}$	E	$|\mathbf{b}| = |\mathbf{d}|$
B	$|\mathbf{b}| = |\mathbf{c}|$	F	$|\mathbf{f}| = |\mathbf{a}|$
C	$\mathbf{b} = \mathbf{d}$	G	$b = c$
D	$|\mathbf{a}| = \mathbf{c}$		

Solution 1.1 C is false because **b** and **d** have different directions. D is meaningless because the left-hand side of the equals sign is a magnitude, i.e. a number, while the right-hand side is a vector. A number cannot be equated to a vector. F is false because the two vectors have different magnitudes. (Statements A, B, E and G are true.)

An isosceles triangle is a triangle with two sides of equal length.

Example 1.2 (*Objective 1*) Fig 1.6 shows an *isosceles* triangle ABC. Point D is the midpoint of the side AC. The displacements **AB**, **BC**, **AD**, **DC** and **AC** are shown as arrows.

(a) Which two displacements are equal?
(b) Which two displacements have the same magnitude but different directions?
(c) Which two displacements have the same directions but different magnitudes?

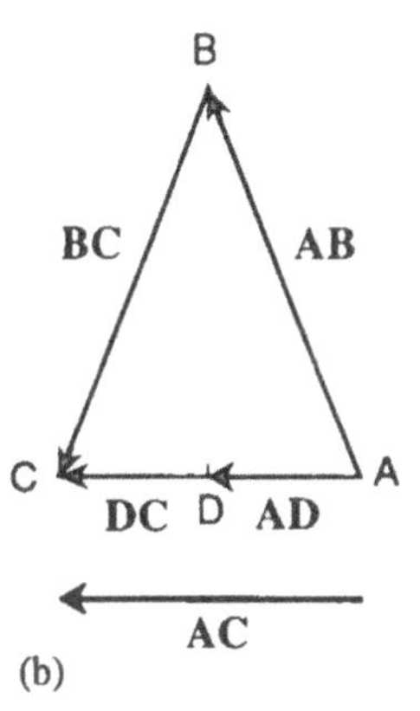

Fig 1.6
The isosceles triangle and displacements in Example 1.2.

Solution 1.2

(a) **AD** = **DC** because they have the same length and the same direction.
(b) **AB** and **BC** have the same length but different directions.
(c) **AD** and **AC** (or **DC** and **AC**) have the same directions but different magnitudes.

Problem 1.1 (*Objective 1*) A point T is located a distance 3 in a northerly direction from point S. Select the *true* statements from options (A)-(H).

A	ST = 3	E	ST denotes a vector
B	$\lvert$**ST**$\rvert$ = 3	F	**ST** is the displacement of T from S
C	**ST** = **TS**	G	$\lvert$**TS**$\rvert$ = –3
D	If **a** = **ST** then **a** = 3	H	If S, T and M lie on a straight line and ST = TM then **ST** = **TM**

Problem 1.2 (*Objective 1*) Fig 1.7 shows a *rhombus* and its two diagonals which bisect one another at O.

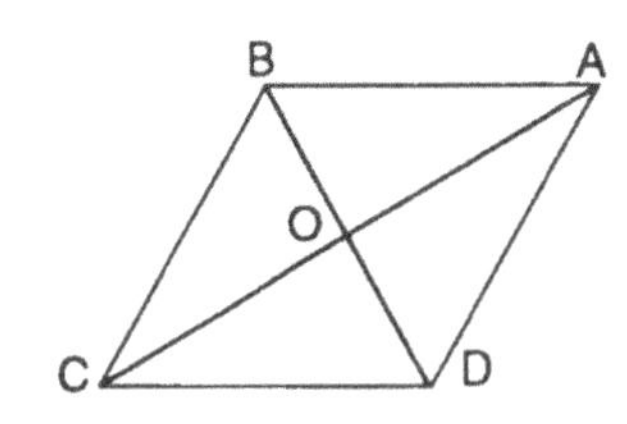

Fig 1.7
A rhombus and its two diagonals. A rhombus is a parallelogram with all four sides of equal length.

(a) Specify four pairs of equal displacements.
(b) State which two of the following displacements have the same magnitude but different directions **CA**, **AO**, **DO**, **BO**, **BD**, **OC**.

1.2 SCALING VECTORS AND UNIT VECTORS

Now that we have defined the set of vectors, the meaning of equality of vectors and the use of symbols to denote vectors, we can introduce the simplest algebraic operation involving vectors, known as *scaling* or *multiplication of a vector by a number*.

1.2.1 Scaling a vector or multiplication of a vector by a number

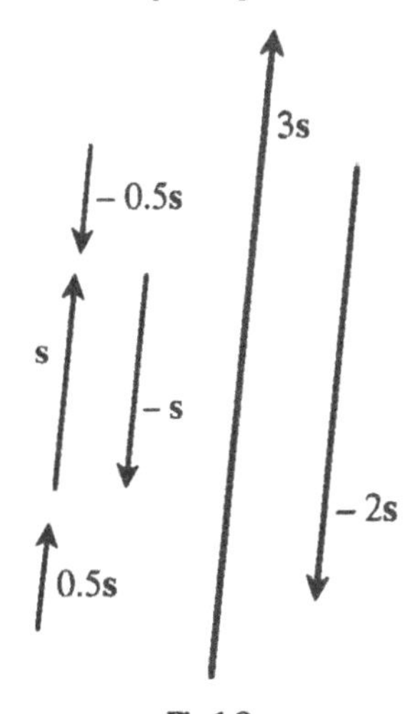

Fig 1.8
Vector **s** and vectors α**s** for $\alpha = 0.5, -0.5, -1, 3$ and -2.

Fig 1.8 shows a vector **s** and some other vectors lying on lines parallel to **s**, or on the straight line defined by **s**. Each of these vectors is related to **s** by the operation known as **scaling**. The scaled vector is denoted by writing α**s** where α is a number specifying the scaling. If α is positive (i.e. $\alpha > 0$) the effect of the scaling is to multiply the magnitude of the vector by α while leaving the direction unchanged. This is illustrated by the vector 3**s** (Fig 1.8) which has a

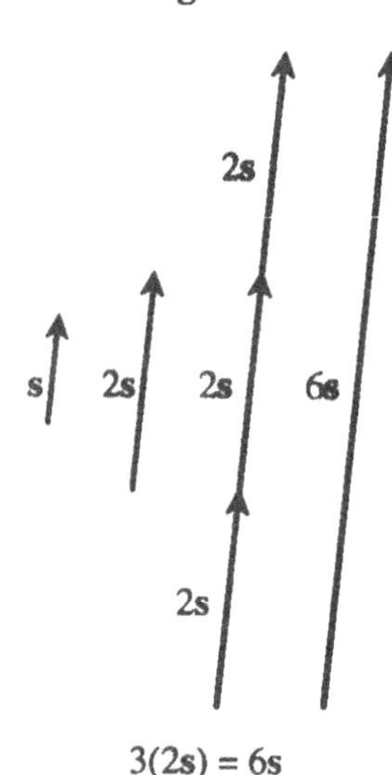

Fig 1.9a
The associative rule with $\alpha = 3$, $\beta = 2$: $3(2\mathbf{s}) = (3 \times 2)\mathbf{s}$.

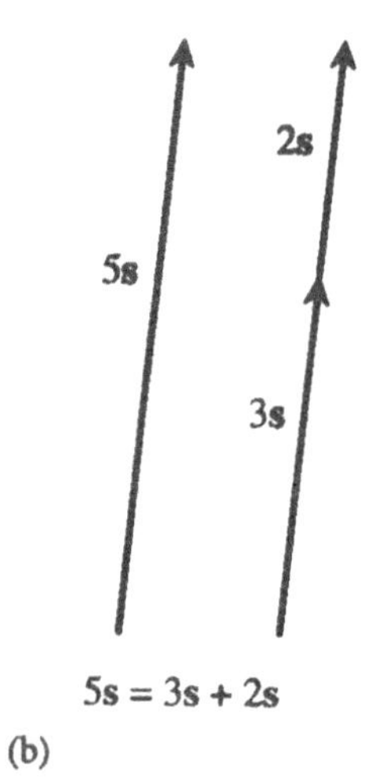

Fig 1.9b
The distributive rule: $(3+2)\mathbf{s} = 3\mathbf{s} + 2\mathbf{s}$.

You should be aware that some technical terms are defined differently in different books. For example, some books use the term collinear to mean parallel. The important thing is to know the meanings of the terms you use and to be consistent.

length three times that of $\mathbf{s}$ and is in the same direction as $\mathbf{s}$. If α is negative ($\alpha < 0$) the effect of the scaling is to multiply the magnitude of the vector by the magnitude of the number, i.e. by $|\alpha|$, while reversing the direction. This is illustrated by the vector $-2\mathbf{s}$ which has a length twice that of $\mathbf{s}$ and is in the opposite direction. If $\alpha = 0$ then we have $0\mathbf{s} = \mathbf{0}$, the zero vector. The effect of scaling on the magnitude of a vector is summarised by

$$|\alpha\mathbf{s}| = |\alpha|\,|\mathbf{s}| \tag{1.3}$$

You should note that the scaled vector is written simply as $\alpha\mathbf{a}$ or as $\mathbf{a}\alpha$, without a dot or a cross between the symbols. To write $\alpha.\mathbf{s}$ or $\alpha \times \mathbf{s}$ is wrong and meaningless. The dot and cross are reserved for other vector operations introduced in Chapter 2.

The following rules and definitions define the algebra of scaling and the notation used ($\mathbf{s}$ is a vector and α and β are numbers)

$\alpha(\beta\mathbf{s}) = (\alpha\beta)\mathbf{s}$	Associative rule
$(\alpha + \beta)\mathbf{s} = \alpha\mathbf{s} + \beta\mathbf{s}$	Distributive rule
$-1\mathbf{s} = -\mathbf{s}$	Definition of $-\mathbf{s}$
$\alpha\mathbf{s} = \mathbf{s}\alpha$	Definition of $\mathbf{s}\alpha$
$(1/\alpha)\mathbf{s} = \mathbf{s}/\alpha$	Definition of $\mathbf{s}/\alpha$

The three definitions simply define what is meant by the symbols $-\mathbf{s}$, $\mathbf{s}\alpha$, and $\mathbf{s}/\alpha$. For example, $\mathbf{s}3$ means the same as $3\mathbf{s}$, and $\mathbf{s}/3$ means $(1/3)\mathbf{s}$. The associative and distributive rules are rather obvious (Fig 1.9). They define the use of brackets in expressions involving scaled vectors; brackets are used in much the same way as in ordinary number algebra.

Some more terminology will be useful at this stage. Any two vectors that are related by a scaling, such as $\mathbf{a}$ and $\alpha\mathbf{a}$, are said to be **collinear**. Thus vectors that lie on the same straight line are collinear vectors. Vectors that lie on parallel straight lines are also collinear vectors; they can be moved onto the same straight line without changing their magnitudes and directions. If α is positive, the two collinear vectors $\mathbf{a}$ and $\alpha\mathbf{a}$ point in the same direction and are said to be **parallel**; if α is negative they point in opposite directions and are **antiparallel**. All vectors shown in Fig 1.8 are collinear; vectors $0.5\mathbf{s}$ and $3\mathbf{s}$ are both parallel to vector $\mathbf{s}$, while $-0.5\mathbf{s}$, $-\mathbf{s}$ and $-2\mathbf{s}$ are parallel to one another but are all antiparallel to $\mathbf{s}$.

Finally we note that any two non-zero vectors that are not collinear, such as vectors $\mathbf{a}$ and $\mathbf{b}$ in Fig 1.10, cannot be related by a scaling. When the two non-collinear vectors are drawn from the same beginning point O they define a plane surface. This plane surface is called the **plane of a and b**. Two or more vectors lying in the same plane are said to be **coplanar**. Vectors that lie in parallel planes are also said to be coplanar because they can be moved to the same plane without changing their magnitudes or directions.

1.2.2 Unit vectors

You have seen that the magnitude $|\mathbf{s}|$ of a vector $\mathbf{s}$ is a non-negative number. Any non-zero vector $\mathbf{s}$ can be scaled by the reciprocal of its magnitude, the number $\frac{1}{|\mathbf{s}|}$, to give a scaled vector,

$$\mathbf{u} = \frac{1}{|\mathbf{s}|}\mathbf{s} = \frac{\mathbf{s}}{|\mathbf{s}|} \qquad (\mathbf{s} \neq \mathbf{0}) \qquad (1.4)$$

The vector $\mathbf{u}$ has a magnitude of unity and the same direction as vector $\mathbf{s}$. $\mathbf{u}$ is called the **unit vector** in the direction of $\mathbf{s}$. A common notation for a unit vector is to use a circumflex (a hat) over the vector symbol, i.e. $\hat{\mathbf{s}} = \mathbf{s}/|\mathbf{s}|$, the unit vector in the direction of $\mathbf{s}$.

Any vector collinear with a vector $\mathbf{s}$ can be expressed in terms of the unit vector $\hat{\mathbf{s}}$ by a scaling. For example, the vector that has magnitude 7 and is in the same direction as $\mathbf{s}$ can be expressed as $7\hat{\mathbf{s}}$. Unit vectors are frequently used to specify direction in space. For example, the direction of the vector $\mathbf{s}$, or of any other vector parallel to $\mathbf{s}$, is specified by $\hat{\mathbf{s}}$ and may be referred to as "the direction $\hat{\mathbf{s}}$".

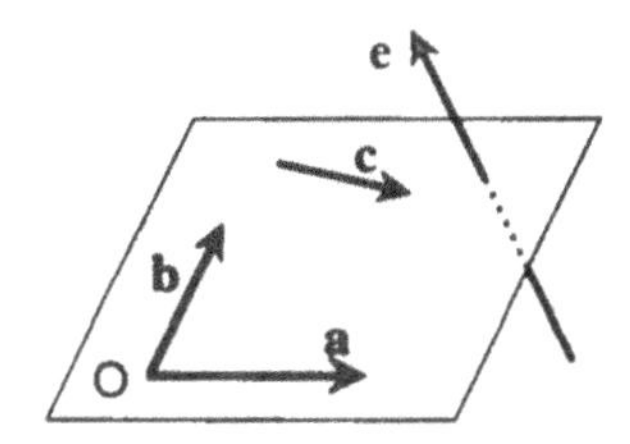

Fig 1.10
The plane of **a** and **b**. Vectors **a**, **b** and **c** are coplanar vectors, all in the plane of **a** and **b**. Vector **e** points out of the plane and is not in the plane of **a** and **b**.

Summary of section 1.2

- A vector $\mathbf{s}$ can be multiplied by a number α to give another vector, $\alpha\mathbf{s}$ or $\mathbf{s}\alpha$, an operation called scaling. Scaling changes the magnitude of the vector by a factor $|\alpha|$, and leaves the direction unchanged for $\alpha > 0$, or reverses the direction for $\alpha < 0$. For $\alpha = 0$, $0\mathbf{s} = \mathbf{0}$.

- Any two vectors related by a scaling, $\mathbf{s}$ and $\alpha\mathbf{s}$, are **collinear**; they are **parallel** for $\alpha > 0$, or **antiparallel** for $\alpha < 0$.

- Two non-collinear vectors $\mathbf{a}$ and $\mathbf{b}$ drawn from the same point define a plane called the **plane of a and b**. Vectors lying in the same plane are **coplanar** vectors.

- A vector of unit magnitude is called a **unit vector**. If $\mathbf{s}$ is any non-zero vector, then the vector $\hat{\mathbf{s}} = \frac{1}{|\mathbf{s}|}\mathbf{s}$, or $\frac{\mathbf{s}}{|\mathbf{s}|}$, is the unit vector in the direction of $\mathbf{s}$.

Example 2.1 (*Objectives 1,2*) Let $\mathbf{p} = 16\mathbf{q}$.

(a) Express $|\mathbf{p}|$ in terms of $|\mathbf{q}|$.

(b) Let $\hat{\mathbf{p}}$ be a unit vector in the direction of $\mathbf{p}$. Specify, in terms of $\hat{\mathbf{p}}$, the unit vector in the direction of $\mathbf{q}$.

(c) Specify the vector that has half the magnitude of $\mathbf{p}$ and is in the direction opposite that of $\mathbf{p}$. Express this vector in terms of $\mathbf{q}$.

Solution 2.1

(a) It follows from Eq (1.3) that $|\mathbf{p}| = 16|\mathbf{q}|$.

(b) Vectors $\mathbf{p}$ and $\mathbf{q}$ have the same direction because $\mathbf{p}$ is obtained from $\mathbf{q}$ by scaling with a positive number. The unit vector in the direction of $\mathbf{q}$ is therefore equal to $\hat{\mathbf{p}}$. (We can derive this result algebraically by writing $\hat{\mathbf{p}} = \mathbf{p}/|\mathbf{p}| = 16\mathbf{q}/16|\mathbf{q}|$ where we have used the given equation and the answer to part (a). Thus we have $\hat{\mathbf{p}} = \mathbf{q}/|\mathbf{q}| = \hat{\mathbf{q}}$.)

(c) The required vector is obtained by scaling $\mathbf{p}$ by $-1/2$. i.e. it is the vector $(-1/2)\mathbf{p}$ or $-\mathbf{p}/2$. In terms of $\mathbf{q}$ it is $-8\mathbf{q}$.

Example 2.2 (*Objectives 1,2*) Fig 1.11 shows a parallelogram ABCD. E is the midpoint of the line AD and F is three quarters of the way from A to B. Let $\mathbf{AE} = \mathbf{a}$ and $\mathbf{AF} = \mathbf{b}$.

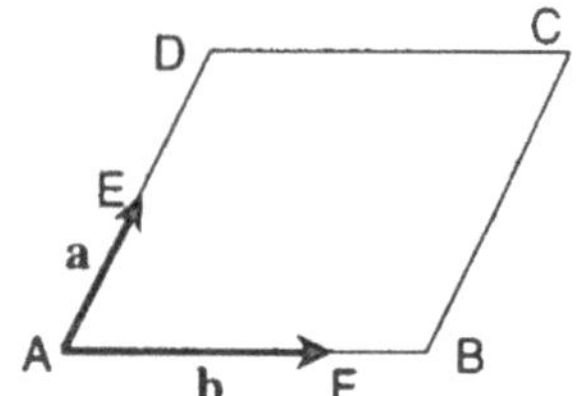

Fig 1.11
The parallelogram for Example 2.2.

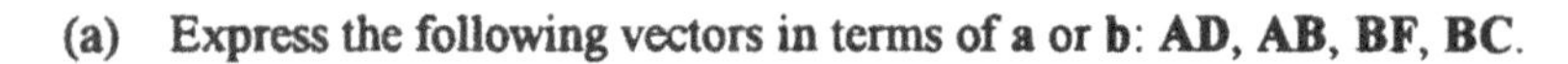

(a) Express the following vectors in terms of $\mathbf{a}$ or $\mathbf{b}$: **AD**, **AB**, **BF**, **BC**.

(b) Suppose $|\mathbf{b}| = 3$.
(i) What is the length of side AB?
(ii) If $\mathbf{AQ} = \hat{\mathbf{b}}$ state where the point Q is.
(iii) Where is P if $\mathbf{CP} = -2\hat{\mathbf{b}}$?

Solution 2.2

(a) The vector **AD** is in the same direction as the vector **AE** but has twice the length. Thus we can obtain **AD** by scaling **AE** by 2: $\mathbf{AD} = 2\mathbf{AE} = 2\mathbf{a}$. Similarly **AB** is obtained by scaling **AF** by 4/3, so $\mathbf{AB} = (4/3)\mathbf{AF} = (4/3)\mathbf{b}$ which can also be written as $4\mathbf{b}/3$. The length of **BF** is 1/3 the length of **b** and is in the opposite direction. Hence $\mathbf{BF} = -(1/3)\mathbf{b}$, or $-\mathbf{b}/3$. The lines BC and AD are opposite sides of a parallelogram. They are therefore parallel to one another and of equal length and so **BC** and **AD** are the same vector. Hence $\mathbf{BC} = \mathbf{AD} = 2\mathbf{a}$.

(b) (i) We have from part (a) that $\mathbf{AB} = (4/3)\mathbf{b}$ and so the length or magnitude of **AB** is $AB = (4/3)|\mathbf{b}| = (4/3)3 = 4$.

(ii) $\hat{\mathbf{b}}$ is a unit vector parallel to **b**, so if **AQ** = $\hat{\mathbf{b}}$ then the point Q must be one unit of distance from A along the line AB, i.e. 1/4 of the way from A to B.

(iii) **CP** is a vector of length 2 units antiparallel to **b**, and so P is a point 2 units from C along the line CD, i.e. P is the midpoint of CD.

Problem 2.1 (*Objective* 2) If **l** is a unit vector directed towards the east, specify

(a) a vector of magnitude 5 directed towards the east,
(b) a vector of magnitude 2 directed towards the west.
(c) Is it possible to obtain a vector directed towards the north by scaling **l**?

Problem 2.2 (*Objectives 1,2*) Select the *incorrect* statements, giving reasons for your choice.

A If **b** = 5**a** then $|\mathbf{b}| = 5|\mathbf{a}|$.
B **w** and -3**w** are collinear and antiparallel.
C If **u** is a unit vector then $|7\mathbf{u}| = 7$.
D Vectors 2**g** and -2**g** have the same magnitude.
E The vector k**q** has the same direction as **q** for any value of k.
F If **r** is any vector and 4**r** is scaled by 1/4 the result is a unit vector.
G If **e** is a unit vector directed towards the north then $-$**e** is a unit vector directed towards the south.

Problem 2.3 (*Objectives 1,2*) The point P divides the straight line AB in the ratio 3:2, the longer length being AP. Express the displacements **AP** and **BP** in terms of **AB**.

1.3 VECTOR ADDITION – THE TRIANGLE ADDITION RULE

The need to consider how two vectors add together to give another vector comes about most simply from considering the effect of making two journeys in succession. This is illustrated in Fig 1.12 which depicts three cities labelled A, B and C. Suppose a journey from B to C is followed by a journey from C to A. The corresponding displacements are shown as **a** = **BC** and **b** = **CA**. The net result of the two displacements is to arrive at A from B, and this can be described by the single displacement **c** = **BA** in Fig 1.12. We denote this algebraically by writing

$$\mathbf{a} + \mathbf{b} = \mathbf{c} \qquad (1.5)$$

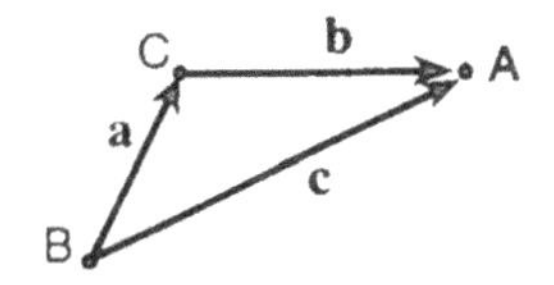

Fig 1.12
Adding displacements - the vector addition rule.

This way of combining displacements defines what is meant by the addition of two vectors. The construction illustrated in Fig 1.12 is called **the vector**

addition rule or the **triangle addition rule**. The single vector $\mathbf{c} = \mathbf{a} + \mathbf{b}$ is called the **resultant** of **a** and **b**.

Vector addition rule (triangle addition rule)

To add **b** to **a**, draw **b** with its beginning point coincident with the end point of **a**. Then the resultant, $\mathbf{c} = \mathbf{a} + \mathbf{b}$, is the vector drawn from the beginning point of **a** to the end point of **b**.

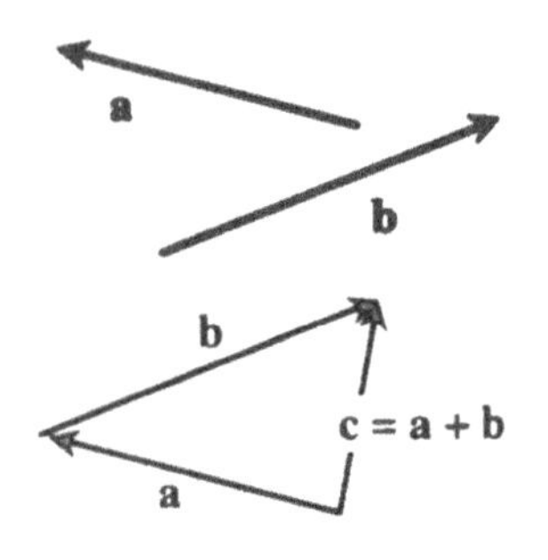

Fig 1.13
The vector addition rule for two arbitrary vectors.

Fig 1.13 illustrates the triangle addition of two arbitrary vectors specified by arrows that are not initially drawn with the beginning point of one at the end point of the other. Recall that an arrow represents the same vector wherever it is drawn on the page provided only that its length and direction are the same. We may therefore draw one of the arrows again, with the same length and direction, so that its beginning point is coincident with the end point of the other; the resultant is then drawn from the free beginning point to the free end point.

Another illustration of the addition rule is shown in Fig 1.14 where two arrows are drawn from the same point O. Vector addition is in this case most easily constructed by completing the parallelogram; the resultant is then represented by the vector $\mathbf{c} = \mathbf{OP}$. This construction is referred to as the **parallelogram addition rule**, but it is clear from Fig 1.14c that the parallelogram rule is entirely equivalent to the triangle rule because the same resultant is obtained.

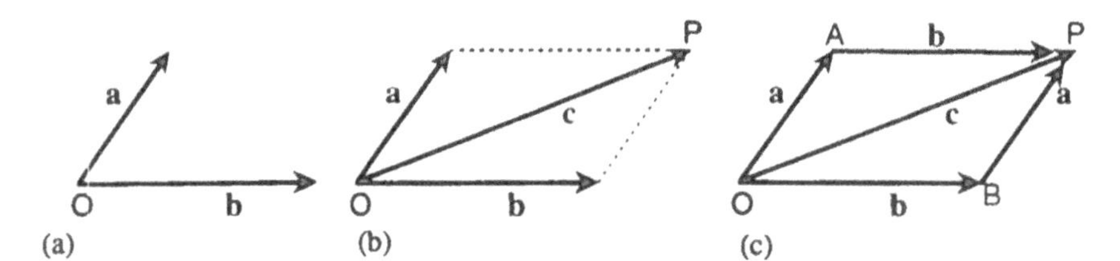

Fig 1.14
The parallelogram addition rule.
(a) Two vectors **a** and **b** with the same beginning point O.
(b) The resultant $\mathbf{c} = \mathbf{OP}$.
(c) The upper triangle OPA or the lower triangle OBP shows the addition of **a** and **b** by the triangle rule to give the resultant **c**.

Vector subtraction can be defined as follows:

$$\mathbf{a} - \mathbf{b} = \mathbf{a} + (-\mathbf{b}) \tag{1.6}$$

To subtract **b** from **a**, we first scale **b** by -1 to obtain $-\mathbf{b}$ and then add $-\mathbf{b}$ to **a** using the triangle rule for addition (Fig 1.15a). An alternative construction which gives the same answer is to draw the two vectors, **a** and **b**, from a common point O; then $\mathbf{a} - \mathbf{b}$ is the vector drawn from the end point of **b** to the end point of **a** (Fig 1.15b).

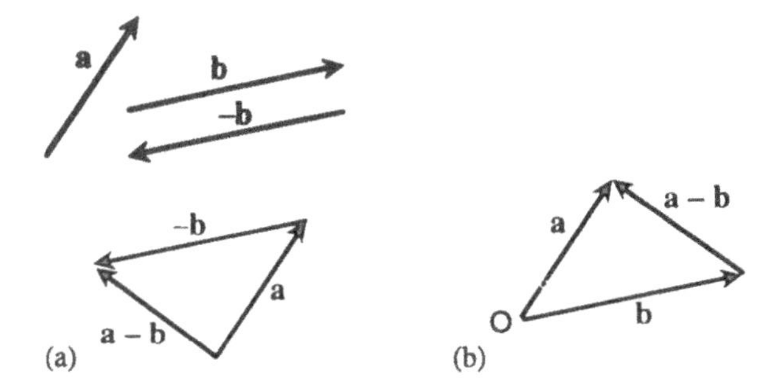

Fig 1.15
Vector subtraction. (**a**) To subtract **b** from **a**, add −**b** to **a** by the triangle rule. (b) Alternatively, draw **a** and **b** from the same point; then **a** − **b** is the vector from the end point of **b** to the end point of **a**.

Any number of vectors can be added by the construction shown in Fig 1.16 known as the **polygon addition rule**, but this is clearly no more than the triangle rule applied successively: the first two vectors **a** and **b** are added by the triangle rule to give the resultant **AC** which is then added to the third vector to give the resultant **AD**, and so on. The polygon rule applies whether or not the vectors are coplanar, but it may be difficult to draw the three-dimensional polygon when the vectors are not coplanar.

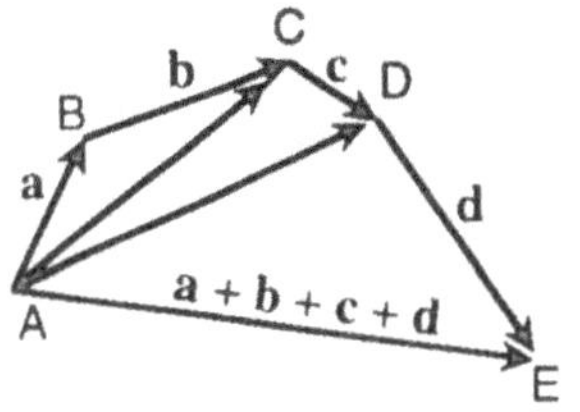

Fig 1.16
The polygon rule.

Vector addition, like ordinary number addition, obeys commutative, associative and distributive rules:

$\mathbf{a} + \mathbf{b} = \mathbf{b} + \mathbf{a}$	commutative rule for addition
$\mathbf{a} + (\mathbf{b} + \mathbf{c}) = (\mathbf{a} + \mathbf{b}) + \mathbf{c}$	associative rule for addition
$\alpha\mathbf{a} + \alpha\mathbf{b} = \alpha(\mathbf{a} + \mathbf{b})$	distributive rule

The commutative rule is illustrated in Fig 1.14c where the same resultant **c** is obtained whether the upper or the lower triangle is used. The associative and distributive rules are illustrated in Fig 1.17. It is seen that these rules express elementary geometrical properties of simple figures.

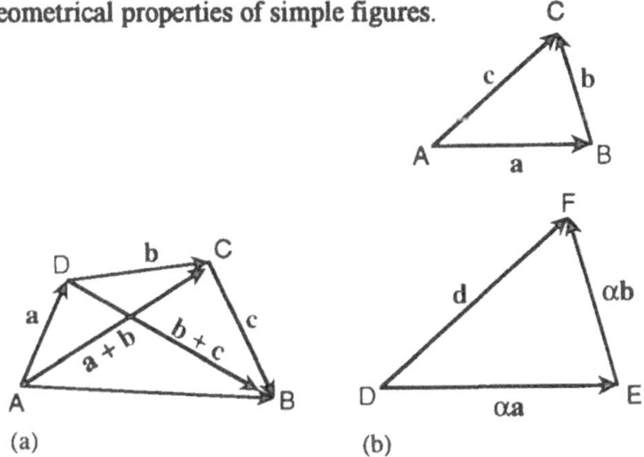

Fig 1.17
(**a**) The associative rule. Triangle ADB shows **a** + (**b** + **c**) = **AB**; triangle ACB shows (**a** + **b**) + **c** = **AB**.
(b) The distributive rule. The vector additions **a** + **b** = **c** and α**a** + α**b** = **d** are shown in triangles ABC and DEF respectively. The triangles are *similar triangles* and so **d** = α**c**. Hence α**a** + α**b** = α(**a** + **b**).

The magnitude and direction of the resultant of two or more vectors can often be found by using Pythagoras's theorem and elementary trigonometry. Note that the magnitude of $\mathbf{a} + \mathbf{b}$ is denoted by $|\mathbf{a} + \mathbf{b}|$ which is not the same as $|\mathbf{a}| + |\mathbf{b}|$ unless the two vectors are parallel, i.e. point in the same direction.

Summary of section 1.3

- The addition of two vectors is achieved by the **vector addition rule** (**triangle addition rule**) illustrated in Figs 1.12 to 1.14. The vector $\mathbf{c} = \mathbf{a} + \mathbf{b}$ is called the **resultant** of **a** and **b**. The **parallelogram addition rule** is equivalent to the triangle addition rule.

- **Vector subtraction** is defined by $\mathbf{a} - \mathbf{b} = \mathbf{a} + (-\mathbf{b})$ (Fig 1.15).

- Three or more vectors can be added by the **polygon addition rule** (Fig 1.16).

- Vector addition is commutative, associative and distributive (Figs 1.14c and 1.17).

Example 3.1 (*Objectives 3,4*) Fig 1.18 shows arrows representing three coplanar vectors **p**, **q** and **r**.

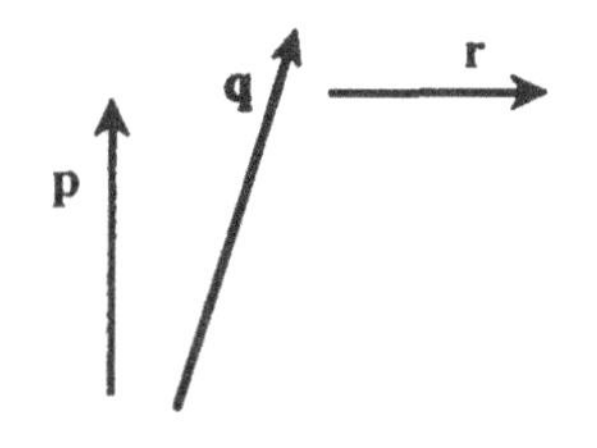

Fig 1.18
Three coplanar vectors for Example 3.1.

(a) Draw rough sketches showing arrows representing
 (i) the resultant of **q** and **r**,
 (ii) the vector $-\mathbf{q}$ and the vector $\mathbf{r} - \mathbf{q}$,
 (iii) $\mathbf{p} + \mathbf{q} + \mathbf{r}$,
 (iv) $\mathbf{p} + \mathbf{q} - \mathbf{r}$.

(b) Is $|\mathbf{p} + \mathbf{q}| = |\mathbf{p}| + |\mathbf{q}|$? State which is the larger of $|\mathbf{p} + \mathbf{q}|$ and $|\mathbf{p} - \mathbf{q}|$.

Solution 3.1 The constructions for part (a) are sketched in Fig 1.19a.

(b) No. For the given vectors $|\mathbf{p} + \mathbf{q}| > |\mathbf{p} - \mathbf{q}|$. See Fig 1.19b.

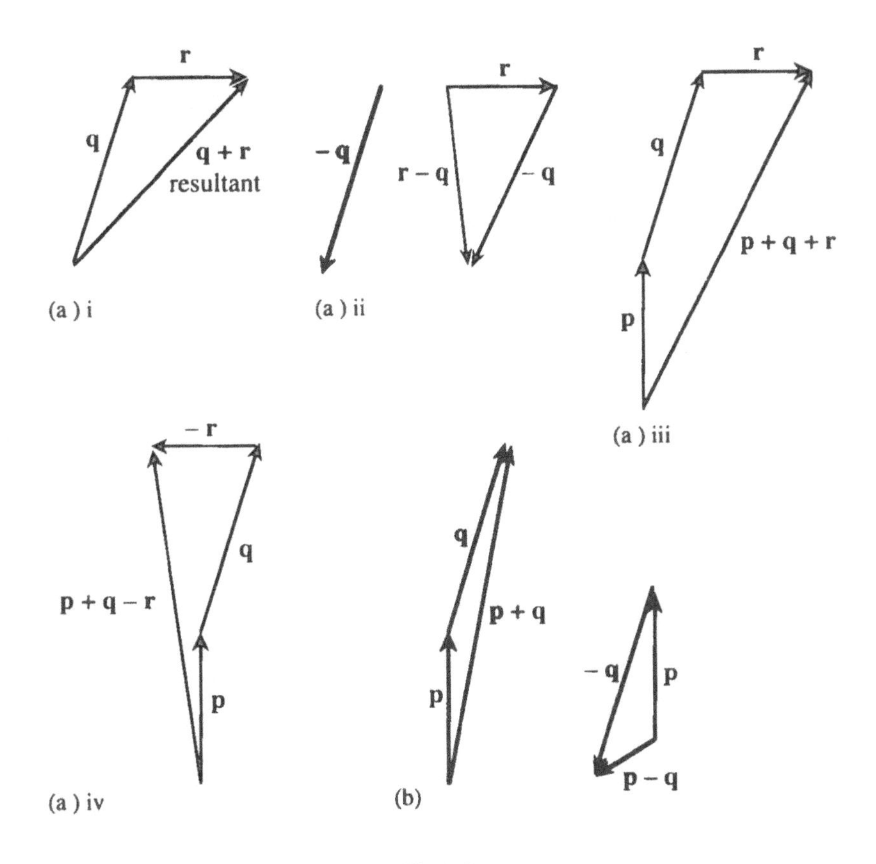

Fig 1.19
Solution 3.1.

Example 3.2 (*Objectives 3,4*)

(a) Sketch an arrow diagram showing the vector addition of three arbitrary vectors **A**, **B** and **C**, assuming none is the zero vector and no two are collinear.

(b) Now suppose vectors **A** and **B** are each of magnitude d, with **A** directed towards the south and **B** towards the west, and suppose that vector **C** is such that $\mathbf{A} + \mathbf{B} + \mathbf{C} = \mathbf{0}$. Sketch a diagram showing the addition of the three vectors. Use Pythagoras's theorem to determine $|\mathbf{C}|$.

Solution 3.2

(a) A sketch is shown in Fig 1.20a where the vectors have been arbitrarily chosen. The resultant of the three vectors is drawn from the beginning point of **A** to the end point of **C**.

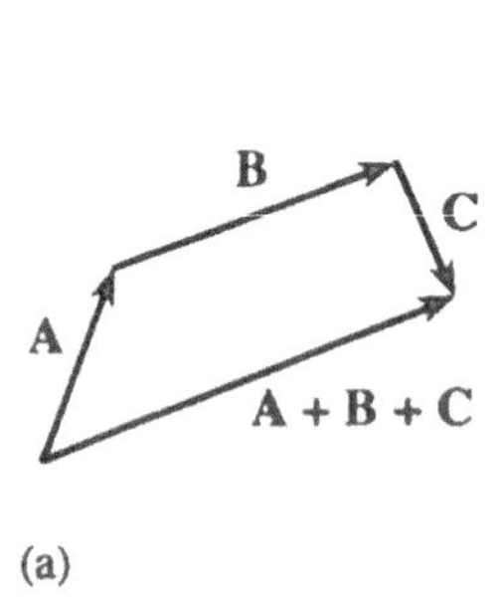

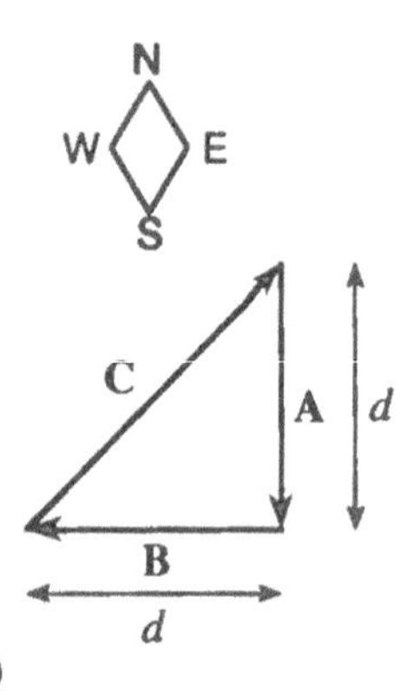

Fig 1.20
Solution 3.2. (a) The addition of three arbitrarily chosen vectors. (b) The three vectors here add to give a zero resultant.

(b) We are told that the three particular vectors specified in the question add to give the zero resultant. This is shown in Fig 1.20b where the end point of **C** meets the beginning point of **A** to give a resultant of zero length, i.e. the zero vector. The magnitude of **C** is found from Pythagoras's theorem: $|\mathbf{C}|^2 = d^2 + d^2 = 2d^2$. Therefore $|\mathbf{C}| = \sqrt{2}d = 1.414d$.

A regular hexagon is a six sided plane figure in which all six sides are of equal length and all internal angles are 120°.

Example 3.3 (*Objectives 1,3*) Fig 1.21(a) shows a *regular hexagon* with displacement vectors **u**, **v** and **w** along three sides.

(a) Express the displacements **DE**, **EF**, **FA**, **CE** and **FC** in terms of **u**, **v** and **w**.
(b) Express **w** in terms of **u** and **v**.

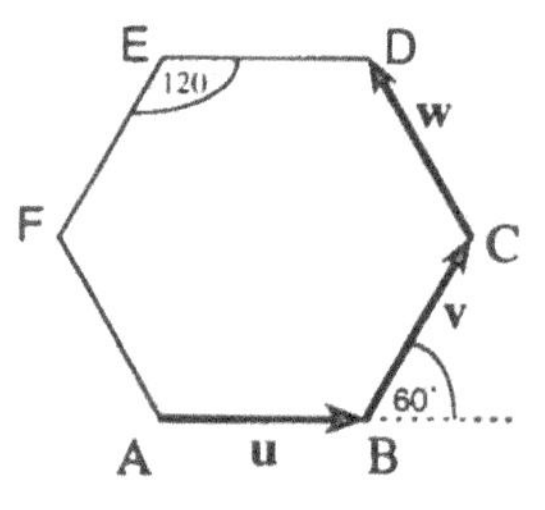

Fig 1.21a
Three displacement vectors along the sides of a regular hexagon.

Solution 3.3

(a) **DE** is antiparallel to and has the same length as **AB** = **u**. Therefore **DE** = –**u**. Similarly **EF** = –**v** and **FA** = –**w**. **CE** = **CD** + **DE** = **w** – **u**. **FC** = **FA** + **AB** + **BC** = –**w** + **u** + **v**.

(b) **w** = **v** – **u**. See the equilateral triangle in Fig 1.21b.

Problem 3.1 (*Objectives 3,4*) A vector **a** is of magnitude k and is directed towards the north. A vector **c**, also of magnitude k, is directed towards the west.

(a) Determine the magnitudes and directions of the vectors **a** + **c** and **a** – **c**.
(b) Specify the magnitude and direction of vector **b** if **a** + **b** + **c** = **0**.

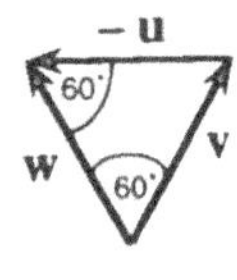

Fig 1.21b
(b) **w** = **v** – **u**.

Problem 3.2 (*Objective 3*) Let **u**, **v** and **w** be vectors of the same magnitude directed towards the north, east and north-west respectively. Draw a sketch to show the addition **u** + **v** + **w**. Draw five other sketches showing the addition of the three vectors in all possible orders. What rules of vector algebra do your sketches illustrate?

Problem 3.3 (*Objective 3*) Draw a sketch to illustrate the distributive rule for vector addition, $\alpha\mathbf{a} + \alpha\mathbf{b} = \alpha(\mathbf{a} + \mathbf{b})$ for the case $\alpha = -1$.

1.4 LINEAR COMBINATIONS OF VECTORS

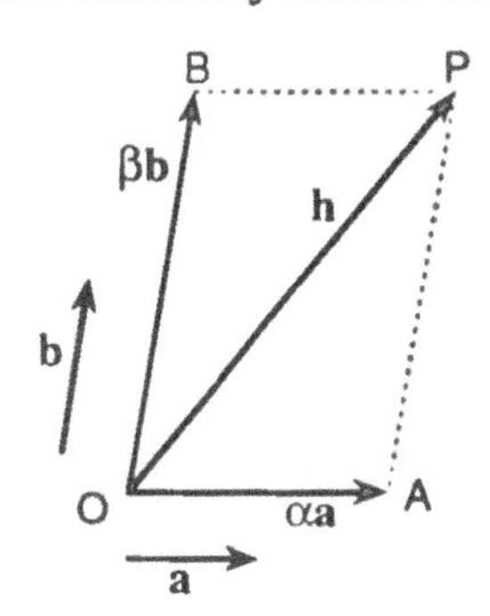

Fig 1.22
Vector **h** is a linear combination of **a** and **b**.

You now know what is meant by the scaling of a vector and the addition of vectors. We can now use these two operations together. Given any two vectors **a** and **b**, we can scale **a** by a number α to give $\alpha\mathbf{a}$, scale **b** by a number β to give $\beta\mathbf{b}$ and then add the two scaled vectors. The resultant **h** is a **linear combination** of **a** and **b** (Fig 1.22). We express this algebraically by writing

$$\mathbf{h} = \alpha\mathbf{a} + \beta\mathbf{b} \tag{1.7}$$

A simple example of a linear combination was seen in the previous section where the subtraction of **b** from **a** was defined to be $\mathbf{a} + (-\mathbf{b})$. This is a linear combination of **a** and **b** with $\alpha = 1$ and $\beta = -1$. Another example is $\mathbf{q} = 3\mathbf{a} - 7\mathbf{b}$. Here $\alpha = 3$ and $\beta = -7$.

If **a** and **b** are nonzero and are not collinear, the vector **h** is always in the plane of **a** and **b** whatever the values of α and β. In fact any vector in the plane of **a** and **b** can be expressed as a linear combination of **a** and **b** for suitably chosen values of the scalars α and β. This is equivalent to saying that any point P in the plane of Fig 1.22 can be reached from O by two successive displacements $\alpha\mathbf{a}$ and $\beta\mathbf{b}$ along, for example, the path OAP or OBP.

We can make linear combinations of any number of vectors. For three vectors **a**, **b** and **c** we have

$$\mathbf{g} = \alpha\mathbf{a} + \beta\mathbf{b} + \gamma\mathbf{c} \tag{1.8}$$

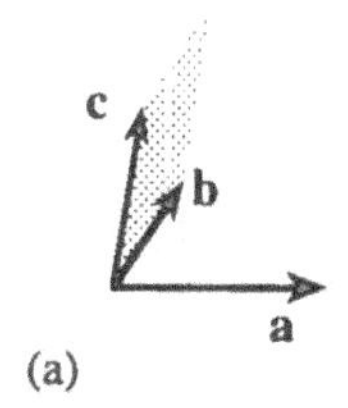

Fig 1.23a
Vectors **a**, **b** and **c** are non-coplanar vectors.

where α, β, and γ are numbers. An example of a linear combination of three vectors is $-3\mathbf{a} + 5\mathbf{b} - \mathbf{c}$. Here $\alpha = -3$, $\beta = 5$ and $\gamma = -1$.

If all three vectors **a**, **b** and **c** are coplanar then vector **g** of Eq 1.8 is also in the same plane, whatever the values of α, β and γ may be. Linear combinations are more useful when the three vectors **a**, **b** and **c** are non-coplanar (Fig 1.23a) because then any vector in three-dimensional space can be expressed as a linear combination of the three non-coplanar vectors. This important fact is shown geometrically in Fig 1.23b where an arbitrary vector **g** is shown as the displacement **OP** along the diagonal of the *parallelepiped* constructed from the scaled vectors $\alpha\mathbf{a} = \mathbf{OA}$, $\beta\mathbf{b} = \mathbf{OB}$ and $\gamma\mathbf{c} = \mathbf{OC}$. The idea of expressing arbitrary vectors as linear combinations of just three non-coplanar vectors is the key to vector analysis. It will be exploited in the next section where the cartesian representation of vectors is introduced.

A **parallelepiped** is a three-dimensional tilted box-shaped figure; each of the six faces is a parallelogram.

We often need to manipulate linear combinations of vectors algebraically. For example, given the linear combination

$$\mathbf{h} = 2\mathbf{a} - 5\mathbf{b} \tag{1.9}$$

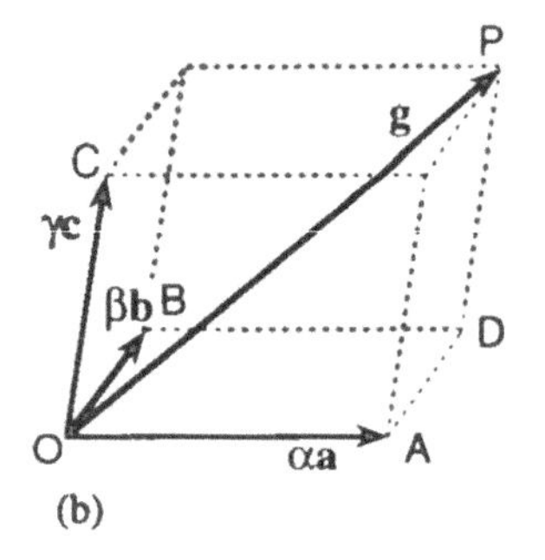

Fig 1.23b
An arbitrary vector $\mathbf{g} = \mathbf{OP}$ is a linear combination of three non-coplanar vectors **a**, **b** and **c**.

we may be required to find **a** in terms of **h** and **b**, i.e. make **a** the subject of the equation. Since a linear combination consists of scaling and vector addition, the rules of algebra for linear combinations are the commutative, associative and distributive rules given in Sections 1.2 and 1.3, which are analogous to the familiar rules of addition and multiplication in ordinary number algebra. Thus we can make vector **a** in Eq (1.9) the subject of the equation, as follows: we add the vector $5\mathbf{b}$ to both sides of the equation to give $\mathbf{h} + 5\mathbf{b} = 2\mathbf{a}$; then scale both sides by 1/2 to obtain $\mathbf{a} = (1/2)(\mathbf{h} + 5\mathbf{b})$, which we can also write as $\mathbf{a} = \mathbf{h}/2 + 5\mathbf{b}/2$.

Summary of section 1.4

- Given any three vectors **a**, **b** and **c** and three numbers α, β and γ, the vector

$$\mathbf{g} = \alpha\mathbf{a} + \beta\mathbf{b} + \gamma\mathbf{c} \tag{1.8}$$

  is a **linear combination** of the three vectors.

- Any vector in space can be expressed as a linear combination of three non-coplanar vectors.

- The rules for the algebraic manipulation of linear combinations of vectors are analogous to the familiar rules of number algebra.

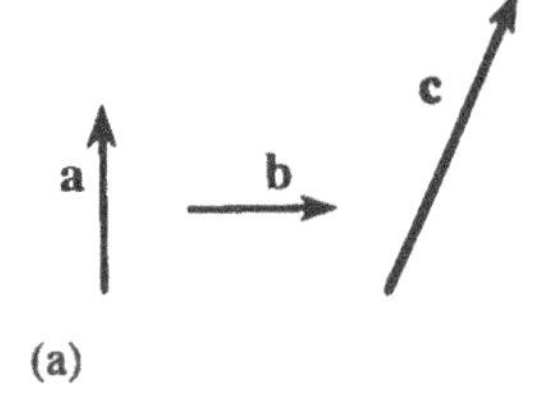

Fig 1.24a
Three coplanar vectors for Examples 4.1 and 4.2

Example 4.1 (*Objective 5*) Refer to the three coplanar vectors represented by the arrows in Figure 1.24a. Make rough sketches to show the following linear combinations: $\mathbf{a} - 2\mathbf{b}$, $\mathbf{c}/2 + \mathbf{a}$, $\mathbf{a} + 3\mathbf{b} - \mathbf{c}$ and $-2\mathbf{b} + \mathbf{a} - 3\mathbf{c}$.

Solution 4.1 See Fig 1.24b.

Example 4.2 (*Objective 5*) Refer to the three coplanar vectors **a**, **b** and **c** in Fig 1.24a. Make a rough sketch showing **b** as a linear combination of **a** and **c**.

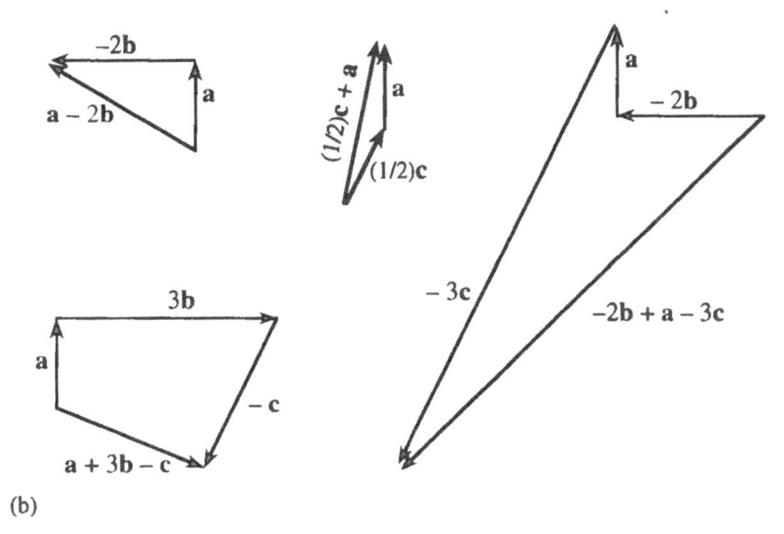

Fig 1.24b
Solution to Example 4.1.

Solution 4.2

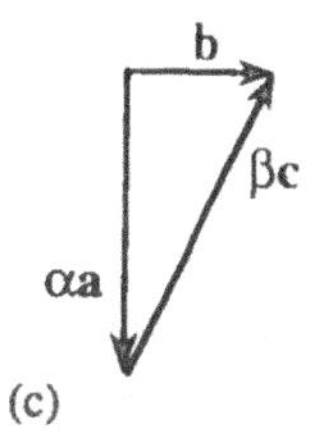

Fig 1.24c
Solution to Example 4.2.

(a) See Fig 1.24c. α and β are scaling constants ($\alpha < 0$; $\beta > 0$).

The term **orthogonal** means perpendicular or at right angles.

Example 4.3 (*Objective 5*) Fig 1.25 shows a box-shaped figure (a cuboid) with three mutually *orthogonal* unit vectors **u**, **v** and **w** directed parallel to the edges. The lengths of the edges are 4, 2 and 3 units as shown. M is the midpoint of DE and G is the midpoint of the upper face. Express the displacement **AG** as a linear combination of the three unit vectors.

Solution 4.3 We can reach G from A by successive displacements along the lines AD, DM and MG. Thus we can write **AG** = **AD** + **DM** + **MG**. But **AD** = 3**w**, **DM** = (1/2)**DE** = (1/2)**AB** = 2**u**, and **MC** = (1/2)**EF** = **v**. Hence **AG** = 3**w** + 2**u** + **v**.

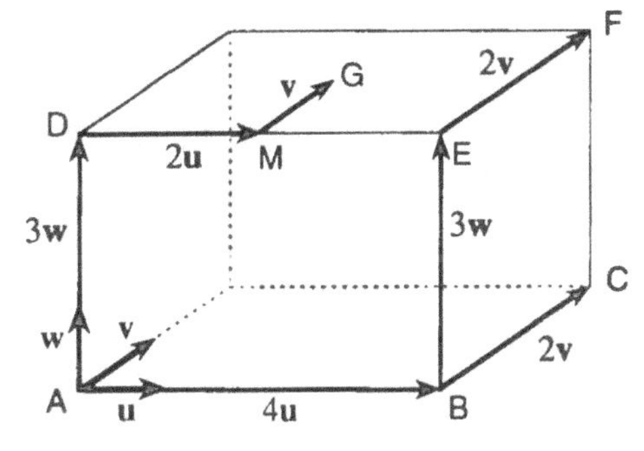

Fig 1.25
The three mutually orthogonal unit vectors **u**, **v** and **w** are parallel to the edges of the cuboid.

Example 4.4 (*Objective 6*) Given **p** = (1/2)(**u** + **v**) and **q** = (1/2)(**u** – **v**), simplify the expressions obtained for **p** + **q** and **p** – **q**.

Solution 4.4

$$\mathbf{p} + \mathbf{q} = \tfrac{1}{2}(\mathbf{u} + \mathbf{v}) + \tfrac{1}{2}(\mathbf{u} - \mathbf{v})$$

$$= \tfrac{1}{2}[(\mathbf{u} + \mathbf{v}) + (\mathbf{u} - \mathbf{v})]$$

$$= \tfrac{1}{2}[\mathbf{u} + \mathbf{v} + \mathbf{u} - \mathbf{v}] = \tfrac{1}{2}[2\mathbf{u}] = \mathbf{u}$$

Similarly,

$$\mathbf{p} - \mathbf{q} = \tfrac{1}{2}[(\mathbf{u} + \mathbf{v}) - (\mathbf{u} - \mathbf{v})] = \tfrac{1}{2}[2\mathbf{v}] = \mathbf{v}$$

Example 4.5 (*Objective 6*) Refer to Eq (1.8). Make vector **a** the subject of the equation.

Solution 4.5 First subtract $\beta\mathbf{b} + \gamma\mathbf{c}$ from both sides of the equation to give $\mathbf{g} - \beta\mathbf{b} - \gamma\mathbf{c} = \alpha\mathbf{a}$; then scale both sides by $1/\alpha$ to give the answer: $\mathbf{a} = (\mathbf{g} - \beta\mathbf{b} - \gamma\mathbf{c})/\alpha$, which can be written as $\mathbf{a} = (1/\alpha)\mathbf{g} - (\beta/\alpha)\mathbf{b} - (\gamma/\alpha)\mathbf{c}$.

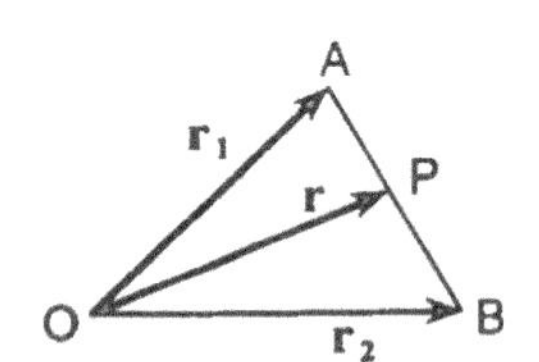

Fig 1.26
Point P is the midpoint of AB.

Example 4.6 (*Objectives 5,6*) Fig 1.26 shows vectors $\mathbf{r}_1 = \mathbf{OA}$, $\mathbf{r}_2 = \mathbf{OB}$ and $\mathbf{r} = \mathbf{OP}$, where P is the midpoint of the line AB.

(a) Express **BP** as a linear combination of $\mathbf{r}_1$ and $\mathbf{r}_2$.

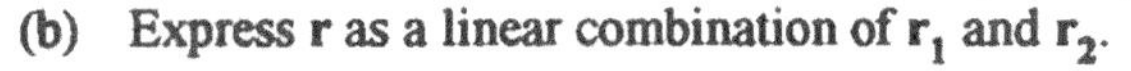

(b) Express **r** as a linear combination of $\mathbf{r}_1$ and $\mathbf{r}_2$.

Solution 4.6

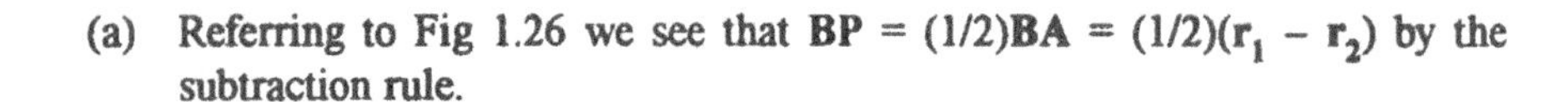

(a) Referring to Fig 1.26 we see that $\mathbf{BP} = (1/2)\mathbf{BA} = (1/2)(\mathbf{r}_1 - \mathbf{r}_2)$ by the subtraction rule.

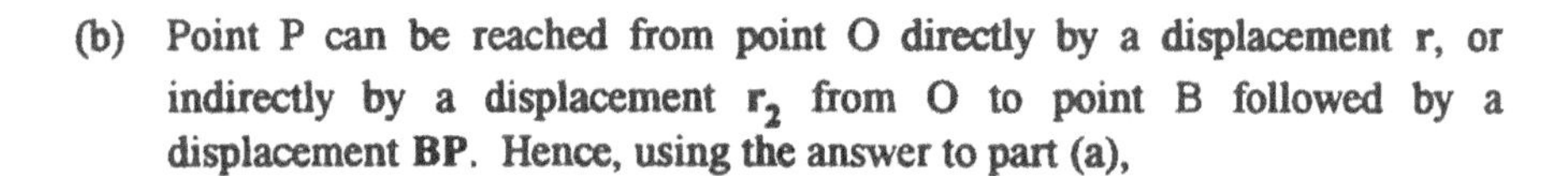

(b) Point P can be reached from point O directly by a displacement **r**, or indirectly by a displacement $\mathbf{r}_2$ from O to point B followed by a displacement **BP**. Hence, using the answer to part (a),

$$\begin{aligned}\mathbf{r} &= \mathbf{r}_2 + \mathbf{BP} \\ &= \mathbf{r}_2 + \tfrac{1}{2}(\mathbf{r}_1 - \mathbf{r}_2) = \mathbf{r}_2 + \tfrac{1}{2}\mathbf{r}_1 - \tfrac{1}{2}\mathbf{r}_2 \\ &= \tfrac{1}{2}\mathbf{r}_1 + \tfrac{1}{2}\mathbf{r}_2 = \tfrac{1}{2}(\mathbf{r}_1 + \mathbf{r}_2)\end{aligned}$$

Thus we have expressed **r** as a linear combination of $\mathbf{r}_1$ and $\mathbf{r}_2$.

Problem 4.1 (*Objective 5*) Given the three vectors **a**, **b** and **c** shown in Fig 1.24a, make rough sketches to show: $\mathbf{b} - 2\mathbf{a}$, $-(2\mathbf{a} - \mathbf{b})$, $\mathbf{c} - 2\mathbf{a} - \mathbf{b}$.

Problem 4.2 (*Objectives 4,5*) Fig 1.27 shows a cuboid with edges constructed from three mutually orthogonal vectors **a**, **b** and **c**.

(a) Express the vector **OP** along the diagonal in terms of **a**, **b** and **c**.
(b) Express the vector **QR** in terms of **a**, **b** and **c**, and find **OP** + **QR**.
(c) If $|\mathbf{b}| = 2|\mathbf{a}|$ and $|\mathbf{c}| = |\mathbf{a}|$, determine $|\mathbf{OP}|$ in terms of $|\mathbf{a}|$.

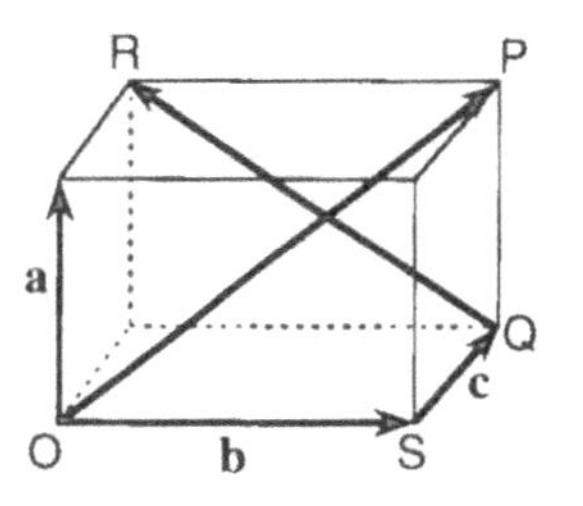

Fig 1.27
The cuboid is constructed from the three mutually orthogonal vectors **a**, **b** and **c**.

Problem 4.3 (*Objectives 4,5*) The directions of vectors **A** and **B** make an angle of 30° with each other. $|\mathbf{A}| = 1$ and $|\mathbf{B}| = 2$. Draw sketches to show the linear combinations 2**A** – **B** and 2**A** + **B**. Determine $|2\mathbf{A} - \mathbf{B}|$ and $|2\mathbf{A} + \mathbf{B}|$. Write down unit vectors in the directions of **B** and 2**A** + **B**.

Problem 4.4 (*Objective 5*) Fig 1.28 shows a parallelepiped with vectors **a** = **AB**, **b** = **AD** and **c** = **AE**.

(a) Express the diagonal displacements **AG** and **HB** as linear combinations of **a**, **b** and **c**.

(b) Let P be the midpoint of the lower face ABCD and Q the midpoint of the face BCGF. (The faces are parallelograms.) Write the displacement **QP** as a linear combination of **a**, **b** and **c**.

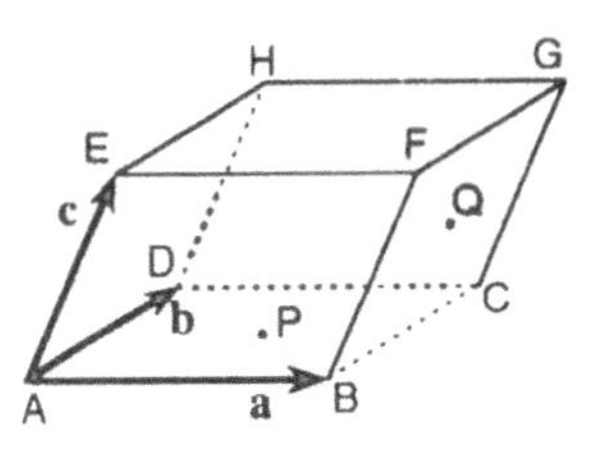

Fig 1.28
The parallelepiped is constructed from three non-coplanar vectors **a**, **b** and **c**.

Problem 4.5 (*Objective 5*) Unit vectors **u** and **v** form two sides of an equilateral triangle. Sketch the triangle and the two vectors $\mathbf{p} = (1/2)(\mathbf{u} + \mathbf{v})$ and $\mathbf{q} = (1/2)(\mathbf{u} - \mathbf{v})$. Identify the two vectors **p** + **q** and **p** – **q** on your sketch.

Problem 4.6 (*Objective 5*) Refer to Fig 1.29. If point F is one third of the way from B to C, express **AF** in terms of **v** and **u**.

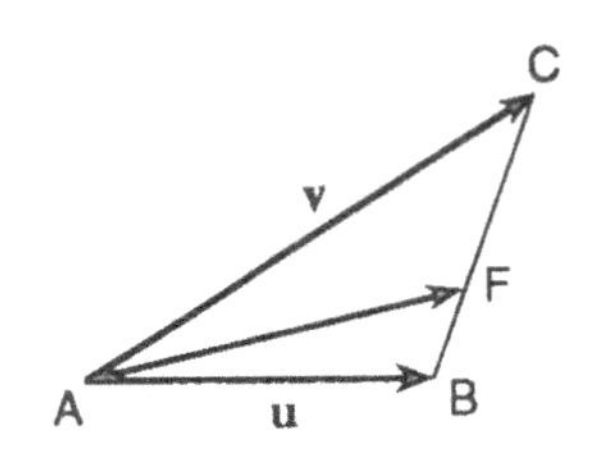

Fig 1.29
The point F is one third of the way from B to C.

Problem 4.7 (*Objective 6*)

(a) Given the linear combination $\mathbf{p} = 2\mathbf{u} - \mathbf{v} + 5\mathbf{w}$, express 10**w** in terms of the other vectors.
(b) Simplify $2(\mathbf{a} - \mathbf{b} + \mathbf{c}) - (3\mathbf{a} + \mathbf{c}) + \mathbf{c} + \mathbf{a}$.

Problem 4.8 (*Objective 6*) If $5\mathbf{a} - 10\mathbf{b} + \mathbf{c} = \mathbf{0}$, find **a** in terms of **b** and **c**.

Problem 4.9 (*Objectives 5,6*) The point Q divides the line AB in the ratio l:m. Let O be a point not lying on the line. Express the vector **OQ** as a linear combination of **OA** and **OB**.

Problem 4.10 (*Objectives 5,6*) Consider three arbitrary points A, B and C in a plane and let O be a reference point. Let the vectors **OA**, **OB** and **OC** be denoted by **a**, **b** and **c**. Then the *centroid* of the three points A, B and C is defined to be (1/3)(**a** + **b** + **c**). Show that if D is the midpoint of BC then the centroid of the three points is located 1/3 of the way from D to A.

The centroid of N points is the sum of the N position vectors multiplied by $1/N$.

1.5 CARTESIAN VECTORS

The previous sections made use of a geometrical picture of vectors based on arrows drawn on the page. Arrows and geometric figures provide us with good visual props for introducing basic vector operations such as scaling and addition. However, for any but the simplest of applications it is difficult to draw or visualise the three-dimensional figures and it becomes necessary to make greater use of algebra. Most applications of vectors in science and engineering use an algebraic description of vectors based on a cartesian coordinate system and the resolution of vectors into their three cartesian components. We begin this section with a review of three-dimensional cartesian coordinate systems and the cartesian coordinates of a point. We then show how the cartesian components of a vector are defined and introduce the associated notation and terminology.

1.5.1 Cartesian coordinates of a point - a review

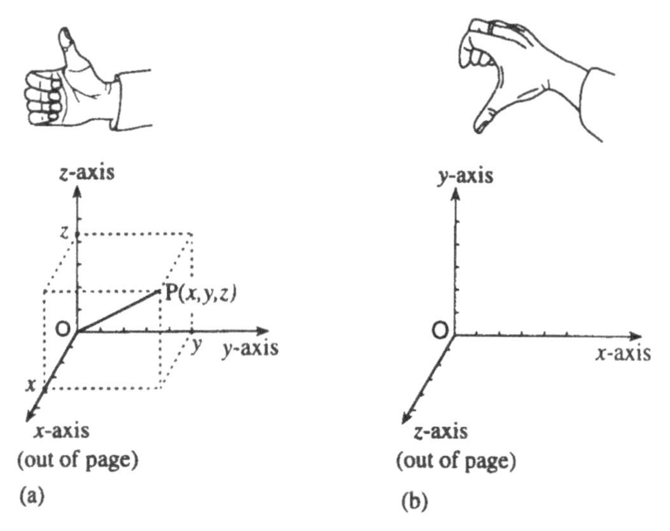

Fig 1.30
(a) A right-handed three-dimensional cartesian coordinate system for specifying the position of a point P. (b) A right-handed coordinate system drawn in a different orientation from that in (a).

Fig 1.30a shows a three-dimensional right-handed **cartesian coordinate system** for specifying the positions of points in space. The three coordinate axes are mutually perpendicular and cross at the **origin point** O. The axes are labelled x-axis, y-axis and z-axis and a definite direction along each axis is chosen to be the positive direction, indicated by an arrowhead. Distances along the axes are measured from the origin point and are positive if in the positive direction and negative if in the negative direction. Any point P in space is specified by its **cartesian coordinates** (x,y,z) shown by the construction in Fig 1.30a. P is called the point $P(x,y,z)$. Thus the origin point is the point $O(0,0,0)$. In a right-handed system the positive directions of the x-, y- and z-axes are given by the **right-hand rule**: hold your right hand in such a way that your fingers curl from the positive x direction to the positive y direction through the smaller (90°) angle

(Fig 1.30a); then your extended thumb points in the positive z direction. An alternative version of the right-hand rule is called the *screw rule* (Fig 1.31a).

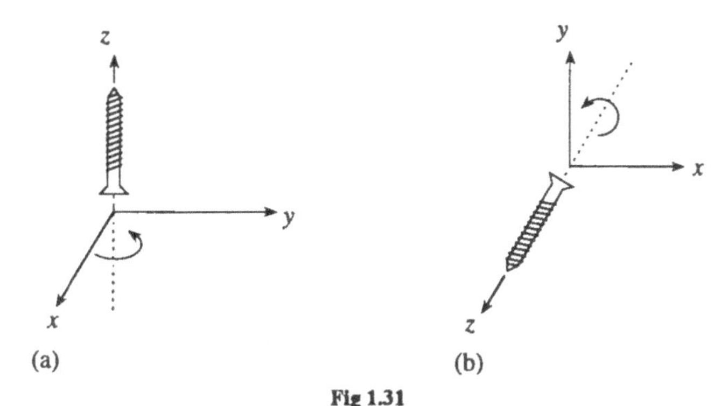

Fig 1.31
(a) The screw rule states that the direction of increasing z is the direction in which a screw lying along the z-axis would advance when the sense in which the screwhead is turned is from the x-axis to the y-axis through the 90° angle. (b) A right-handed system shown in a different orientation.

A left-handed system is shown in Fig 1.32. No rotation of a left-handed system can bring it into a right-handed system, but the image of a left-handed system seen in a plane mirror is a right-handed system. By convention only right-handed systems are used routinely in physics and engineering. Sometimes a right-handed system is shown in a rotated orientation, as in Figs 1.30b and 1.31b. We are free to orient our system in any convenient way but the orientation shown in Fig 1.30a is the one that will be routinely used in this book.

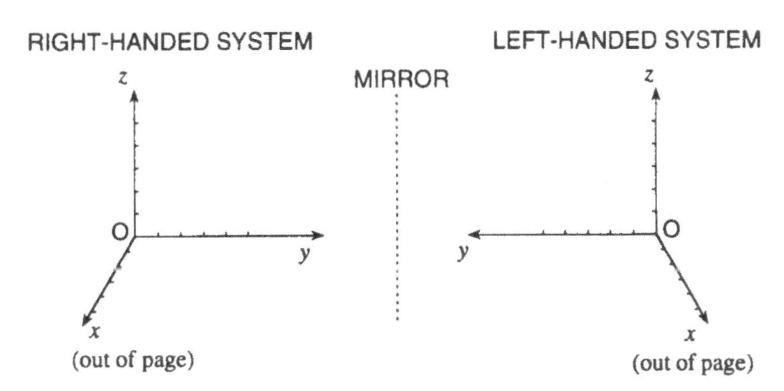

Fig 1.32
A left-handed system is the mirror image of a right-handed system.

1.5.2 Cartesian unit vectors and cartesian components of a vector

Fig 1.33 shows three mutually orthogonal unit vectors denoted by the symbols **i**, **j** and **k**, directed parallel to the positive x, y and z directions respectively; they are called **cartesian unit vectors**. We can express a displacement parallel to one of the cartesian axes as a scaling of the corresponding unit vector. For example, a displacement along the x-axis from the origin O to the point A a distance x

from O is **OA** = x**i**. Any vector in space can be expressed as a linear combination of the cartesian unit vectors corresponding to successive displacements parallel to the three coordinate axes. Thus the vector **r** = **OP** in Fig 1.34 is the linear combination

$$\mathbf{r} = x\mathbf{i} + y\mathbf{j} + z\mathbf{k} \tag{1.10}$$

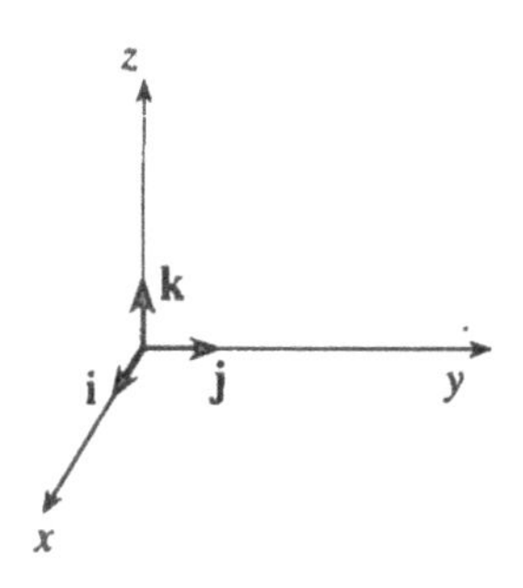

Fig 1.33
The cartesian unit vectors **i**, **j** and **k**.

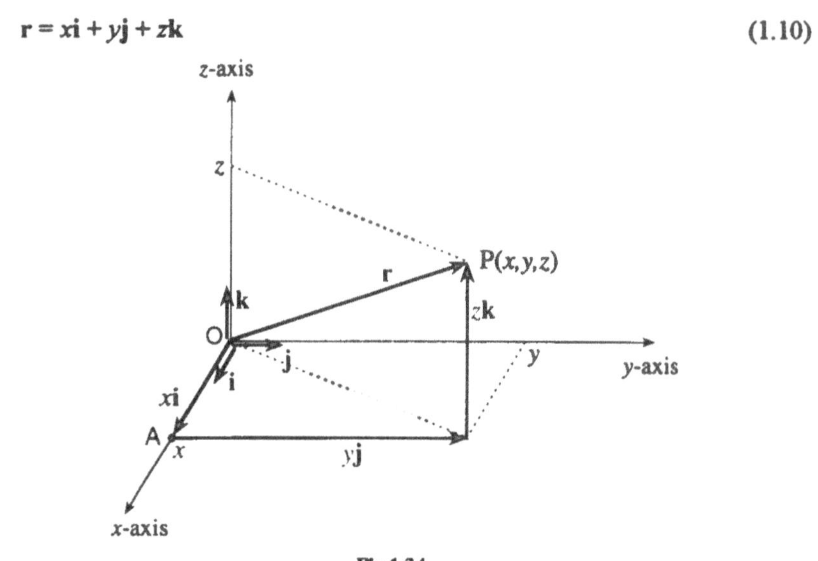

Fig 1.34
The position vector **r** is a linear combination of the cartesian unit vectors.

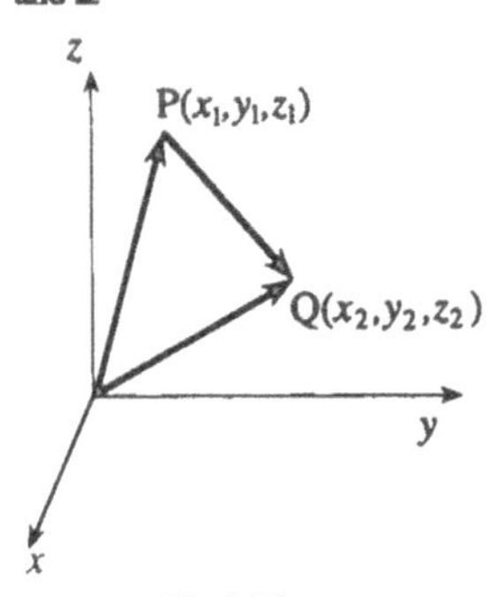

Fig 1.35
The vector **PQ**.

The vectors x**i**, y**j** and z**k** are the **cartesian component vectors** or **resolutes of r**, and the vector **r** is said to be **resolved** into its vector components. The numbers x, y and z are the **cartesian components** of **r**, or simply the **components of r**. When a vector is expressed in terms of cartesian unit vectors, as in Eq (1.10), it is called a **cartesian vector**. Of course, a cartesian vector is not physically different from any other vector; the term refers only to its mathematical representation, Eq (1.10).

There is no universal agreement about the use of the term *component*. Some books refer to the component vectors as the components of **r**. In this book the *components* are numbers not vectors.

The vector **r** = **OP** in Fig 1.34 is the displacement of the point P from the origin point O and is often called the **position vector** of point P. The components of the position vector of P are simply the cartesian coordinates x, y and z of P. More generally, a vector is drawn from some arbitrary point. Consider the vector **PQ** in Fig 1.35 where the beginning and end points are $P(x_1,y_1,z_1)$ and $Q(x_2,y_2,z_2)$. We can find the components of **PQ** in terms of the coordinates of P and Q by using the vector subtraction rule

$$\mathbf{PQ} = \mathbf{OQ} - \mathbf{OP}$$

and expressing **OQ** and **OP** as cartesian position vectors using Eq (1.10)

$$\mathbf{OQ} = x_2\mathbf{i} + y_2\mathbf{j} + z_2\mathbf{k} \quad \text{and} \quad \mathbf{OP} = x_1\mathbf{i} + y_1\mathbf{j} + z_1\mathbf{k}$$

Hence

$$\mathbf{PQ} = (x_2 - x_1)\mathbf{i} + (y_2 - y_1)\mathbf{j} + (z_2 - z_1)\mathbf{k} \qquad (1.11)$$

The x-component of the vector is $x_2 - x_1$, the y-component is $y_2 - y_1$ and the z-component is $z_2 - z_1$.

We can express the vector **PQ** of Eq (1.11) as

$$\mathbf{PQ} = (x_2 - x_1, y_2 - y_1, z_2 - z_1) \qquad (1.12)$$

where the cartesian unit vectors **i**, **j** and **k** are not shown; only the components are written down, in sequence, separated by commas and enclosed by brackets. This ordered set of three numbers is called a **cartesian ordered triple**; it provides us with a complete specification of a cartesian vector.

Ordered triples of numbers enclosed within brackets have two different usages: to denote the cartesian *coordinates* of *a point* in space, and to denote the cartesian *components* of a *vector*. The context will always make the usage clear.

There is a very useful notation for the cartesian components of a vector. The components of a vector **PQ** are denoted by putting brackets around the vector symbol and attaching a subscript to specify the component. For example, the x-component of the vector **PQ** is denoted by $(\mathbf{PQ})_x$. When the vector is denoted by a symbol such as **a**, a more common notation is a_x rather than $(\mathbf{a})_x$. Thus we could put $\mathbf{a} = \mathbf{PQ}$ and write Eq (1.11) as

$$\mathbf{a} = a_x\mathbf{i} + a_y\mathbf{j} + a_z\mathbf{k} \qquad (1.13)$$

or as an ordered triple

$$\mathbf{a} = (a_x, a_y, a_z) \qquad (1.14)$$

where $a_x = (\mathbf{PQ})_x = x_2 - x_1$, etc. This notation will be used frequently from hereon. Note that Eqs (1.11) to (1.14) illustrate four different notations for a cartesian vector; all are in common use so you must become familiar with all of them.

In some problems all vectors lie in the same plane which we can take to be the x-y plane. We can then ignore the z coordinates which are all zero. Thus if points P and Q are in the x-y plane with coordinates (x_1, y_1) and (x_2, y_2), then a displacement **PQ** can be expressed as $\mathbf{PQ} = (x_2 - x_1)\mathbf{i} + (y_2 - y_1)\mathbf{j}$, or as an **ordered pair,** $\mathbf{PQ} = (x_2 - x_1, y_2 - y_1)$.

In one-dimensional problems, where all vectors point in the positive or negative x direction for example, the ordered triple notation $(a_x, 0, 0)$ is often abbreviated simply to a_x. Thus **one-dimensional vectors** can be represented by positive or negative numbers.

The idea of resolving a vector into its components is an important milestone in the development of the subject. The three numerical components of a vector completely specify the vector in a given coordinate system. This allows us to move from a geometric picture of a vector based on arrows to a numerical one based on an ordered triple of numbers. To illustrate the change of view that this entails, consider what it means to say that two vectors are equal. In the geometric picture, two vectors are equal if the representative arrows have the same length and the same direction (Section 1.1). In the cartesian view, two vectors **a** and **b** are equal if they have the same cartesian components, i.e.

$$(a_x, a_y, a_z) = (b_x, b_y, b_z) \tag{1.15}$$

if $\quad a_x = b_x,\quad a_y = b_y,\quad a_z = b_z \qquad (1.16)$

Here we have used the notation introduced in Eqs (1.13) and (1.14) for the components of a vector. Eq (1.16) shows that a single vector equation **a** = **b** is equivalent to three scalar equations for the components.

In the next section we show how magnitudes and directions of vectors can be calculated from the components, and then in Section 1.7 we express the operations of scaling and vector addition in component form.

Summary of section 1.5

- A point P in space is specified by its cartesian coordinates (x,y,z) in a right-handed **cartesian coordinate system** in which the positive directions of the axes are given by the **right-hand rule**.

- Three unit vectors **i**, **j** and **k** directed parallel to the positive directions of the x, y and z axes respectively are called **cartesian unit vectors**.

- The **position vector r** of a point P(x,y,z) is the displacement from the origin, $\mathbf{r} = x\mathbf{i} + y\mathbf{j} + z\mathbf{k}$. The numbers x, y and z are the **cartesian components** of **r**. A vector expressed in this way is called a **cartesian vector**. The vectors $x\mathbf{i}$, etc. are called the **cartesian component vectors** or **resolutes** of **r**.

- When a vector has beginning point P(x_1,y_1,z_1) and end point Q(x_2,y_2,z_2), its cartesian components are the numbers $x_2 - x_1$, $y_2 - y_1$ and $z_2 - z_1$, i.e. $\mathbf{PQ} = (x_2 - x_1)\mathbf{i} + (y_2 - y_1)\mathbf{j} + (z_2 - z_1)\mathbf{k}$.

- A vector is specified completely by its cartesian components and can be written as an **ordered triple of numbers**, i.e. $\mathbf{PQ} = (x_2 - x_1, y_2 - y_1, z_2 - z_1)$. Vectors restricted to a plane surface can be represented by **ordered pairs** of numbers.

- The x-component of a vector **PQ** can be denoted by $(\mathbf{PQ})_x$, etc. The x-component of a vector **a** can be denoted by $(\mathbf{a})_x$ or more commonly by a_x, etc. Thus $\mathbf{a} = a_x\mathbf{i} + a_y\mathbf{j} + a_z\mathbf{k} = (a_x, a_y, a_z)$.

- If two vectors are equal then they have the same cartesian components.

Example 5.1 (*Objectives 7,8*) Point P(1,2,0) has position vector **r** and point Q(0,1,3) has position vector **s**. Sketch the cartesian coordinate system showing the cartesian unit vectors and the position vectors **r** and **s**. Express **r** and **s** as linear combinations of the cartesian unit vectors, and specify **r** and **s** as ordered triples.

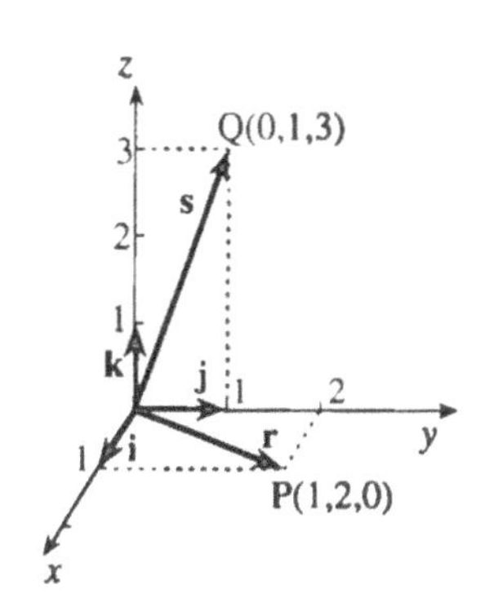

Fig 1.36
Vectors **r** and **s** in Solution 5.1.

Solution 5.1 The vectors and the coordinate system are shown in Fig 1.36. The position vector **r** can be expressed as $\mathbf{r} = \mathbf{i} + 2\mathbf{j} + 0\mathbf{k}$, and so **r** can be represented by the ordered triple (1,2,0). Similarly $\mathbf{s} = 0\mathbf{i} + \mathbf{j} + 3\mathbf{k} = (0,1,3)$.

Example 5.2 (*Objectives 1-11*) Consider the displacement **BA** of the point A(3,–6,–2) from the point B(6,9,–1). Determine the *x*-, *y*- and *z*-components of **BA** and write down the ordered triple representing **BA**.

Solution 5.2 Use Eq (1.11). The *x*-component of **BA** is $(\mathbf{BA})_x = 3 - 6 = -3$, the *y*-component is $(\mathbf{BA})_y = -6 - 9 = -15$ and $(\mathbf{BA})_z = -2 - (-1) = -1$. The ordered triple representation is therefore $\mathbf{BA} = (-3, -15, -1)$.

Example 5.3 (*Objectives 7,8*) Fig 1.37 shows some vectors drawn as arrows in the *x-y* plane.

(a) Express the vectors **c**, **d**, **u** and **v** as linear combinations of the unit vectors **i** and **j**.
(b) Specify the components c_x and d_y.
(c) Write down the ordered pairs representing **c**, **d**, **u**, **v**.

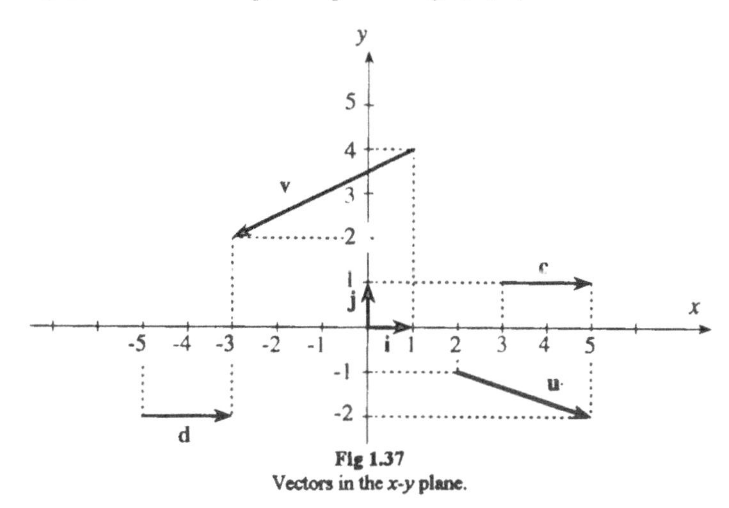

Fig 1.37
Vectors in the *x-y* plane.

Solution 5.3

(a) From inspection of Fig 1.37 we have $\mathbf{c} = 2\mathbf{i}$, $\mathbf{d} = 2\mathbf{i}$, $\mathbf{u} = 3\mathbf{i} - \mathbf{j}$, $\mathbf{v} = -4\mathbf{i} - 2\mathbf{j}$.

(b) c_x, the *x*-component of vector **c**, is seen to be $c_x = 2$. Similarly, $d_y = 0$.

(c) From the answers to part (a) it can be seen that the x and y components of vector **c** are 2 and 0 respectively. Thus **c** = (2,0). Similarly, **d** = (2,0), **u** = (3,–1), **v** = (–4,–2).

Example 5.4 (*Objective 9*) Given that

$$\mathbf{a} = 5\mathbf{i} - 3\mathbf{j} \quad \text{and} \quad \mathbf{b} = (\alpha - \beta)\mathbf{i} - (\alpha + \beta)\mathbf{j} - \gamma\mathbf{k},$$

and that **a** = **b**, determine the scalar constants α, β and γ.

Solution 5.4 If two vectors are equal then they have the same cartesian components (Eq (1.16)). Hence we have the three scalar equations

$$5 = \alpha - \beta, \quad -3 = -(\alpha + \beta) \quad \text{and} \quad 0 = -\gamma.$$

The first two equations give $\alpha = 4$ and $\beta = -1$. The third equation gives $\gamma = 0$.

Problem 5.1 (*Objectives 7,8*) Consider the vector **ST** = (9,–1,–7).

(a) Determine the coordinates of point T when the beginning point S is
(i) the origin,
(ii) the point (–6,0,2).

(b) Express **ST** in terms of the cartesian unit vectors.

Problem 5.2 (*Objective 8*) Express the cartesian unit vectors and the zero vector as ordered triples.

Problem 5.3 (*Objective 7*) What condition has to be satisfied if the cartesian coordinates of the end point of a vector are to be equal to the cartesian components of the vector?

Problem 5.4 (*Objective 9*) Given that $\mathbf{P} = (\alpha + \beta + 2)\mathbf{i} + (\alpha - \beta - 1)\mathbf{j} + (\alpha + 2\beta - \gamma)\mathbf{k}$ and that $\mathbf{P} = \mathbf{0}$, determine the scalars α, β and γ.

1.6 MAGNITUDES AND DIRECTIONS OF CARTESIAN VECTORS

The positive square root of a positive number A is denoted by $A^{1/2}$ or $\sqrt{A}$. Thus Eq 1.17(a) can be written as $|\mathbf{OP}| = |\mathbf{r}| = \sqrt{x^2 + y^2 + z^2}$. We shall generally use the *index notation* $A^{1/2}$ rather than $\sqrt{A}$.

The magnitude of a vector can be calculated from its cartesian components by using Pythagoras's theorem. Fig 1.38 shows the position vector $\mathbf{r} = \mathbf{OP}$ of the point P(x,y,z). It can be seen from the right-angled triangle OAN in the x-y plane that $\text{ON} = (x^2 + y^2)^{1/2}$. It then follows from the right-angled triangle ONP that the magnitude of **OP** is

$$|\mathbf{OP}| = |\mathbf{r}| = (x^2 + y^2 + z^2)^{1/2} \tag{1.17a}$$

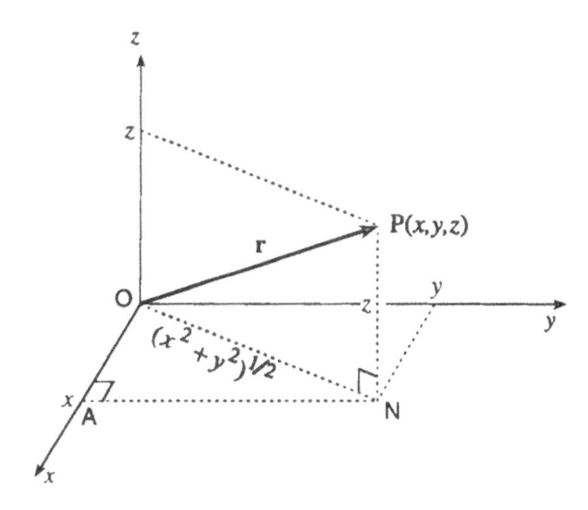

Fig 1.38
The magnitude of vector $\mathbf{r} = \mathrm{OP}$ is $(x^2 + y^2 + z^2)^{1/2}$.

More generally, when the vector is drawn from an arbitrary point $\mathrm{P}(x_1,y_1,z_1)$ to point $\mathrm{Q}(x_2,y_2,z_2)$, the magnitude is given by

$$|\mathbf{PQ}| = |(x_2 - x_1, y_2 - y_1, z_2 - z_1)|$$

$$= [(x_2 - x_1)^2 + (y_2 - y_1)^2 + (z_2 - z_1)^2]^{1/2} \qquad (1.17b)$$

Using the notation for components introduced in Section 1.5 we can write the **magnitude of a cartesian vector a** as

$$|\mathbf{a}| = (a_x^2 + a_y^2 + a_z^2)^{1/2} \qquad (1.17c)$$

Eqs (1.17a), (b) and (c) are equivalent formulae for the magnitude of a cartesian vector. Eq (1.17a) is useful when the vector has its beginning point at the origin; otherwise Eq (1.17b) or (1.17c) should be used.

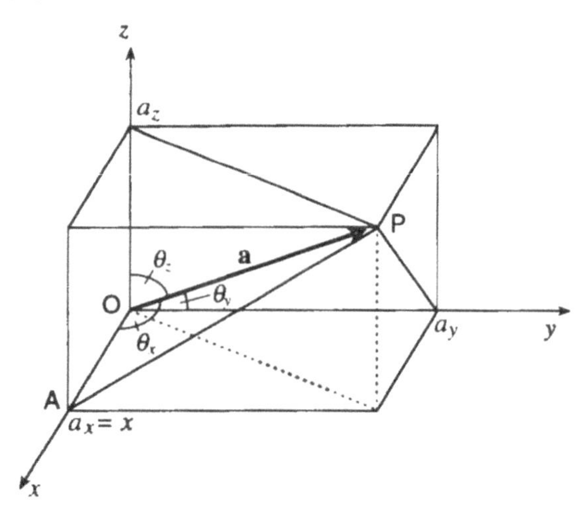

Fig 1.39
The angles between the direction of vector **a** and the coordinate axes.

The angles between the direction of a vector and the positive directions of the coordinate axes are found from simple trigonometry. Let the vector **a** be drawn from the origin (Fig 1.39). Triangle OAP is a right-angled triangle with the right-angle at point A at a distance $a_x = x$ from O. You can see that the angle θ_x between the vector and the positive direction of the x-axis is given by $\cos\theta_x = a_x/|\mathbf{a}|$, and similarly for the angles θ_y and θ_z between the vector and the y- and z-axes. Thus we have

$$\cos\theta_x = \frac{a_x}{|\mathbf{a}|}, \quad \cos\theta_y = \frac{a_y}{|\mathbf{a}|}, \quad \cos\theta_z = \frac{a_z}{|\mathbf{a}|} \tag{1.18a}$$

By convention, the angles θ_x etc. are in the range from 0 to 180°; they are acute angles when the cosines and the corresponding components are positive, and obtuse angles when the cosines and the components are negative. The three angles or the three cosines of Eq (1.18a) completely specify the direction of the vector. The cosines are called **direction cosines**.

The direction cosines are commonly denoted by l, m and n. Thus

$$l = \cos\theta_x$$
$$m = \cos\theta_y$$
$$n = \cos\theta_z$$

Eqs (1.18a) are also useful for the inverse problem of finding the components of a vector when the magnitude and direction are given. Thus we write Eqs (1.18a) as

$$a_x = |\mathbf{a}|\cos\theta_x, \quad a_y = |\mathbf{a}|\cos\theta_y, \quad a_z = |\mathbf{a}|\cos\theta_z \tag{1.18b}$$

Given any two cartesian vectors you could in principle work out the angle between them using trigonometry, but this could be quite difficult. A general way of calculating the magnitudes and directions of cartesian vectors is given in Chapter 2 after the scalar product of two vectors has been introduced.

Summary of section 1.6

- The **magnitude of a cartesian vector a** is

$$|\mathbf{a}| = (a_x^2 + a_y^2 + a_z^2)^{1/2} = \sqrt{x^2 + y^2 + z^2}$$

and its direction is specified by the **direction cosines**

$$\cos\theta_x = \frac{a_x}{|\mathbf{a}|}, \quad \text{etc.}$$

The angles θ_x, etc. are in the range from 0° to 180°.

- A given vector can be resolved into its cartesian components by using

$$a_x = |\mathbf{a}| \cos\theta_x, \text{ etc.}$$

Example 6.1 (*Objective 10*) Consider the vectors **r** and **s** of Example 5.1 and Fig 1.36. Determine the magnitude of each vector and its direction relative to the coordinate axes.

Solution 6.1 Use Eq (1.17a) for the magnitudes. Thus

$$|\mathbf{r}| = (1^2 + 2^2 + 0^2)^{1/2} = 5^{1/2} = 2.236$$

The three angles specifying the direction or **r** are found from the direction cosines (Eq (1.18a)):

$$\cos\theta_x = \frac{1}{5^{1/2}}, \quad \cos\theta_y = \frac{2}{5^{1/2}}, \quad \cos\theta_z = \frac{0}{5^{1/2}}$$

Thus $\theta_x = 63.4°$, $\theta_y = 26.6°$, $\theta_z = 90°$.

For vector **s**,

$$|\mathbf{s}| = (0^2 + 1^2 + 3^2)^{1/2} = 10^{1/2} = 3.16$$

The direction of **s** is specified by the direction cosines (Eq (1.18a)):

$$\cos\theta_x = \frac{0}{10^{1/2}}, \quad \cos\theta_y = \frac{1}{10^{1/2}}, \quad \cos\theta_z = \frac{3}{10^{1/2}}$$

Thus $\theta_x = 90°$, $\theta_y = 71.6°$, $\theta_z = 18.4°$.

Example 6.2 (*Objective 10*) A vector **F** has magnitude 5 and lies in the x-y plane at an angle of 30° from the x-axis, as shown in Fig 1.40. Determine the x- and y-components of **F**.

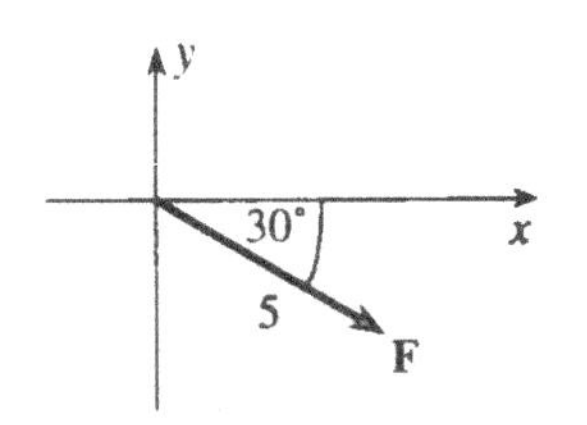

Fig 1.40
The vector **F** for Example 6.2.

Solution 6.2 Refer to Fig 1.40 and use Eq (1.18b). The components are:

$$F_x = 5\cos 30° = 5 \times \frac{3^{1/2}}{2} = 4.33$$

$$F_y = 5\cos(90° + 30°) = -2.5$$

Problem 6.1 (*Objective 10*) Determine the magnitude and direction of the vector **p** = (–5,3,–4).

Problem 6.2 (*Objective 10*) Show that the sums of the squares of the direction cosines are equal to 1.

Problem 6.3 (*Objectives 7,8,10*) Consider the vector **P** = –5**i** – **j** + 2**k**. Select the statements that are incorrect or meaningless.

A $|\mathbf{P}| = 5.477$

B $P_z = 2$

C $\mathbf{P} = (-5,-1,2)$

D $|\mathbf{P}| = P_x + P_y + P_z$

E The direction of **P** makes an acute angle with the x-axis

F $\mathbf{P}/|\mathbf{P}|$ is parallel to **P**

G $|1/\mathbf{P}| = 0.1826$

Problem 6.4 (*Objectives 7,10*) The cartesian coordinates of points P and Q are (2,1,0) and (–1,0,3) respectively. Let **a** be the position vector of P and **b** that of Q.

(a) Give the values of a_x and b_z.
(b) Calculate $|\mathbf{b}| - |\mathbf{a}|$.
(c) Determine the angle between the direction of **a** – **b** and the x-axis.

Problem 6.5 (*Objectives 7,8,10*) Points S and T have coordinates (–1,0,0) and (0,–7,–1) respectively.

(a) Write down ordered triples representing the position vectors of points S and T, and the vector **ST**.

(b) State the values of $(\mathbf{ST})_x$, $(\mathbf{ST})_y$ and $(\mathbf{ST})_z$, and determine $|\mathbf{ST}|$.

1.7 SCALING AND ADDING CARTESIAN VECTORS

In Sections 1.2 and 1.3, the operations of scaling and vector addition were defined geometrically in terms of arrows. We now show how these operations can be expressed in cartesian form.

Consider a cartesian vector $\mathbf{a} = a_x\mathbf{i} + a_y\mathbf{j} + a_z\mathbf{k}$. Scaling $\mathbf{a}$ by a number α gives the vector

$$\alpha\mathbf{a} = \alpha(a_x\mathbf{i} + a_y\mathbf{j} + a_z\mathbf{k}) = \alpha a_x\mathbf{i} + \alpha a_y\mathbf{j} + \alpha a_z\mathbf{k} \qquad (1.19a)$$

The removal of brackets here is justified by the distributive rule (Section 1.3).

Thus when a vector is scaled by a number α, each component is multiplied by α. Expressing the vectors as ordered triples we have the equivalent statement

$$\alpha\mathbf{a} = \alpha(a_x, a_y, a_z) = (\alpha a_x, \alpha a_y, \alpha a_z) \qquad (1.19b)$$

Now consider vector addition. For two vectors $\mathbf{a}$ and $\mathbf{b}$ we have

$$\begin{aligned}\mathbf{a} + \mathbf{b} &= (a_x\mathbf{i} + a_y\mathbf{j} + a_z\mathbf{k}) + (b_x\mathbf{i} + b_y\mathbf{j} + b_z\mathbf{k}) \\ &= (a_x + b_x)\mathbf{i} + (a_y + b_y)\mathbf{j} + (a_z + b_z)\mathbf{k} \qquad (1.20a)\end{aligned}$$

The regrouping and reordering of terms is justified by the associative and commutative rules of vector addition (Section 1.3).

Thus the addition of two vectors is accomplished simply by adding the corresponding components. Eq (1.20a) can be written as

$$\mathbf{a} + \mathbf{b} = (a_x + b_x, a_y + b_y, a_z + b_z) \qquad (1.20b)$$

The vector addition law expressed by Eqs (1.20a and 1.20b) is entirely equivalent to triangle addition.

The addition law is readily generalised to the addition of any number of vectors in an obvious way. Thus the *x*-component of the sum of any number of vectors is the sum of the *x*-components, etc.

Summary of section 1.7

- The scaling of a cartesian vector $\mathbf{a}$ by a number α gives

$$\begin{aligned}\alpha\mathbf{a} &= \alpha a_x\mathbf{i} + \alpha a_y\mathbf{j} + \alpha a_z\mathbf{k} \\ &= (\alpha a_x, \alpha a_y, \alpha a_z)\end{aligned}$$

- The addition of two cartesian vectors **a** and **b** gives the resultant

$$\mathbf{a}+\mathbf{b}=(a_x+b_x)\mathbf{i}+(a_y+b_y)\mathbf{j}+(a_z+b_z)\mathbf{k}$$

$$=(a_x+b_x, a_y+b_y, a_z+b_z)$$

Example 7.1 (*Objectives 7,8,11*) Consider the three vectors $\mathbf{u}=(2,-1,0)$, $\mathbf{v}=(-3,-1,3)$ and $\mathbf{w}=(2,2,-1)$.

(a) Write down ordered triples representing $3\mathbf{u}$, $\mathbf{u}+\mathbf{v}$, $3\mathbf{u}-\mathbf{v}$, $\mathbf{h}=(1/2)\mathbf{u}-(3/2)\mathbf{w}$ and $\mathbf{l}=\mathbf{u}-\mathbf{i}$.

(b) Evaluate $|\mathbf{l}|$, $|\mathbf{u}-\mathbf{v}|$, $(\mathbf{u}+\mathbf{w})_z$.

(c) Specify a unit vector in the direction of **v**.

Solution 7.1

(a) Use Eqs (1.19b) and (1.20b).

$$3\mathbf{u}=(6,-3,0)$$

$$\mathbf{u}+\mathbf{v}=(2+(-3),\ -1+(-1),\ 0+3)=(-1,-2,3)$$

$$3\mathbf{u}-\mathbf{v}=(6-(-3),\ -3-(-1),\ 0-3)=(9,-2,-3)$$

$$\mathbf{h}=\tfrac{1}{2}(2,-1,0)-\tfrac{3}{2}(2,2,-1)$$

$$=\left(1-3,\ -\tfrac{1}{2}-3,\ 0+\tfrac{3}{2}\right)=\left(-2,-\tfrac{7}{2},\tfrac{3}{2}\right)$$

$$\mathbf{l}=(2,-1,0)-(1,0,0)=(1,-1,0)$$

(b) $|\mathbf{l}|=(1^2+(-1)^2+0^2)^{1/2}=2^{1/2}=1.414$

$$|\mathbf{u}-\mathbf{v}|=|(2-(-3),-1-(-1),0-3)|$$

$$=|(5,0,-3)|=(5^2+0^2+(-3)^2)^{1/2}=34^{1/2}=5.831$$

$$(\mathbf{u}+\mathbf{w})_z=u_z+w_z=0+(-1)=-1$$

(c) The unit vector is

$$\hat{\mathbf{v}} = \frac{\mathbf{v}}{|\mathbf{v}|} = \frac{(-3,-1,3)}{[(-3)^2 + (-1)^2 + 3^2]^{1/2}}$$

$$= \frac{(-3,-1,3)}{19^{1/2}} = 0.2294(-3,-1,3)\ .$$

Example 7.2 (*Objective 11*) Find the distance from the origin to the midpoint of the line joining A(3,–5,7) and B(3,–4,2).

Solution 7.2 The position vector of the **midpoint** of the line AB is $\mathbf{OM} = \frac{1}{2}(\mathbf{OA} + \mathbf{OB})$ where **OA** and **OB** are the position vectors of points A and B. (See Example 4.6.) Therefore

$$\mathbf{OM} = \frac{1}{2}[(3,-5,7) + (3,-4,2)] = \frac{1}{2}(6,-9,9)\ .$$

The required distance is the magnitude of this vector:

$$|\mathbf{OM}| = \tfrac{1}{2}[6^2 + (-9)^2 + 9^2]^{1/2} = \tfrac{1}{2} \times 198^{1/2} = 7.046\ .$$

Problem 7.1 (*Objectives 7,10,11*) Consider the three vectors **u**, **v** and **w** of Example 7.1.

(a) Express **u**, **v** and **w** explicitly in terms of the cartesian unit vectors.

(b) Write down the vector $\mathbf{q} = \mathbf{u} - \mathbf{w} - \mathbf{k}$ in terms of the cartesian unit vectors.

(c) Evaluate $|\mathbf{q}|$ and write down the unit vector $\hat{\mathbf{q}}$.

(d) What is the angle between the unit vector $\hat{\mathbf{q}}$ and the cartesian unit vector **k**?

Problem 7.2 (*Objectives 7,8,10,11*) Given two vectors

$$\mathbf{r} = 3\mathbf{i} + 2\mathbf{j} - \mathbf{k},\quad \mathbf{s} = -2\mathbf{i} + 5\mathbf{j} + 7\mathbf{k}$$

(a) Determine $\mathbf{r} + \mathbf{s}$, $\mathbf{r} - \mathbf{s}$ and $\mathbf{r} + \mathbf{j} + \mathbf{k}$, and give your answers as ordered triples.

(b) Determine r_x, s_z, $(\mathbf{r} - \mathbf{s})_y$ and $|\mathbf{r} + \mathbf{s}|$. Find the number λ such that $\lambda\mathbf{s}$ is a unit vector.

(c) Give a geometrical interpretation of the statement $|\mathbf{r} + \mathbf{s}| < |\mathbf{r}| + |\mathbf{s}|$.

Problem 7.3 (*Objectives 7,10,11*) Let OAPB be a parallelogram where O is the origin point and A and B are the points specified in Example 7.2. Determine the position vector of the point P and the direction this vector makes with the z-axis.

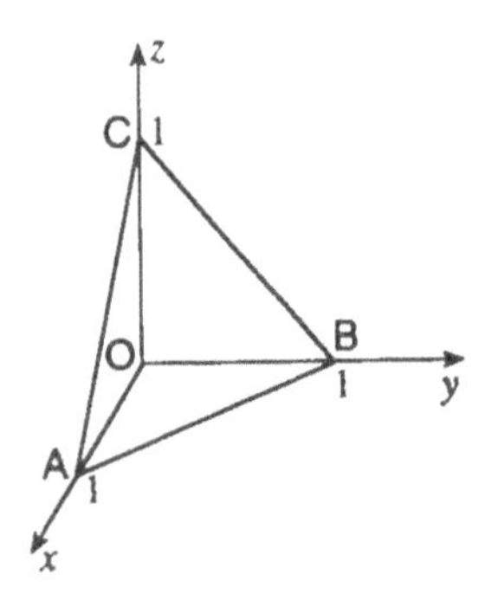

Fig 1.41
The tetrahedron for Problem 7.4.

Problem 7.4 (*Objectives 7,11*) Fig 1.41 shows a tetrahedron with apices (i.e. corners) at the four points O, A, B and C. Given that **OA** = **i**, **OB** = **j** and **OC** = **k**, determine:

(a) the displacement **CA**, and the position vector of the mid-point M of the line CA,

(b) the centroid of the four points O, A, B and C. (The definition of centroid is given in Problem 4.10.)

1.8 VECTORS IN SCIENCE AND ENGINEERING

We have based the properties of vectors on the displacement vector. We now turn to science and engineering where the scaling and adding of vectors is applied to problems involving other vector quantities such as forces, accelerations and velocities, as well as displacements. These other vector quantities obey the same scaling and addition laws as displacements. In this section we look at some of the evidence supporting this assertion. This is followed by Examples and Problems on the scaling and adding of forces, velocities and other vector quantities.

1.8.1 Definition of a vector and evidence for vector behaviour

All vectors obey the laws that we have established for displacements; in particular, vectors obey the triangle addition rule. The triangle rule is in fact taken to be the defining feature of a vector and we have the following **definition of a vector**:

> A vector is any quantity that has a magnitude and a direction in space and combines according to the triangle addition rule.

This definition provides us with a practical criterion for deciding whether or not a particular physical quantity is a vector. We now apply the criterion to two physical quantities – force and velocity.

Force as a vector

Force obviously has magnitude, measured in newtons (N), and a direction in space – the direction in which the force is pushing or pulling. A force can therefore be represented in a diagram by an arrow pointing in the direction of the force and of length proportional to the magnitude of the force. To show that force is in fact a vector we must demonstrate that two forces applied simultaneously at a point are equivalent to a single force equal to the resultant of the two as given by the triangle addition rule. In practice it is easier to apply three forces $\mathbf{F}_1$, $\mathbf{F}_2$ and $\mathbf{F}_3$ at the same point O and to adjust one of them, $\mathbf{F}_3$ say, until they balance. An apparatus for doing this is sketched in Fig 1.42a. The magnitudes and directions of $\mathbf{F}_1$ and $\mathbf{F}_2$ can be set by adding weights to the scale pans P_1 and P_2 and choosing the angles α and β. The magnitude of the force $\mathbf{F}_3$, acting vertically downwards, is then adjusted by adding weights to scale pan P_3 until the system balances when released. The three balancing forces are said to be in **equilibrium** and are together equivalent to a zero force. The vector nature of force is confirmed if, in all arrangements that give equilibrium, the arrows representing the three forces form a triangle (Fig 1.42b) expressing the vector equation

$$\mathbf{F}_1 + \mathbf{F}_2 + \mathbf{F}_3 = \mathbf{0} \tag{1.21}$$

Confidence in the vector nature of force is confirmed directly in experiments of this kind and more compellingly by the achievements of engineers – bridges stay up, courtesy of the triangle rule!

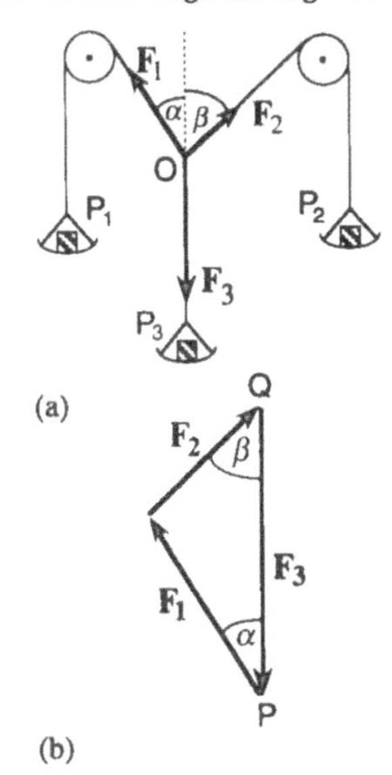

Fig 1.42
(a) An apparatus for investigating how forces add. Scale pans P_1 and P_2 are attatched to light strings that pass over frictionless pulleys. The three forces act at point O.
(b) When the three forces are in equilibrium the representative arrows form a triangle.

Velocity as a vector

Here we restrict our attention to constant velocities, i.e. velocities that do not vary in time. Velocity has a magnitude which we call speed, measured in metres per second (m s^{-1}), and a direction in space given by the direction of motion. A velocity can therefore be represented in a diagram by an arrow showing the direction of motion and of length proportional to the speed.

The need to add velocities comes about when we consider the velocity of a body as measured by two observers who are themselves in relative motion. Fig 1.43(a) depicts a coastguard labelled O, a passenger ferry labelled O′ and an oil tanker T. The coastguard is a stationary shore-bound observer who measures the velocities of the two ships using radar and obtains velocities **u** and **v** for the ferry and tanker respectively. The ferry captain is also interested in the tanker's movements. His ship-bound radar indicates the tanker's velocity to be **v′**; this is the velocity of the tanker relative to the ferry. How are **v** and **v′** related? From the point of view of the coastguard O, the measurement **v′** obtained by the ferry captain fails to take account of the ferry's own velocity **u** which should therefore be added to **v′** to agree with his measurement

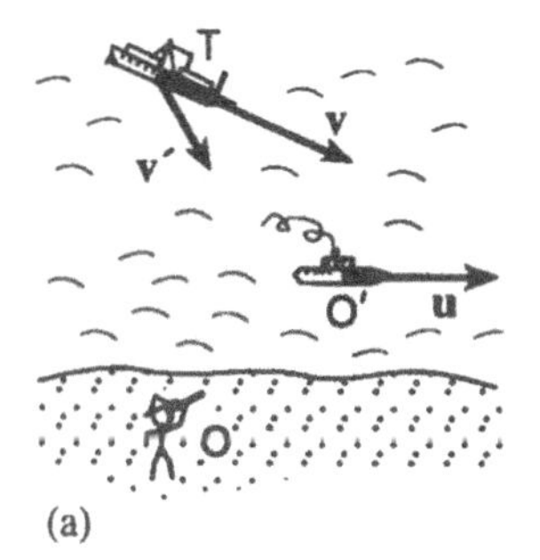

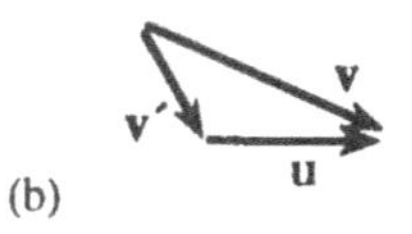

Fig 1.43
(a) A coastguard O and a ferry captain O′ both measure the velocity of the tanker T.
(b) $\mathbf{v} = \mathbf{v}' + \mathbf{u}$.

Eq (1.22) will be derived more generally in Chapter 3.

$$\mathbf{v} = \mathbf{v}' + \mathbf{u} \tag{1.22}$$

The vector nature of velocity is expressed by the fact that in all such measurements the three arrows representing the three velocities always form a triangle (Fig 1.43b).

Similar problems arise in air traffic control where the controller in the airport control tower (Observer O) and the navigator in an airliner (Observer O′) both make radar measurements of the velocities of other aircraft in the vicinity. Their measurements are related by the triangle law.

Problems involving the addition of velocities are called **relative velocity** problems. We can base a systematic approach to such problems on the example considered above. We first identify two observers: a stationary observer O and a observer O′ moving with velocity **u** relative to O. A third body is observed to move with velocity **v** relative to O and **v′** relative to O′. The three velocities are then related by the triangle addition law, Eq (1.22).

The vector nature of forces and velocities, and many of the other vector quantities that we shall discuss in later chapters, has been confirmed to great accuracy in many measurements over a wide range of magnitudes. In fact the whole of Newtonian mechanics and mechanical engineering rest on the vector nature of force, displacement, velocity, momentum, torque, acceleration and many other vector quantities. There are, however, discrepancies from the triangle addition law in extreme circumstances. You will probably know that there is an upper limit to the speed of any material particle. This limiting speed is the speed of electromagnetic waves in free space, $c = 2.998 \times 10^{8}\ \mathrm{m\,s^{-1}}$, usually referred to as the **speed of light**. However it is obvious from the preceding sections that there is no limit to the magnitudes that can be achieved mathematically by adding vectors. It follows therefore that when speeds approach the speed of light, velocities can no longer add by the triangle rule, and so velocity is not a vector quantity under these conditions. Such extreme speeds are well beyond those encountered in everyday applications. For an airliner flying at a speed of $v = 200\ \mathrm{m\,s^{-1}}$ (about 450 miles per hour), the ratio v/c is less than 10^{-6}; and even for a space probe travelling at say $10\ \mathrm{km\,s^{-1}}$, v/c is less than 10^{-4} Actually the discrepancies from the triangle addition of velocities, Eq 1.22, depend on the magnitude of the quantity $v'u/c^2$ which is normally extremely small, and so we can safely regard velocity as a vector quantity in everyday engineering applications. This isn't the case in all applications however. In atomic physics and in elementary particle physics, speeds approaching the speed of light are actually quite common. Some cosmic ray protons, for example, enter the Earth's atmosphere with speeds in excess of $2.9 \times 10^{8}\ \mathrm{m\,s^{-1}}$ and only very slightly less than the speed of light. In such circumstances velocities no longer behave as ordinary vectors. Similar remarks can be made about displacement, force, acceleration etc. However we shall always assume that normal conditions prevail and that these quantities are vectors. We shall introduce many vector quantities in this book but we shall not always indicate the experimental evidence for triangle addition.

1.8.2 Vector problems in science and engineering

The Examples and Problems in this section are designed to give practice in applying the algebra of scaling and vector addition to problems involving forces, velocities and accelerations, as well as displacements. No additional vector analysis is needed to solve these problems but additional skills of a different kind are involved. For example, the problems to be solved and the physical laws used to solve them have to be expressed in the language of scalars and vectors. Also a consistent set of units of measurement has to be used. The SI system of units (see Appendix A) is used throughout this book. Although it is good practice to show the physical units at every stage of a calculation, it is sometimes convenient to suppress them, especially when algebraic manipulations are being carried out. Vigilance is needed however, especially when factor-of-ten prefixes are used in the specification of units (as in cm, mm, etc.). An answer that is wrong by just one factor of 10 is completely wrong and could lead to disastrous results if put into practice!

When a sketch or diagram is asked for in a problem, it is sufficient to give a freehand sketch labelled with all relevant information. Even when such a sketch is not asked for, it is often a good idea to make one for your own reference to help visualise the important aspects of the problem and to assist in a "check" on the credibility of your answer.

Summary of section 1.8

- The **definition of a vector** is based on the properties of displacement: A vector is any quantity that has magnitude and direction in space and obeys the triangle addition rule.

- Experiments show that force, velocity and many other quantities having magnitude and direction in space, are vectors under normal conditions.

Example 8.1 *(Objectives 3, 9, 10)* Fig 1.44 shows three forces in equilibrium, all in the x-y plane.

(a) Determine F_{1x} and F_{3x}. Hence deduce the value of F_{2x}.

(b) Determine F_{1y} and F_{3y}. Hence deduce the value of F_{2y}.

(c) Determine $|\mathbf{F}_2|$ and the angle α.

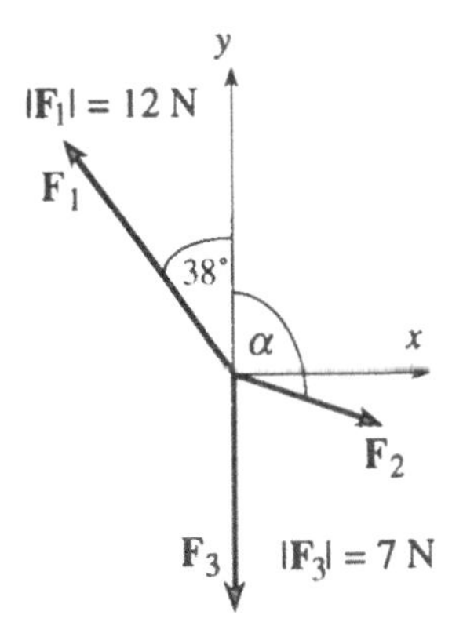

Fig 1.44
Three forces in equilibrium.

Solution 8.1

(a) The angle between $\mathbf{F}_1$ and the positive x-axis is seen from Fig 1.44 to be $\theta_{1x} = 90° + 38°$. Thus using the first of Eqs (1.18b) we obtain

$$F_{1x} = (12 \text{ N})\cos(128°) = -7.388 \text{ N}$$

Similarly

$$F_{3x} = |\mathbf{F}_3|\cos(90°) = 0$$

We deduce F_{2x} from the equilibrium condition, Eq (1.21). This vector equation is equivalent to two scalar equations for the components

$$F_{1x} + F_{2x} + F_{3x} = 0; \qquad F_{1y} + F_{2y} + F_{3y} = 0 \qquad (1.23)$$

The two scalar equations (1.23) express the fact that when the forces are in equilibrium the x and y components of the forces are separately in equilibrium. This greatly simplifies the problem because we can forget about the y components when considering the equilibrium of the x components, and vice versa.

We use the first of Eqs (1.23) with the values of F_{1x} and F_{3x} found above. Thus

$$-7.388 \text{ N} + 0 + F_{2x} = 0$$

Hence $F_{2x} = 7.388$ N.

(b) Using Eqs (1.18b) we obtain $F_{1y} = (12 \text{ N})\cos(38°) = 9.456$ N; and $F_{3y} = (7 \text{ N})\cos(180°) = -7$ N. We now use the second of Eqs (1.23) to give

$$9.456 \text{ N} + F_{2y} - 7 \text{ N} = 0$$

Hence $F_{2y} = -2.456$ N.

(c) $|\mathbf{F}_2| = (F_{2x}^2 + F_{2y}^2)^{1/2} = (7.388^2 + (2.456)^2)^{1/2} \text{ N} = 7.786 \text{ N}$.

Now use the second of Eqs (1.18a) to obtain

$$\cos\alpha = \frac{F_{2y}}{|\mathbf{F}_2|} = -\frac{2.456 \text{ N}}{7.786 \text{ N}} = -0.3154$$

Therefore $\alpha = 108.4°$.

Example 8.2 (*Objectives 3,4*) A motorboat is to cross a river from a point A on one bank to a point B directly opposite A on the other bank. The motorboat moves relative to the water with a speed of 2.5 m s^{-1}. The river is 500 m wide and a steady current of 0.5 m s^{-1} flows. Give a sketch showing how the velocity $\mathbf{v}$ of the boat relative to the bank can be calculated and determine $|\mathbf{v}|$. Show also a similar sketch for the return journey from B to A and calculate the total time for the round trip.

Solution 8.2 This is a relative velocity problem and so we use Eq (1.22). The crucial first step is to identify the two observers. The stationary observer O is obviously the observer on the river bank according to whom the motor boat moves with velocity $\mathbf{v}$ straight across the river. The question tells us that the motor boat moves with a velocity of magnitude $2.5\ \mathrm{m\,s^{-1}}$ relative to the water. The moving observer O′ is therefore the water itself, or if you like, an observer floating in the water and moving with the current.

The velocity $\mathbf{u}$ of O′ is of magnitude $|\mathbf{u}| = 0.5\ \mathrm{m\,s^{-1}}$ and is directed parallel to the river bank. The velocity $\mathbf{v}'$ of the motor boat relative to O′ (the water) has magnitude $|\mathbf{v}'| = 2.5\ \mathrm{m\,s^{-1}}$.

We now use Eq (1.22) which tells us that the velocity $\mathbf{v}$ of the motor boat relative to the bank is equal to the velocity $\mathbf{v}'$ relative to the water plus the velocity $\mathbf{u}$ of the water. Knowing the magnitude and direction of $\mathbf{u}$ and the magnitude of $\mathbf{v}'$, and knowing that $\mathbf{v}$ must point directly across the river from A to B, we can draw a sketch (Fig 1.45a) representing the triangle addition rule, Eq (1.22). The vectors $\mathbf{u}$, $\mathbf{v}$ and $\mathbf{v}'$ make a right-angled triangle and so we use Pythagoras's theorem to obtain

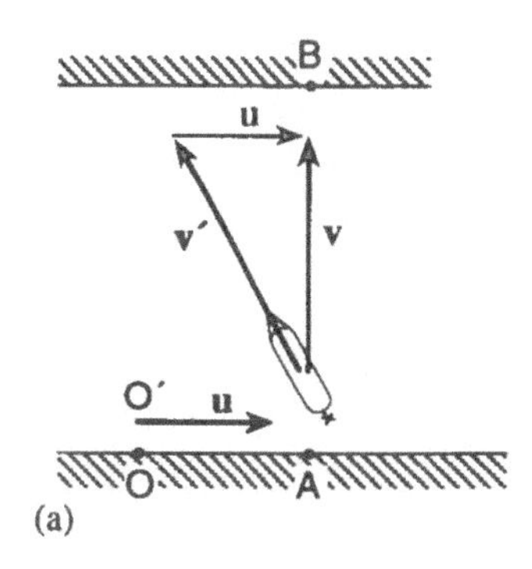

$$|\mathbf{v}| = (2.5^2 - 0.5^2)^{1/2}\ \mathrm{m\,s^{-1}} = 2.449\ \mathrm{m\,s^{-1}}$$

For the return journey the direction of $\mathbf{v}$ is reversed and the velocity addition is shown in Fig 1.45b. Thus $|\mathbf{v}|$ is the same as before and so the time for a round trip is

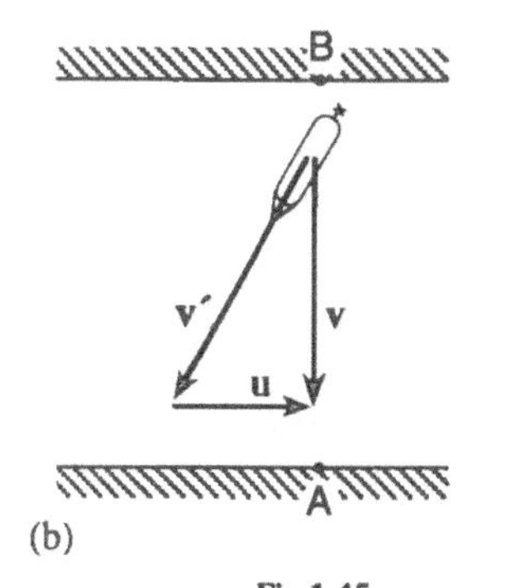

Fig 1.45
(a) The motorboat crosses the river from A to B.
(b) The return journey.

$$2\times\frac{500\ \mathrm{m}}{2.499\ \mathrm{ms^{-1}}} = 408\ \mathrm{s}$$

Example 8.3 (*Objectives 6,10*) The position vector $\mathbf{R}$ of the **centre of mass** of n particles of masses m_i at positions $\mathbf{r}_i$ ($i = 1, 2, \ldots, n$) is given by

$$M\mathbf{R} = m_1\mathbf{r}_1 + m_2\mathbf{r}_2 + \ldots m_i\mathbf{r}_i + \ldots m_n\mathbf{r}_n \qquad (1.24)$$

where $M = m_1 + m_2 + \ldots m_n$ is the total mass. Determine the position vector of the centre-of-mass of three particles: one of mass 2 kg at the origin O, one of mass 4 kg on the positive x-axis at a distance of 2 m from O and the third of mass 1kg on the positive y-axis at a distance of 4 m from O. How far is the centre of mass from the origin and what is its direction of the centre of mass from the origin relative to the direction of the x-axis?

Solution 8.3 The position vector $\mathbf{R}$ of the centre-of-mass is found by using Eq (1.24) with $\mathbf{r}_1 = \mathbf{0}$, $\mathbf{r}_2 = 2\mathbf{i}$ and $\mathbf{r}_3 = 4\mathbf{j}$:

$$(2+4+1)\mathbf{R} = 2(\mathbf{0}) + 4(2\mathbf{i}) + 1(4\mathbf{j})$$

which gives $\mathbf{R} = \frac{1}{7}(8\mathbf{i} + 4\mathbf{j})$. Hence the distance in metres from the origin is $|\mathbf{R}| = \frac{1}{7}(8^2 + 4^2)^{1/2} = 1.278$. The direction is given by the first of Eqs (1.18a):

$$\cos\theta_x = \frac{R_x}{|\mathbf{R}|} = \frac{8}{7} \times \frac{1}{1.278} = 0.8943$$

from which $\theta_x = 26.6°$. It is necessary to decide whether **R** is directed above or below the x-axis. We note that $R_y = 4/7$ which is positive. **R** is therefore directed above the x-axis by 26.6°.

Example 8.4 (*Objectives 12*) Coulomb's law of electrostatics states that the electrostatic force **F** acting on a point electric charge q at a distance d from another point electric charge Q has a magnitude

$$F = \frac{|qQ|}{4\pi\epsilon_0 d}$$

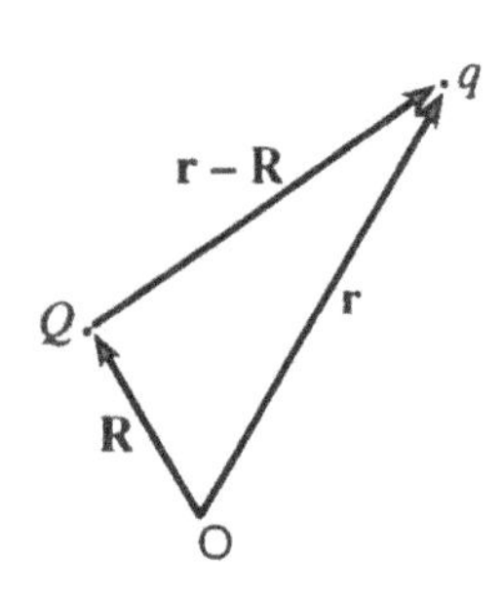

Fig 1.46
The position of charge q relative to charge Q is given by the displacement **r** − **R**.

where $4\pi\epsilon_0$ is a scalar constant (Appendix A). The direction of **F** is along the line joining the charges and is directed away from Q (repulsive) when the charges have the same sign (like charges) and towards Q (attractive) when they have opposite signs (unlike charges). Express the force vector **F** in terms of the position vectors **r** of q and **R** of Q.

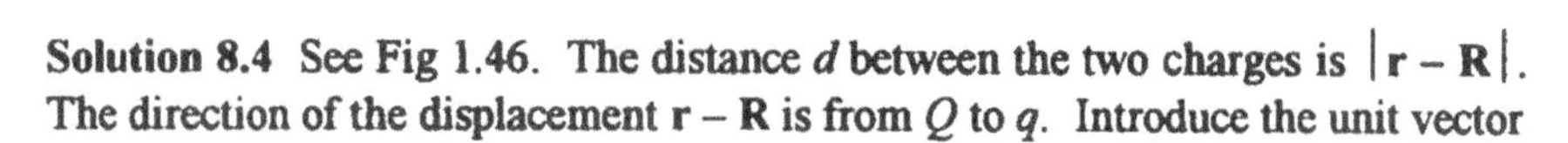

Solution 8.4 See Fig 1.46. The distance d between the two charges is $|\mathbf{r} - \mathbf{R}|$. The direction of the displacement **r** − **R** is from Q to q. Introduce the unit vector

$$\mathbf{u} = \frac{\mathbf{r} - \mathbf{R}}{|\mathbf{r} - \mathbf{R}|}$$

Then the vector quantity $qQ\mathbf{u}$ is directed from Q to q when q and Q have the same signs and from q to Q when q and Q have opposite signs, and so we can write

$$\mathbf{F} = \frac{qQ\mathbf{u}}{4\pi\epsilon_0 |\mathbf{r} - \mathbf{R}|^2} = \frac{qQ(\mathbf{r} - \mathbf{R})}{4\pi\epsilon_0 |\mathbf{r} - \mathbf{R}|^3} .$$

Problem 8.1 (*Objectives 3,8,10*) Find the resultant of the three forces $\mathbf{f} = (2,1,0)$ N, $\mathbf{s} = (1,-1,2)$ N and $\mathbf{t} = (-4,-2,2)$ N, and specify the unit vector in the direction of this resultant.

Problem 8.2 (*Objective 3*)

(a) Show that if a particle is held in equilibrium under the action of three non-collinear and non-zero forces then the forces must be coplanar.

(b) Is the statement in part (a) also true for four forces?

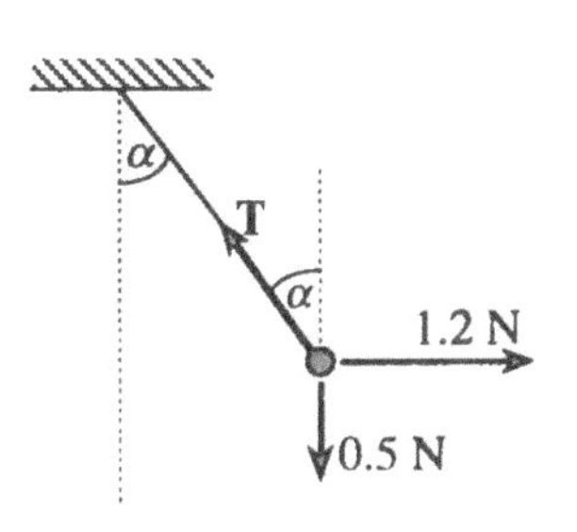

Fig 1.47
The pendulum bob is held in equilibrium by three forces.

Problem 8.3 (*Objectives 3,4*) The pendulum bob in Fig 1.47 is held in equilibrium by three forces: its weight of magnitude 0.5 N, a horizontally directed force of magnitude 1.2 N and the tension force **T** of the string. Determine the magnitude and the direction of **T**.

Problem 8.4 (*Objectives 3,4*) Determine the magnitude and direction of the force **F** in Fig 1.48 given that the four forces are in equilibrium.

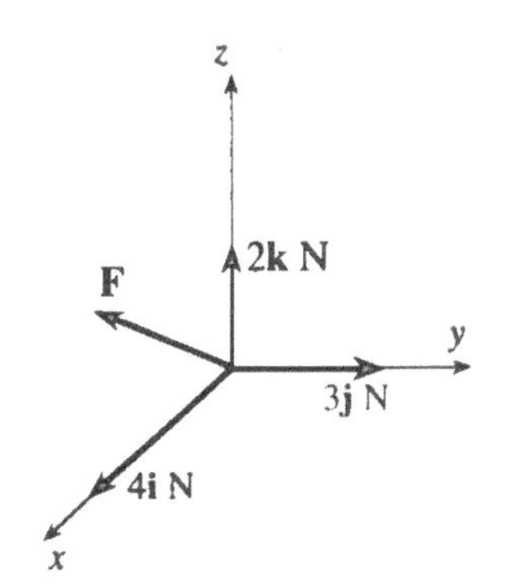

Fig 1.48
The four forces are in equilibrium.

Problem 8.5 (*Objectives 3,4*) ABCD is a square. Forces of magnitudes 1 N, 2 N and 3 N act parallel to **AB**, **BC** and **CD** respectively.

(a) Determine the magnitude and direction of the resultant force.

(b) Suppose the three forces are applied at a point. What is the magnitude and direction of the additional force that should be applied at that point to produce equilibrium?

Problem 8.6 (*Objectives 3,4*) Refer to Example 8.2. Suppose the boat starts at A and points directly towards the opposite bank.

(a) Determine the velocity of the boat relative to the river bank.

(b) Determine how far the boat is carried downstream from B when it reaches the other side.

Problem 8.7 (*Objectives 3,4*) A pilot wishes to fly from London to Birmingham, a distance of 185 km in a direction 45° west of north. His aeroplane has an air speed of 240 km per hour and the wind is blowing at 30 km per hour from the east.

(a) Sketch a velocity addition diagram showing the wind velocity **u**, the velocity **v**′ of the aeroplane relative to the air, and the velocity **v** of the aeroplane relative to the ground. Write down two equations expressing the relationship between

(i) the northerly components of these velocities,
(ii) the easterly components.

Use these two equations and the known direction of **v** to show that the pilot should follow a compass bearing of α degrees west of north where α satisfies the equation $\cos\alpha - \sin\alpha = 1/8$.

(b) Given that this equation is equivalent to $\sin(45° - \alpha°) = \sqrt{2}/16$, determine α.

Problem 8.8 (*Objectives 4,6*) Given that the masses of the Earth and the Moon are 5.975×10^{24} kg and 7.343×10^{22} kg respectively, and the mean Earth-Moon distance is 3.844×10^{5} km, estimate how far the centre-of-mass of the system is from the centre of the Earth, stating any assumptions that you make. (See Example 8.3 for the definition of centre-of-mass.)

Problem 8.9 (*Objectives 3,4*) A car moves with a constant speed $|\mathbf{u}| = 45$ km per hour in a direction 20° west of north. Determine the components of the car's velocity **u** in the northerly and easterly directions. The wind blows from the west with wind speed $|\mathbf{v}| = 20$ km per hour. Determine the magnitude and direction of the wind velocity **v'** relative to the car.

Problem 8.10 (*Objectives 3,8,9*) Forces **P** and **Q** are represented by the ordered pairs (3 N,–2 N) and (x,y) respectively. Determine x and y given that three forces 3**P**, 2**Q** and **P** – **Q** are in equilibrium.

Problem 8.11 (*Objectives 3*) Six forces each of magnitude F act along the sides of a regular hexagon. All six forces are in directions that would tend to cause an anticlockwise rotation in the plane of the hexagon. Determine the resultant force. Suppose the direction of one of the forces were reversed. What would the magnitude of the resultant be then?

Problem 8.12 (*Objectives 3,4*) ABCD is a square. The point F is two-thirds of the way from D to C and E is half way from C to B. Obtain the displacements **DE** and **AF** in terms of **a** = **AD** and **b** = **AB**. Determine $|\mathbf{DE}|$ in terms of the side length a of the square.

Problem 8.13 (*Objectives 3,4*) Determine the resultant **F** of the three forces $\mathbf{f} = 2\mathbf{i} + 3\mathbf{j} + 4\mathbf{k}$, $\mathbf{g} = \mathbf{j} - \mathbf{k}$, and $\mathbf{h} = -2\mathbf{i}$. Determine the magnitude $F = |\mathbf{F}|$ and the unit vector **u** if $\mathbf{F} = F\mathbf{u}$.

Problem 8.14 (*Objective 12*) A hydrogen molecule ion consists of two protons each carrying an electric charge e, and an electron carrying an electric charge $-e$. Suppose the two protons are at points P_1 and P_2 with $\mathbf{R} = \mathbf{P_1P_2}$ and the electron is at point Q with $\mathbf{r} = \mathbf{P_1Q}$. Express the resultant electrostatic force **F** acting on the electron in terms of the vectors **R** and **r**. (Coulomb's law giving the electrostatic force between point electric charges is stated in Example 8.4.)

Problem 8.15 (*Objectives 3,4*) The mast of a yacht is held by three cables each attached to the top of the mast and secured at points on the deck around the bottom of the mast. Each cable makes an angle with the vertical of 20° and the tension in each cable has a magnitude of 4500 N. Determine the magnitude of the resultant force of the cables on the mast.

Problem 8.16 (*Objective 6*) **Newton's second law of motion** relates the acceleration **a** of a body of mass m to the resultant force **F** acting on the body, by the vector equation $\mathbf{F} = m\mathbf{a}$. Determine the magnitude and direction of the acceleration of a 0.1 kg mass acted on simultaneously by the three forces specified in Problem 8.1.

2

Vector algebra II
Scalar and vector products

After you have studied this chapter you should be able to

Objectives

- Use the geometric definition of the scalar product to calculate the scalar product of two given vectors (*Objective 1*).
- Use the scalar product to determine the projection of a vector onto another vector (*Objective 2*).
- Manipulate algebraic expressions involving the scalar product (*Objective 3*).
- Test two given vectors for orthogonality (*Objective 4*).
- Determine the scalar product of two cartesian vectors (*Objective 5*).
- Use the scalar product to determine the magnitude of a given vector and the angle between two given vectors (*Objective 6*).
- Use the geometric definition of the vector product to calculate the vector product of two given vectors (*Objective 7*).
- Use the vector product to determine the areas of parallelograms and triangles (*Objective 8*).
- Manipulate expressions containing vector products (*Objective 9*).
- Determine the vector product of two cartesian vectors (*Objective 10*).
- Use the vector product to test for collinear vectors (*Objective 11*).
- Recognise and calculate scalar triple products and vector triple products of given vectors (*Objective 12*).
- Manipulate expressions involving triple products of vectors (*Objective 13*).
- Use scalar products and vector products in a scientific and engineering context (*Objective 14*).

Chapter 1 was concerned with multiplying a vector by a scalar (scaling) and the addition of vectors. This chapter is concerned with products of vectors. There are two useful ways in which a product can be formed from two vectors. One is called the *scalar product* because the product so formed is a scalar quantity; the other is called a *vector product* because the product is itself another vector. Both kinds of product have wide applications in science and technology, as well as in mathematics. We begin with the scalar product.

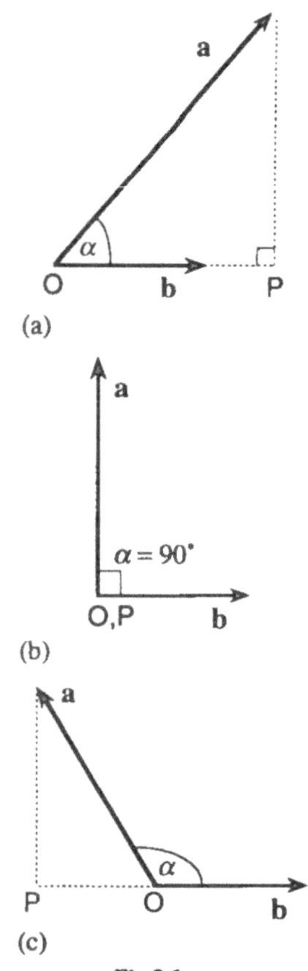

Fig 2.1
The length OP = |**a**|cosα is the projection of **a** onto **b**; it can be (a) positive, (b) zero or (c) negative.

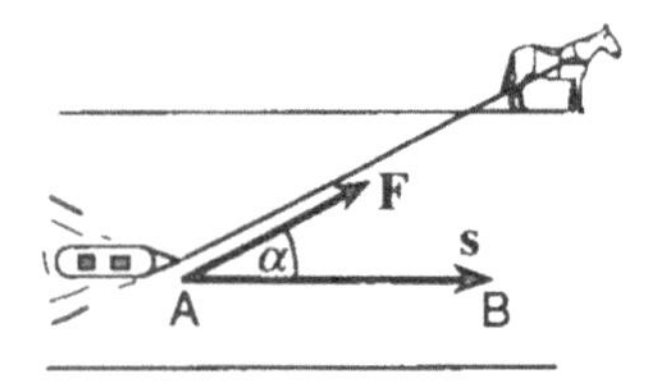

Fig 2.2
The horse pulls the canal boat with a force **F** directed at an angle α to the displacement **s** = AB. The work done by the force when the boat moves from A to B is |**F**| |**s**| cosα. Other forces acting on the boat are not shown.

2.1 THE SCALAR PRODUCT

The scalar product is associated with the idea of projection. An everyday example of projection is the shadow of a solid body cast on a flat surface by a source of light such as the Sun. This is an example of the projection of a solid body onto a plane. We are interested in the projection of one vector onto another vector. This is illustrated in Fig 2.1a which shows a vector **a** directed at an angle α from the direction of a vector **b**. The point P is obtained by dropping a perpendicular from the end point of **a** onto the line of **b**. The length OP = $|\mathbf{a}|\cos\alpha$ is the *projection* of vector **a** onto vector **b**. We often refer to this as the projection of **a** onto the direction of **b** or onto the unit vector $\hat{\mathbf{b}}$. Note that the projection of **a** onto **b** depends on the direction but not the magnitude of **b**. The projection is a positive number when the angle α is an acute angle as in Fig 2.1a; it can be zero or negative as illustrated in Figs 2.1b and c.

If you have studied mechanics you may have come across projections in the context of the *work done by a force*. Fig 2.2 shows a force **F** applied to a body which undergoes a displacement **s**. The work done by the force is the projection of the force onto the displacement, $|\mathbf{F}|\cos\alpha$, times the magnitude $|\mathbf{s}|$ of the displacement

$$\text{work done} = |\mathbf{F}|\,|\mathbf{s}|\cos\alpha$$

Note that this expression treats the two vectors on an equal footing; it depends on the magnitudes of both vectors and can equally well be interpreted as the projection of the displacement onto the force, i.e. $|\mathbf{s}|\cos\alpha$, times the magnitude $|\mathbf{F}|$ of the force. The work done by a force is an example of a scalar product of two vectors. You can see that it is indeed a scalar quantity, positive, negative or zero depending on the angle α. Other applications of the scalar product will be taken up in the final section of this chapter.

2.1.1 Definition of the scalar product and projections

Consider two vectors **a** and **b** and let α be the angle between their directions (Fig 2.3a). The **scalar product** of **a** and **b** is denoted by writing **a . b** and is defined to be the scalar quantity

$$\mathbf{a}\cdot\mathbf{b} = |\mathbf{a}|\,|\mathbf{b}|\cos\alpha \tag{2.1}$$

The dot symbol in **a . b** is reserved for this use in vector algebra; a dot between two vectors always means the scalar product of the two vectors. In fact the scalar product is often called the **dot product**, and in speech we say "**a** dot **b**".

We are free to take α to be either the larger or the smaller of the two angles between the vectors (Fig 2.3b) because $\cos\alpha = \cos(360° - \alpha)$, but it is usually more convenient to take the smaller angle, i.e. $0 \le \alpha \le 180°$. A common error is illustrated in Fig 2.3c where two vectors are drawn head-to-tail. The angle between the directions of the vectors is α, not $\gamma = 180 - \alpha$, as is clear from Fig 2.3d where both vectors are drawn from the same point. If the angle between the vectors is taken wrongly to be γ rather than α then the wrong sign will be obtained for the scalar product because $\cos\gamma = -\cos\alpha$.

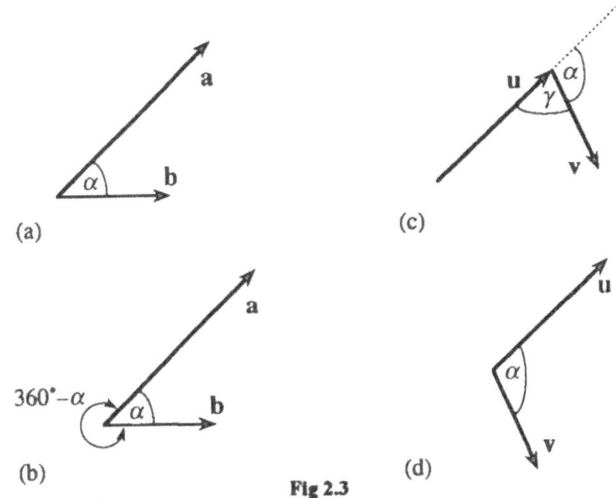

Fig 2.3
(a) $\mathbf{a} \cdot \mathbf{b} = |\mathbf{a}|\,|\mathbf{b}|\cos\alpha$. (b) The angle α is usually taken to be the smaller of the two. (c) and (d) The angle between **u** and **v** is α not γ.

The scalar product is zero when one (or both) of the vectors is the zero vector, or when the two vectors are orthogonal (i.e. at right-angles) to each other for then $\cos\alpha = \cos 90° = 0$. Thus we have the **orthogonality condition** for non-zero vectors,

$$\mathbf{a} \cdot \mathbf{b} = 0 \qquad \text{(orthogonality condition)} \tag{2.2}$$

Another special case is when $\mathbf{b} = \mathbf{a}$. Then $\alpha = 0$, $\cos 0 = 1$ and the scalar product is the square of the magnitude of **a**,

$$\mathbf{a} \cdot \mathbf{a} = |\mathbf{a}|^2 \tag{2.3}$$

The **projection of a onto b** is the number $|\mathbf{a}|\cos\alpha$ where α is the angle between the directions of **a** and **b**; it is obtained formally by taking the scalar product of **a** with $\hat{\mathbf{b}}$, the unit vector in the direction of **b**,

Some books define the projection of **a** onto **b** to be the **projected vector** $(|\mathbf{a}|\cos\alpha)\hat{\mathbf{b}}$, i.e. a vector of magnitude $|\mathbf{a}|\,|\cos\alpha|$ in the direction of **b** or opposite to **b** depending on the sign of $\cos\alpha$.

$$\mathbf{a} \cdot \hat{\mathbf{b}} = |\mathbf{a}|\,|\hat{\mathbf{b}}|\cos\alpha = |\mathbf{a}|\cos\alpha \tag{2.4}$$

For example, the projection of a vector **a** onto the cartesian unit vector **i** is

$$\mathbf{a} \cdot \mathbf{i} = |\mathbf{a}|\cos\theta_x = a_x \tag{2.5}$$

where θ_x is the angle between **a** and the x-axis. Thus the projections of **a** onto the cartesian unit vectors are the cartesian components of **a**.

The projection of a vector onto a surface is defined in Problem 2.4.

2.1.2 The scalar product in vector algebra

We need to know the rules of algebra for manipulating expressions involving scalar products.

It is easy to see from the definition (Eq (2.1)) that the scalar product is commutative (i.e. the order of the two vectors doesn't matter),

$$\mathbf{a}\cdot\mathbf{b} = |\mathbf{a}|\,|\mathbf{b}|\cos\alpha = |\mathbf{b}|\,|\mathbf{a}|\cos\alpha = \mathbf{b}\cdot\mathbf{a} \tag{2.6}$$

and for any scalar λ

$$\mathbf{a}\cdot\lambda\mathbf{b} = \lambda\mathbf{a}\cdot\mathbf{b} = \lambda(\mathbf{a}\cdot\mathbf{b}) \tag{2.7}$$

Moreover, the scalar product is distributive over vector addition, i.e.

$$(\mathbf{b}+\mathbf{c})\cdot\mathbf{a} = \mathbf{b}\cdot\mathbf{a} + \mathbf{c}\cdot\mathbf{a} \tag{2.8}$$

This rule expresses the fact that the projection of the sum of two vectors is equal to the sum of the projections. This is easily demonstrated geometrically (see Fig 2.4).

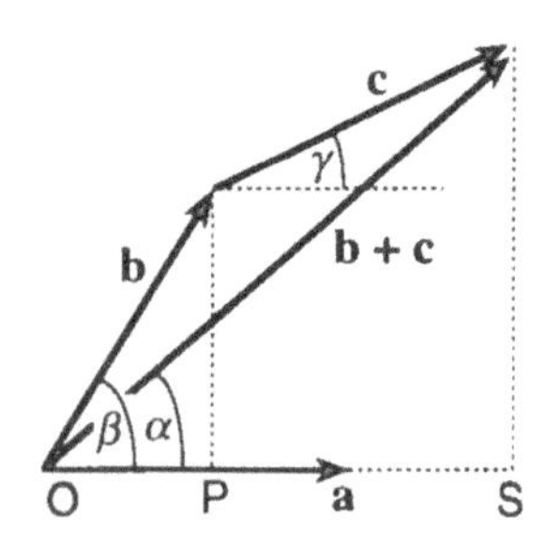

Fig 2.4
The distributive rule.
OS = OP + PS
But OS = $(\mathbf{b}+\mathbf{c})\cdot\hat{\mathbf{a}}$
OP = $\mathbf{b}\cdot\hat{\mathbf{a}}$
PS = $\mathbf{c}\cdot\hat{\mathbf{a}}$
Hence
$(\mathbf{b}+\mathbf{c})\cdot\hat{\mathbf{a}} = \mathbf{b}\cdot\hat{\mathbf{a}} + \mathbf{c}\cdot\hat{\mathbf{a}}$
Multiplying both sides by $|\mathbf{a}|$ yields Eq 2.8.

The above three rules allow us to manipulate expressions involving scalar products in the same way as we manipulate ordinary algebraic expressions. However, it is not possible to divide by a vector.

Summary of section 2.1

- The **scalar product** or **dot product** of two vectors is a scalar quantity defined by

$$\mathbf{a}\cdot\mathbf{b} = |\mathbf{a}|\,|\mathbf{b}|\cos\alpha \tag{2.1}$$

  where α is the angle between the directions of **a** and **b**.

- When $\alpha = 90°$

$$\mathbf{a}\cdot\mathbf{b} = 0 \qquad \textbf{(orthogonality condition)} \tag{2.2}$$

  and when $\mathbf{b} = \mathbf{a}$,

$$\mathbf{a}\cdot\mathbf{a} = |\mathbf{a}|^2 \tag{2.3}$$

- The **projection of a onto b** is the number $\mathbf{a}\cdot\hat{\mathbf{b}} = |\mathbf{a}|\cos\alpha$. The **projected vector** is $(|\mathbf{a}|\cos\alpha)\hat{\mathbf{b}}$.

- The rules of algebra for the scalar product (Eqs 2.6-2.8) are similar to those of ordinary number algebra, but there is no division by a vector.

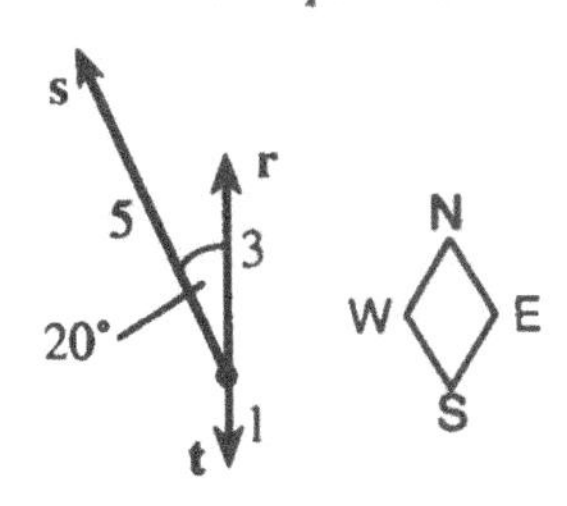

Fig 2.5
The vectors **s**, **r** and **t**.

Example 1.1 (*Objectives 1,3*) Let **r** be of magnitude 3 and directed towards the north, **s** of magnitude 5 directed 20° west of north and **t** a unit vector directed south (Fig 2.5).

(a) Evaluate $\mathbf{r} \cdot \mathbf{s}$, $\mathbf{r} \cdot \mathbf{t}$, and $3\mathbf{s} \cdot (\mathbf{r} - \mathbf{t})$.
(b) Evaluate $|(\mathbf{s} \cdot \mathbf{t})\mathbf{s}|$.

Solution 1.1

(a) Using Eq (2.1), we have

$$\mathbf{r} \cdot \mathbf{s} = 3\times5\cos(20°) = 14.1$$

$$\mathbf{r} \cdot \mathbf{t} = 3\times1\cos(180°) = -3$$

Using Eqs (2.1), (2.7) and (2.8), we find

$$3\mathbf{s} \cdot (\mathbf{r} - \mathbf{t}) = 3\mathbf{s} \cdot \mathbf{r} - 3\mathbf{s} \cdot \mathbf{t} = 3\times5\times3\cos(20°) - 3\times5\times1\cos(180° - 20°) = 56.4.$$

(Alternatively, note that $\mathbf{r} - \mathbf{t}$ is directed towards the north and $|\mathbf{r} - \mathbf{t}| = 4$, and so $3\mathbf{s} \cdot (\mathbf{r} - \mathbf{t}) = 3\times5\times4\cos(20°) = 56.4$.)

(b) $\mathbf{s} \cdot \mathbf{t} = 5\times1\cos(180° - 20°) = -4.70$ and so $(\mathbf{s} \cdot \mathbf{t})\mathbf{s} = -4.70\mathbf{s}$. Hence

$$|(\mathbf{s} \cdot \mathbf{t})\mathbf{s}| = |-4.70\mathbf{s}| = |-4.70|\,|\mathbf{s}| = 4.70\times5 = 23.5$$

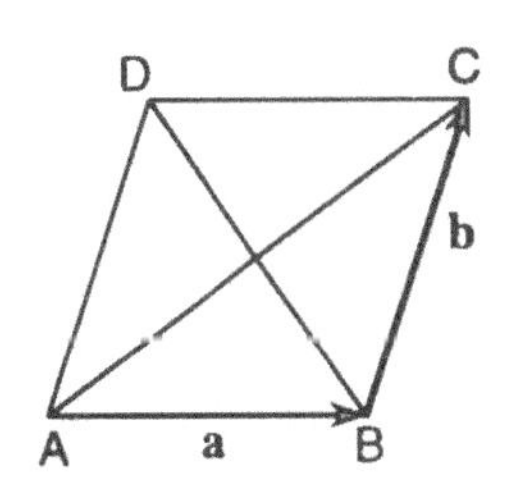

Fig 2.6
A rhombus.

Example 1.2 (*Objectives 1,3,4*) Fig 2.6 shows a rhombus ABCD and the displacements $\mathbf{a} = \mathbf{AB}$ and $\mathbf{b} = \mathbf{BC}$.

(a) Express **AC** and **DB** in terms of **a** and **b**, and
(b) hence show that the diagonals of a rhombus are at right-angles to one another.

Solution 1.2

(a) $\mathbf{AC} = \mathbf{a} + \mathbf{b}$;
$\mathbf{DB} = \mathbf{a} - \mathbf{AD} = \mathbf{a} - \mathbf{b}$.

(b) Vectors **AC** and **DB** lie on the diagonals of the rhombus. The scalar product of **AC** and **DB** is

We have multiplied out the brackets and simplified the resulting expression using the rules of algebra, Eqs (2.6) to (2.8).

$$(\mathbf{a} + \mathbf{b}) \cdot (\mathbf{a} - \mathbf{b}) = \mathbf{a} \cdot \mathbf{a} + \mathbf{a} \cdot (-\mathbf{b}) + \mathbf{b} \cdot \mathbf{a} + \mathbf{b} \cdot (-\mathbf{b})$$

$$= |\mathbf{a}|^2 - \mathbf{a} \cdot \mathbf{b} + \mathbf{a} \cdot \mathbf{b} - |\mathbf{b}|^2 = |\mathbf{a}|^2 - |\mathbf{b}|^2$$

$$= 0$$

because $|\mathbf{a}| = |\mathbf{b}|$ for a rhombus. Thus we have shown that the scalar product of two vectors on the diagonals of a rhombus is zero. It follows from Eq (2.2) that the diagonals are orthogonal.

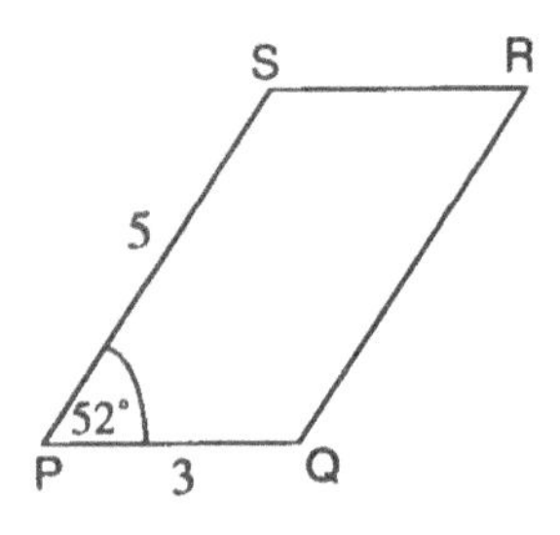

Fig 2.7
The parallelogram PQRS.

Problem 1.1 (*Objective 1*) The parallelogram shown in Fig 2.7 has sides of length PQ = 3 and PS = 5. The angle between sides PQ and PS is 52°. Determine the scalar products **PQ . PS** and **PQ . PR**.

Problem 1.2 (*Objective 2*) Refer to Fig 2.7. Determine the projection of **PS** onto **PQ** and the projection of **PR** onto **PQ**. Determine also the projection of **RS** onto **QR**.

Problem 1.3 (*Objectives 1,3*) ABC is a triangle with **BC** = **a**, **CA** = **b** and β is the angle between the sides CA and CB. Express the scalar product $(\mathbf{a} + \mathbf{b}) \cdot (\mathbf{a} + \mathbf{b})$ in terms of the lengths AC and BC and the angle β, and hence obtain the **cosine rule**:

$$AB^2 = AC^2 + BC^2 - 2AC \times BC \cos\beta$$

where AB^2 denotes the square of the length AB, etc.

Problem 1.4 Select the meaningless expressions, giving reasons for your answers:

A	$\mathbf{P} \cdot \mathbf{Q} + 6$	E	$\mathbf{q} \cdot \mathbf{r} + (\mathbf{q} + \mathbf{r})$
B	$\mathbf{a}(\mathbf{b} \cdot \mathbf{c}) + \mathbf{c}$	F	$\mathbf{q} \cdot \mathbf{r} + (\mathbf{q} - \mathbf{r}) \cdot \hat{\mathbf{r}}$
C	$\mathbf{a} \cdot (\mathbf{b} \cdot \mathbf{c})$	G	$|\mathbf{p}| \cdot \mathbf{q}$
D	$3 \cdot \mathbf{q}$	H	$\mathbf{a} \cdot \mathbf{b}/\mathbf{c}$

2.2 CARTESIAN FORM OF THE SCALAR PRODUCT

Eq (2.1) defines the scalar product in geometric terms. We now derive an expression for the scalar product which is equivalent to Eq (2.1) but which involves only the cartesian components of the two vectors.

Consider two cartesian vectors $\mathbf{a} = a_x\mathbf{i} + a_y\mathbf{j} + a_z\mathbf{k}$ and $\mathbf{b} = b_x\mathbf{i} + b_y\mathbf{j} + b_z\mathbf{k}$. Using the rules of algebra given in Section 2.1.2, we can expand the scalar product $\mathbf{a} \cdot \mathbf{b}$ to obtain nine terms,

$$\mathbf{a} \cdot \mathbf{b} = (a_x\mathbf{i} + a_y\mathbf{j} + a_z\mathbf{k}) \cdot (b_x\mathbf{i} + b_y\mathbf{j} + b_z\mathbf{k})$$

$$= a_xb_x\mathbf{i} \cdot \mathbf{i} + a_xb_y\mathbf{i} \cdot \mathbf{j} + a_xb_z\mathbf{i} \cdot \mathbf{k}$$

$$+ a_yb_x\mathbf{j} \cdot \mathbf{i} + a_yb_y\mathbf{j} \cdot \mathbf{j} + a_yb_z\mathbf{j} \cdot \mathbf{k}$$

$$+ a_zb_x\mathbf{k} \cdot \mathbf{i} + a_zb_y\mathbf{k} \cdot \mathbf{j} + a_zb_z\mathbf{k} \cdot \mathbf{k} \qquad (2.9)$$

This formidable-looking expression simplifies considerably when the scalar products of the cartesian unit vectors are evaluated. These are: $\mathbf{i} \cdot \mathbf{i} = |\mathbf{i}|\,|\mathbf{i}|\cos 0 = 1$, $\mathbf{i} \cdot \mathbf{j} = |\mathbf{i}|\,|\mathbf{j}|\cos 90° = 0$, and so on. Collecting them together we have

$$\begin{aligned} \mathbf{i} \cdot \mathbf{i} &= \mathbf{j} \cdot \mathbf{j} = \mathbf{k} \cdot \mathbf{k} = 1 \\ \mathbf{i} \cdot \mathbf{j} &= \mathbf{j} \cdot \mathbf{k} = \mathbf{k} \cdot \mathbf{i} = 0 \end{aligned} \qquad (2.10)$$

The word **orthonormality** is a contraction of the two words: **orthogonal** meaning perpendicular and **normalised** meaning of unit magnitude.

Eqs (2.10) are the **orthonormality relations** for the cartesian unit vectors. Note that all possible scalar products among the cartesian unit vectors are implicitly included in Eqs (2.10) because the scalar product is commutative and so $\mathbf{j} \cdot \mathbf{i} = \mathbf{i} \cdot \mathbf{j} = 0$, etc.

Using Eqs (2.10) we find that Eq (2.9) reduces to the extremely simple and enormously important result

$$\mathbf{a} \cdot \mathbf{b} = a_xb_x + a_yb_y + a_zb_z \qquad (2.11)$$

Eq (2.11) is the **cartesian form of the scalar product**.

We now revisit some earlier results. We have seen that the projections of a vector onto the cartesian unit vectors are the cartesian components of the vector (Eq (2.5)). This result also follows from Eq (2.11). The projection of **a** onto **i** is found by putting **b** = **i** in Eq (2.11). This gives $\mathbf{a} \cdot \mathbf{i} = a_x \times 1 + a_y \times 0 + a_z \times 0 = a_x$, and so we have

$$\mathbf{a} \cdot \mathbf{i} = a_x \quad \mathbf{a} \cdot \mathbf{j} = a_y \quad \mathbf{a} \cdot \mathbf{k} = a_z \qquad (2.12)$$

Now suppose we have a vector equation

$$\mathbf{a} = \mathbf{b} \qquad (2.13)$$

We can project both sides of this equation onto each of the cartesian unit vectors in turn, and so obtain the three scalar equations

$$a_x = b_x, \quad a_y = b_y, \quad a_z = b_z. \qquad (2.14)$$

Thus a single vector equation is equivalent to three scalar equations for the components. This result has many applications. For example, when forces are in equilibrium the corresponding components of the forces are separately in equilibrium (see Example 8.1 of Chapter 1).

Summary of section 2.2

- The **orthonormality relations** for the cartesian unit vectors are:

$$\mathbf{i}\,.\,\mathbf{i}=\mathbf{j}\,.\,\mathbf{j}=\mathbf{k}\,.\,\mathbf{k}=1$$
$$\mathbf{i}\,.\,\mathbf{j}=\mathbf{j}\,.\,\mathbf{k}=\mathbf{k}\,.\,\mathbf{i}=0 \qquad (2.10)$$

- The cartesian form of the scalar product is:

$$\mathbf{a}\,.\,\mathbf{b}=a_xb_x+a_yb_y+a_zb_z \qquad (2.11)$$

  where a_x and b_x are the x-components of **a** and **b**, etc.

- The projections of a vector onto the cartesian unit vectors give the cartesian components of the vector: $\mathbf{a}\,.\,\mathbf{i}=a_x$, etc.

- A vector equation is equivalent to three scalar equations for the components (Eqs (2.13) and (2.14)).

Example 2.1 (*Objective 5*)

(a) Determine the scalar product **p . q** where **p** = 2**i** – **k** and **q** = **i** + 4**j** – **k**.
(b) Determine (3,2,–5) . (–4,0,2).

Solution 2.1

(a) Use Eq (2.11). $\mathbf{p}\,.\,\mathbf{q}=2\times1+0\times4+(-1)\times(-1)=3$.

(b) Here the two vectors are given as ordered triples. Thus with **a** = (3,2,–5) and **b** = (–4,0,2) we use Eq (2.11) to obtain

$$\mathbf{a}\,.\,\mathbf{b}=(3,2,-5)\,.\,(-4,0,-2)$$
$$=3\times(-4)+2\times0+(-5)\times(-2)=-12+0+10=-2.$$

Example 2.2 (*Objectives 2,5*) Find the projections of the vector $\mathbf{A}=3\mathbf{i}-2\mathbf{j}+4\mathbf{k}$ onto the three mutually orthogonal unit vectors $\mathbf{u}=(\mathbf{i}+\mathbf{j})/\sqrt{2}$, $\mathbf{v}=(\mathbf{i}-\mathbf{j})/\sqrt{2}$ and **k**.

Solution 2.2 The projections are equal to the scalar products of **A** with the unit vectors. Using Eq (2.11), and taking the factor $1/\sqrt{2}$ outside brackets, we have

$$\mathbf{A}\cdot\mathbf{u} = \tfrac{1}{\sqrt{2}}[3\times1 + (-2)\times1 + 4\times0] = \tfrac{1}{\sqrt{2}}$$

$$\mathbf{A}\cdot\mathbf{v} = \tfrac{1}{\sqrt{2}}[3\times1 + (-2)\times(-1) + 4\times0] = \tfrac{5}{\sqrt{2}}$$

$$\mathbf{A}\cdot\mathbf{k} = (3\times0 + (-2)\times0 + 4\times1) = 4$$

Problem 2.1 (*Objective 5*) Given $\mathbf{a} = 3\mathbf{i} - 2\mathbf{j} + 5\mathbf{k}$ and $\mathbf{b} = \mathbf{i} - \mathbf{j}$, determine: $\mathbf{a}\cdot\mathbf{b}$, $\mathbf{a}\cdot\mathbf{i}$, $(\mathbf{a}+\mathbf{b})\cdot(\mathbf{a}-\mathbf{b})$, $\mathbf{j}\cdot\mathbf{b}$, $|(\mathbf{a}\cdot\mathbf{j})\mathbf{i}|$ and the x-component of $(\mathbf{b}-\mathbf{a})$.

Problem 2.2 (*Objective 5*) Given $\mathbf{A} = (1,2,3)$ and $\mathbf{B} = (-1,0,5)$, determine $\mathbf{A}\cdot\mathbf{B}$, $\mathbf{i}\cdot\mathbf{B}$, $(\mathbf{i}+\mathbf{j})\cdot(\mathbf{A}-\mathbf{B})$ and $\mathbf{A}\cdot(A_x + B_y)\mathbf{B}$.

Problem 2.3 (*Objective 5*) Find the scalar product of $\mathbf{p} = (1,6,5)$ and $\mathbf{q} = (-3,-7,2)$. Find also the unit vector in the direction of $\mathbf{p}$ and the projection of $\mathbf{q}$ onto $\mathbf{p}$.

Problem 2.4 (*Objective 5*) Let $\mathbf{n}$ be a unit vector in a direction at right angles to a plane surface. Then the **projection of a vector b onto the surface** is defined to be the vector $\mathbf{b} - (\mathbf{n}\cdot\mathbf{b})\mathbf{n}$.

(a) Determine the projection of $\mathbf{b} = 3\mathbf{i} - 2\mathbf{j} + 4\mathbf{k}$ onto the x-y plane.

(b) Determine the magnitude of the projection of the acceleration of gravity vector $\mathbf{g}$ ($|\mathbf{g}| = g = 10\ \text{ms}^{-2}$) onto a house roof that slopes at 20° from the horizontal.

2.3 THE ANGLE BETWEEN TWO VECTORS

In Chapter 1 magnitudes and directions of vectors were calculated using Pythagoras's theorem and trigonometry. That approach is feasible when simple figures can be sketched and right-angled triangles readily identified. We now show how to calculate magnitudes and directions algebraically. The scalar product provides us with the tools for doing this. The definition (Eq (2.1)) gives the scalar product in terms of the magnitudes of the vectors and the angle between them, while the cartesian form (Eq (2.11)) gives the scalar product in terms of the cartesian components of the two vectors. We now put these two results together to obtain some extremely useful formulas for the magnitudes and directions of vectors.

Substitute Eq (2.11) into the left-hand side of Eq (2.1) to give

$$a_x b_x + a_y b_y + a_z b_z = |\mathbf{a}|\,|\mathbf{b}|\cos\alpha \tag{2.15}$$

Now put $\mathbf{b} = \mathbf{a}$ and take the square root of both sides of Eq (2.15), and we obtain

$$(a_x^2 + a_y^2 + a_z^2)^{1/2} = |\mathbf{a}| \tag{2.16}$$

which you will recognise as Eq (1.17c) of Chapter 1 for the magnitude of a cartesian vector; there it was obtained by application of Pythagoras's theorem, here it comes from the scalar product.

A new and extremely useful result is obtained by making $\cos\alpha$ the subject of Eq (2.15) and then using Eq (2.16). Thus

$$\cos\alpha = \frac{(a_x b_x + a_y b_y + a_z b_z)}{(a_x^2 + a_y^2 + a_z^2)^{1/2}(b_x^2 + b_y^2 + b_z^2)^{1/2}} \tag{2.17}$$

This equation can be used to calculate the angle α between any two cartesian vectors since the right-hand side is expressed entirely in terms of cartesian components. When using Eq (2.17) remember that with α restricted to the range $0 \le \alpha \le 180°$, $\cos\alpha$ determines α uniquely.

The direction cosines of Chapter 1 (Eq (1.18a)) are included in Eq (2.17) as special cases. For example, if $\mathbf{b} = \mathbf{i}$ then α is the angle θ_x between $\mathbf{a}$ and the x-axis, and Eq (2.17) simplifies to give

$$\cos\theta_x = a_x/(a_x^2 + a_y^2 + a_z^2)^{1/2} = a_x/|\mathbf{a}|$$

Another special case is when the two vectors $\mathbf{a}$ and $\mathbf{b}$ are at right-angles to each other. The orthogonality condition (Eq (2.2)) together with Eq (2.11) leads directly to the orthogonality condition for cartesian vectors

$$a_x b_x + a_y b_y + a_z b_z = 0 \qquad \text{(orthogonality condition)} \tag{2.18}$$

which can also be obtained by putting $\alpha = 90°$ in Eq (2.17).

Eqs (2.16) and (2.17) form the basis of the cartesian analysis of distances and angles in three-dimensional geometrical figures.

Summary of section 2.3

- The magnitude of a cartesian vector and the angle α between the directions of two cartesian vectors can be calculated from

$$(\mathbf{a} \cdot \mathbf{a})^{1/2} = (a_x^2 + a_y^2 + a_z^2)^{1/2} = |\mathbf{a}| \tag{2.16}$$

$$\cos\alpha = \frac{(a_x b_x + a_y b_y + a_z b_z)}{(a_x^2 + a_y^2 + a_z^2)^{1/2}(b_x^2 + b_y^2 + b_z^2)^{1/2}} \tag{2.17}$$

- The orthogonality condition for cartesian vectors is

$$a_xb_x + a_yb_y + a_zb_z = 0 \qquad \text{(orthogonality condition)} \qquad (2.18)$$

Example 3.1 (*Objective 6*) Given the following vectors

$$\mathbf{f} = 2\mathbf{i} - \mathbf{j} + 5\mathbf{k}, \quad \mathbf{g} = -\mathbf{i} - 3\mathbf{j} - 2\mathbf{k}, \quad \mathbf{h} = \mathbf{i} - 4\mathbf{j} + 3\mathbf{k}$$

(a) Calculate $|\mathbf{f}|$, $|\mathbf{g}|$ and $|\mathbf{h}|$.

(b) Calculate the angle between the directions of
(i) **f** and **g**, (ii) **g** and **h**, (iii) **h** and **f**.

(c) Do the three given vectors make a triangle?

Solution 3.1

(a) Use Eq (2.16).
$|\mathbf{f}| = (2^2 + (-1)^2 + 5^2)^{1/2} = 30^{1/2} = 5.48,$
$|\mathbf{g}| = ((-1)^2 + (-3)^2 + (-2)^2)^{1/2} = 14^{1/2} = 3.74,$
$|\mathbf{h}| = (1^2 + (-4)^2 + 3^2)^{1/2} = 26^{1/2} = 5.10.$

(b) Using Eq (2.17), with an obvious notation for the angles, we obtain

(i) $\cos\alpha_{fg} = \dfrac{2(-1) + (-1)(-3) + 5(-2)}{30^{1/2} \times 14^{1/2}} = -0.439$, and so $\alpha_{fg} = 116.0°$.

(ii) $\cos\alpha_{gh} = \dfrac{(-1)(1) + (-3)(-4) + (-2)(3)}{14^{1/2} \times 26^{1/2}} = 0.262$, and so $\alpha_{gh} = 74.8°$.

(iii) $\cos\alpha_{hf} = \dfrac{(1)(2) + (-4)(-1) + (3)(5)}{26^{1/2} \times 30^{1/2}} = 0.752$, and so $\alpha_{hf} = 41.2°$.

(c) Yes. Fig 2.8 shows the angles between the vectors and the triangle formed by them. α_{fg} is the angle between the two vectors **f** and **g** and so the corresponding interior angle of the triangle is $180° - \alpha_{fg} = 64.0°$. The interior angles are seen to add to 180°. (The fact that the vectors form a triangle can also be seen from inspection of the vectors: **h** is the resultant of **f** and **g**.)

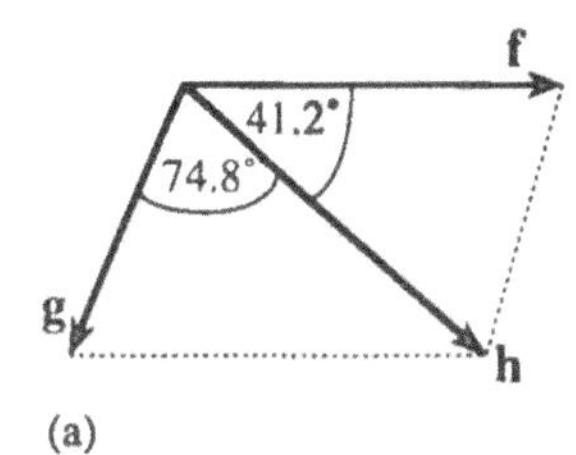

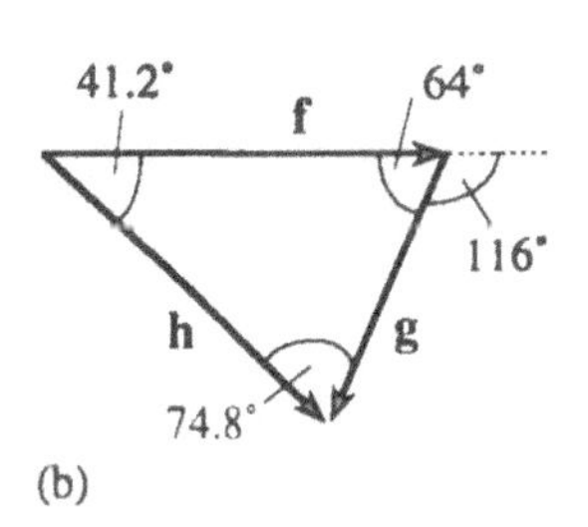

Fig 2.8
(a) The three vectors f, g and h. (b) The three vectors form a triangle.

Example 3.2 (*Objective 6*) The four points O(0,0,0), X(3,0,0), Y(0,3,0) and P(1,1,3) define a *tetrahedron*, see Fig 2.9. M is the mid-point of XY.

(a) Determine the angles α, β and γ shown in the figure.

(b) Find the centroid of triangle OXY and show that it is directly below point P. (The centoid of the triangle is the centroid of the points O, X and Y.)

A tetrahedron is a four-sided solid figure, e.g. a triangular pyramid.

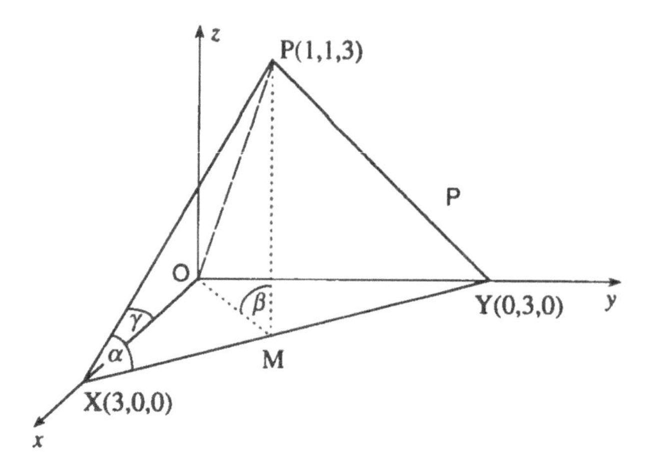

Fig 2.9
A tetrahedron.

Solution 3.2

(a) α is the angle between the directions of displacements **XP** and **XY**. We can use Eq (1.12) of Chapter 1 to obtain **XP** = (–2,1,3) and **XY** = (–3,3,0) where for convenience we are using the cartesian ordered triple representation of the vectors. Now use Eq (2.17) to obtain

$$\cos\alpha = \frac{(-2)(-3)+(1\times 3)+(3\times 0)}{((-2)^2+1^2+3^2)^{1/2}((-3)^2+3^2+0^{1/2})}$$

$$= \frac{9}{14^{1/2}\times 18^{1/2}} = 0.567$$

Hence $\alpha = 55.5°$.

Similarly, to find β we need **MO** and **MP**. Refer to Example 4.6 of Chapter 1 for the position vector of a mid-point, and we have

MO = –**OM** = –(1/2)(**OX** + **OY**) = –(3/2,3/2,0)

The coordinates of M are (3/2,3/2,0) and so

MP = (1,1,3) – (3/2,3/2,0) = (–1/2,–1/2,3)

Thus

$$\cos\beta = \frac{(-3/2)(-1/2)+(-3/2)(-1/2)+0\times 3}{((-3/2)^2+(-3/2)^2+0^2)^{1/2}((-1/2)^2+(-1/2)^2+3^2)^{1/2}}$$

$$= \frac{3/2}{(9/2)^{1/2} \times (19/2)^{1/2}} = 0.229.$$

Hence $\beta = 76.7°$.

To find γ we need **XP** = (–2,1,3) and **XO** = (–3,0,0). Thus

$$\cos\gamma = \frac{(-2)(-3)+0+0}{((-2)^2+1^2+3^2)^{1/2}((-3)^2+0+0)^{1/2}}$$

$$= \frac{6}{14^{1/2} \times 3} = 0.535.$$

Hence $\gamma = 57.7°$.

(b) The centroid of OXY (see Problem 4.10 of Chapter 1) is the point N given by

ON = (1/3)[(0,0,0) + (3,0,0) + (0,3,0)] = (1,1,0)

Thus N is the point N(1,1,0) which is clearly 3 units directly below the point P(1,1,3). (We can use Eq (2.18) to confirm that **ON** and **NP** are orthogonal, i.e. **ON** . **NP** = (1,1,0) . (0,0,3) = 0.)

Problem 3.1 (*Objective 4*) Select the vectors that are orthogonal to **r** = 2**i** – **j** –3**k**.

A **i** – 4**j** + 2**k**
B 4**i** + 2**j** – 2**k**
C **i** + **j** + **k**
D 2**i** + **j** + 3**k**
E **i** – **j** + **k**
F **i** + 2**j** – **k**

Problem 3.2 (*Objective 6*) Consider the following points: A(1,3,1), B(–2,–1,0) and C(1,1,4). Find the angles between the directions of the pairs of vectors

(i) **c** = **AB** and **b** = **AC**,
(ii) **a** = **BC** and **b**,
(iii) **a** and **c**.

Problem 3.3 (*Objective 6*) Find the angle between two diagonals passing through the centre of a cube.

Problem 3.4 (*Objective 6*) The great Egyptian pyramid "Cheops" has a square base of side length 230 m and is of height 147 m. Make a sketch of the pyramid with one corner of the base at the origin of a cartesian coordinate system. Write down the cartesian coordinates of the five vertices (corners) and determine the angle of ascent to the summit (a) along an edge and (b) directly up one face.

(The angle of ascent is the angle between the line of ascent and the horizontal line that lies vertically below it.)

Problem 3.5 (*Objective 4*) Show that the two vectors **u** and **v** of Example 2.2 are orthogonal.

Problem 3.6 (*Objective 6*) Show that the sum of the squares of the projections in Example 2.2 is equal to the square of the modulus of **A** and comment on why this should be so.

Problem 3.7 (*Objective 6*) A laser beam propagating in the direction defined by unit vector $\mathbf{u} = (\mathbf{i} + \mathbf{j} + \mathbf{k})/\sqrt{3}$ is reflected by a plane mirror. The orientation of the mirror is defined by unit vector $\mathbf{n} = -(\mathbf{i} + 2\mathbf{j})/\sqrt{5}$ pointing normal to the plane surface of the mirror from the back silvered face towards the front face.

(a) Determine the angle α of incidence, i.e. the angle between $-\mathbf{u}$ and the mirror normal **n**.

(b) Determine the angle of reflection, i.e. the angle between the reflected beam and the mirror normal **n**. Assume that the reflection reverses the normal component of the velocity $c\mathbf{u}$ associated with the incident laser beam, while leaving the other components unchanged, where c is the speed of light.

2.4 THE VECTOR PRODUCT

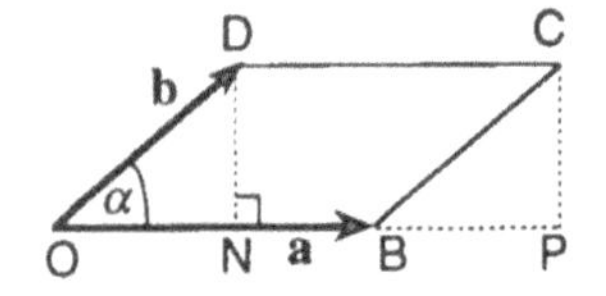

Fig 2.10
(a) The area of parallelogram OBCD is equal to the area of the rectangle NPCD because triangles OND and BPC are congruent and therefore have the same area. Thus the area of the rectangle and that of the parallelogram are both equal to DN × DC. But DN = $|\mathbf{b}|\sin\alpha$ and DC = OB = $|\mathbf{a}|$ and so the area is DN × DC = $|\mathbf{a}||\mathbf{b}|\sin\alpha$.

The vector product of two vectors is itself a vector. We begin with two examples of how vector products arise in applications. We then give a definition of the vector product and obtain the rules of algebra for using vector products.

The two displacements **a** and **b** shown in Fig 2.10a define the parallelogram OBCD. The area A of the parallelogram is equal to the base length $|\mathbf{a}|$ multiplied by its height $|\mathbf{b}|\sin\alpha$, where α is the interior angle. Thus

$$A = |\mathbf{a}||\mathbf{b}|\sin\alpha$$

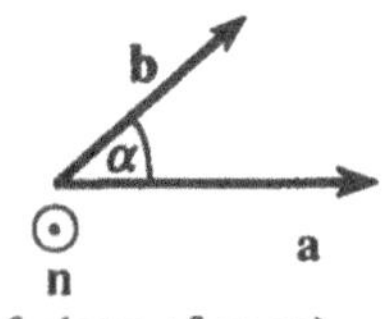

(b) The unit vector **n** is directed at right-angles to the plane of **a** and **b**, out of the plane of the paper as indicated by the dot in a circle. (You can think of this symbol as representing the head end of an approaching arrow.)

We can associate a direction with this scalar area A by taking a unit vector **n** directed at right-angles to the plane of **a** and **b** (Fig 2.10b). Putting magnitude and direction together we obtain a directed area, i.e. an area vector

$$\mathbf{A} = (|\mathbf{a}||\mathbf{b}|\sin\alpha)\mathbf{n}$$

This area vector is the vector product of the two displacements **a** and **b**. There are of course two unit vectors at right-angles to the plane of **a** and **b**; they are **n** and $\mathbf{n}' = -\mathbf{n}$. We choose **n** rather than **n'** by invoking the right-hand rule which we describe in detail in the next section. Meanwhile we consider another example of a vector product familiar to scientists and engineers.

Fig 2.11 shows a horizontal force **F** applied to a gate. The force is applied at a point P specified by the horizontal displacement **s** from a point O on the hinge. It is common experience that the "turning effect" or *torque* of the force increases with the distance $|\mathbf{s}|$ to the hinge and the magnitude $|\mathbf{F}|$ of the force, and is largest when the force is at right-angles to the displacement **s** and the line of the hinge, i.e. when the angle α in Fig 2.11 is 90°. These observations are encompassed by the formula $|\mathbf{s}|\,|\mathbf{F}|\sin\alpha$ for the magnitude of the torque.

The torque also has a direction associated with it. The torque in Fig 2.11 tends to rotate the gate about a vertical axis, i.e. the gate hinge. The direction of this vertical axis is specified by the unit vector **n** directed vertically downwards (Fig 2.11), and this is taken to be the direction of the torque vector. Thus

$$\text{torque} = (|\mathbf{s}|\,|\mathbf{F}|\sin\alpha)\mathbf{n}$$

The torque is the vector product of the two vectors **s** and **F**. Once again we use the right-hand rule (described below) to define the direction of the torque to be that of unit vector **n** (Fig 2.11) rather than $\mathbf{n}' = -\mathbf{n}$.

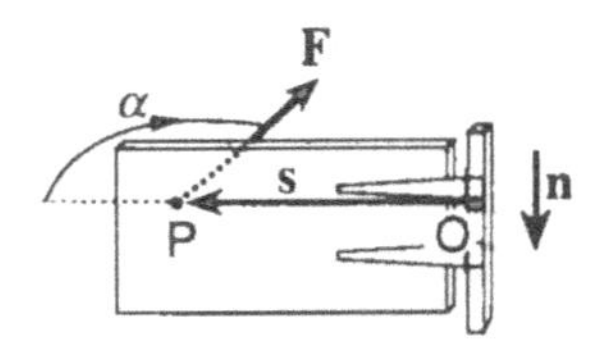

Fig 2.11
The horizontal force **F** applied to the gate at point P exerts a torque about point O on the hinge equal to $(|\mathbf{F}|\,|\mathbf{s}|\sin\alpha)\mathbf{n}$. The direction of unit vector **n** is vertically downwards.

2.4.1 Definition of the vector product

Consider two vectors **a** and **b** and let α be the angle between them (Fig 2.12). The **vector product** of **a** and **b** is denoted by $\mathbf{a} \times \mathbf{b}$ and is defined to be the vector

$$\mathbf{a} \times \mathbf{b} = (|\mathbf{a}|\,|\mathbf{b}|\sin\alpha)\mathbf{n} \qquad (0 \le \alpha \le 180°) \tag{2.19}$$

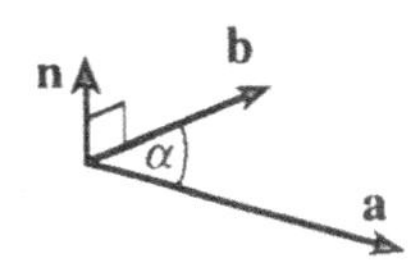

Fig 2.12

$\mathbf{a} \times \mathbf{b} = (|\mathbf{a}|\,|\mathbf{b}|\sin\alpha)\mathbf{n}$.

The vector product $\mathbf{a} \times \mathbf{b}$ is sometimes denoted by $\mathbf{a} \wedge \mathbf{b}$.

The cross symbol between two vectors always denotes the vector product. For this reason the vector product is sometimes called the **cross product**, and in speech we refer to $\mathbf{a} \times \mathbf{b}$ as "**a** cross **b**". Note that α is the smaller of the two angles between **a** and **b** and so $\sin\alpha$ is always non-negative. The direction of the vector product $\mathbf{a} \times \mathbf{b}$ is the direction of the unit vector **n** which we define to be at right angles to the plane of **a** and **b** in the direction given by **the right-hand rule for vector products** (Fig 2.13a): hold your right hand such that your fingers curl from the first vector **a** towards the second vector **b** through the angle α; then your extended thumb points in the direction of the vector product. This is essentially the same rule that was used in Chapter 1 to define the direction of the *z*-axis of a right-handed cartesian coordinate system. The screw rule (Fig 2.13b) is an alternative version of the right-hand rule.

The magnitude of the vector product is, from Eq (2.19), simply

$$|\mathbf{a} \times \mathbf{b}| = |\mathbf{a}|\,|\mathbf{b}|\sin\alpha \qquad (0 \le \alpha \le 180°) \tag{2.20}$$

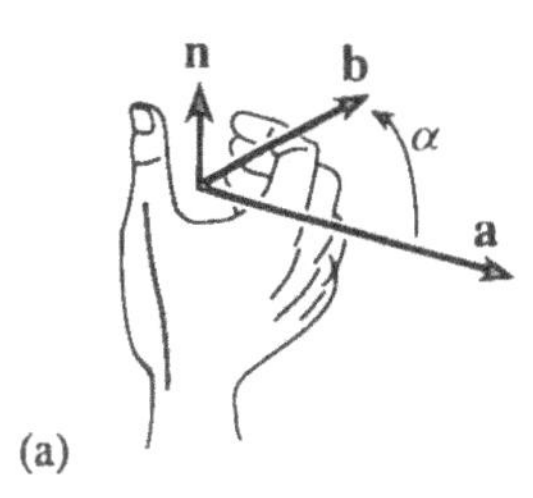

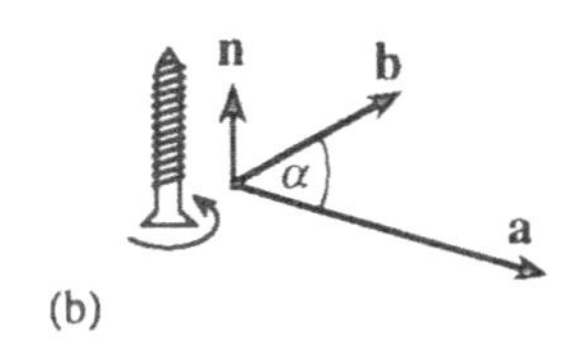

Fig 2.13
(a) The right-hand rule. (b) An equivalent rule is the screw rule: the screw advances in the direction **n** of $\mathbf{a} \times \mathbf{b}$ when the screwhead is turned in the sense from the first vector **a** to the second vector **b** through α.

We have seen (Fig 2.10) that $|\mathbf{a} \times \mathbf{b}|$ is the area of the parallelogram defined by vectors **a** and **b**. It follows that the area of a triangle with two sides formed from **a** and **b** is $|\mathbf{a} \times \mathbf{b}|/2$ (Fig 2.14).

It follows from the definition, Eq (2.19), that the vector product is zero if any one of the two vectors is the zero vector, or whenever the angle α is 0 or

180°, i.e. whenever the two vectors are collinear (parallel or antiparallel). This property provides us with the **collinearity condition** for non-zero vectors

We have defined collinear vectors to mean vectors that lie on the same straight line (or parallel lines) and therefore point in the same or in opposite directions.

$$\mathbf{a} \times \mathbf{b} = \mathbf{0} \qquad \text{(for } \mathbf{a} \text{ and } \mathbf{b} \text{ collinear)} \qquad (2.21)$$

A special case is the vector product of any vector with itself:

$$\mathbf{a} \times \mathbf{a} = \mathbf{0} \qquad (2.22)$$

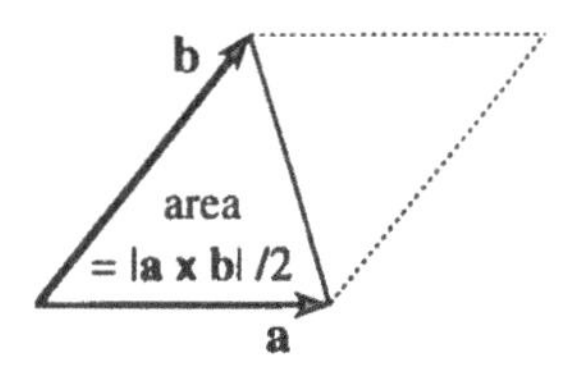

Fig 2.14
The area of the triangle is $|\mathbf{a} \times \mathbf{b}|/2$.

2.4.2 The vector product in vector algebra

You can see from the right-hand rule or the screw rule (Fig 2.13) that the direction of the vector product is reversed when the order in which the two vectors appear in the vector product is changed. Thus the vector product is not commutative; it changes sign when the order is changed

$$\mathbf{b} \times \mathbf{a} = -\mathbf{a} \times \mathbf{b} \qquad \text{(anticommutative rule)} \qquad (2.23)$$

Other rules of algebra are

$$\mathbf{a} \times (\mathbf{b} + \mathbf{c}) = \mathbf{a} \times \mathbf{b} + \mathbf{a} \times \mathbf{c} \qquad \text{(distributive rule)} \qquad (2.24)$$

$$\mathbf{a} \times (\lambda\mathbf{b}) = (\lambda\mathbf{a}) \times \mathbf{b} = \lambda(\mathbf{a} \times \mathbf{b}) \qquad (2.25)$$

where λ is any scalar. The brackets in Eq (2.25) are optional.

Because of the non-commutative property (Eq (2.23)) extra care is needed when manipulating expressions involving vector products. If the order of the vectors in a vector product is changed a minus sign must be introduced. Apart from that, we may manipulate vector products by rules (Eqs (2.24) and (2.25)) similar to those used in ordinary number algebra. Recall also that we cannot divide by a vector.

Summary of section 2.4

- The **vector product**, also called the **cross product**, is defined by

  $$\mathbf{a} \times \mathbf{b} = (|\mathbf{a}|\,|\mathbf{b}|\sin\alpha)\mathbf{n} \qquad (0 \le \alpha \le 180°) \qquad (2.19)$$

  where the direction of unit vector $\mathbf{n}$ is specified by the right-hand rule or the screw rule (Fig 2.13).

- The magnitude of the vector product $\mathbf{a} \times \mathbf{b}$ is the area of the parallelogram defined by displacements $\mathbf{a}$ and $\mathbf{b}$ (Fig 2.10a).

- The **collinearity condition** for non-zero vectors is $\mathbf{a} \times \mathbf{b} = \mathbf{0}$. An example is $\mathbf{a} \times \mathbf{a} = \mathbf{0}$.

- Vector products can be manipulated algebraically according to rules that are similar to those of ordinary number algebra except that the vector product is non-commutative (Eq (2.23)) and there is no division by a vector.

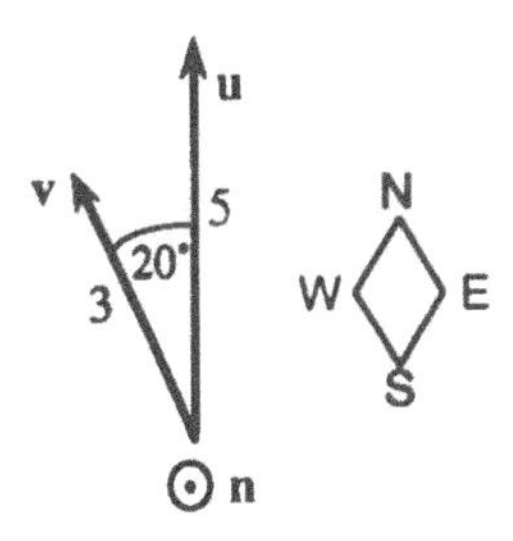

Fig 2.15
The direction n of $\mathbf{u} \times \mathbf{v}$ is out of the plane of the page towards you.

Example 4.1 (*Objective 7*) Let **u** be of magnitude 5 directed towards the north and **v** of magnitude 3 directed 20° west of north (Fig 2.15). Specify the vectors $\mathbf{u} \times \mathbf{v}$, $\mathbf{v} \times \mathbf{u}$, $\mathbf{u} \times \mathbf{u}$, $\mathbf{u} \times (\mathbf{v} + \mathbf{u})$, $-2\mathbf{u} \times 7\mathbf{v}$.

Solution 4.1 The magnitudes are found from Eq (2.20) and the directions are specified by the right-hand rule or the screw rule. The magnitude $|\mathbf{u} \times \mathbf{v}| = 5 \times 3 \sin 20° = 5.13$, and the direction of $\mathbf{u} \times \mathbf{v}$ is vertically upwards, i.e. in the direction of unit vector **n** pointing out of the plane of Fig 2.15 towards you. Thus $\mathbf{u} \times \mathbf{v} = 5.13\mathbf{n}$.

$\mathbf{v} \times \mathbf{u} = -\mathbf{u} \times \mathbf{v} = -5.13\mathbf{n}$.

$\mathbf{u} \times \mathbf{u} = 0$ (Eq (2.22)).

We can write $\mathbf{u} \times (\mathbf{v} + \mathbf{u}) = \mathbf{u} \times \mathbf{v} + \mathbf{u} \times \mathbf{u} = \mathbf{u} \times \mathbf{v}$ since $\mathbf{u} \times \mathbf{u} = 0$. $\mathbf{u} \times \mathbf{v}$ has already been specified as $5.13\mathbf{n}$.

Finally, $-2\mathbf{u} \times 7\mathbf{v} = -14\mathbf{u} \times \mathbf{v} = -71.8\mathbf{n}$.

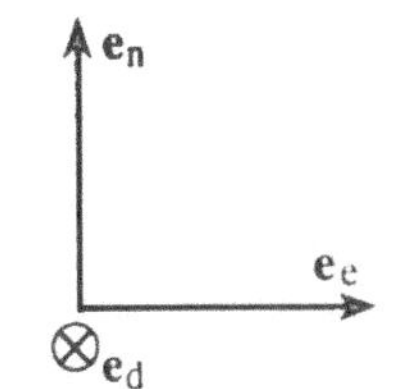

Fig 2.16
The vectors $\mathbf{e}_n$, $\mathbf{e}_e$ and $\mathbf{e}_d$ are mutually orthogonal. The cross in a circle indicates that the direction of $\mathbf{e}_d$ is into the plane of the page. You can think of this symbol as representing the tail of an arrow receding from you.

Example 4.2 (*Objective 7*) Consider the mutually orthogonal unit vectors $\mathbf{e}_n$, $\mathbf{e}_e$ and $\mathbf{e}_d$ directed towards the north, towards the east and vertically downwards respectively. Determine all possible vector products $\mathbf{e}_i \times \mathbf{e}_j$ where i and j can each be n, e or d.

Solution 4.2 In cases where subscripts i and j are the same we have $\mathbf{e}_i \times \mathbf{e}_i = 0$ (Eq (2.22)).

When i = **n** and j = **e** we have (Fig 2.16) $\mathbf{e}_n \times \mathbf{e}_e = (1 \times 1 \times \sin 90°)\mathbf{e}_d = \mathbf{e}_d$. Similarly, $\mathbf{e}_e \times \mathbf{e}_d = \mathbf{e}_n$ and $\mathbf{e}_d \times \mathbf{e}_n = \mathbf{e}_e$.

When the order of the vectors in a vector product is changed, the sign is reversed, and so we have $\mathbf{e}_e \times \mathbf{e}_n = -\mathbf{e}_d$, etc.

Example 4.3 (*Objectives 7,8*) Vectors **a**, **b**, **c** and **b** + **c** are shown in Fig 2.17. **a** is normal to the plane of **b** and **c**, and the angle between **b** and **c** is an acute angle.

(a) Specify the geometrical figures that have areas represented by

(i) $|\mathbf{a} \times \mathbf{b}|$,
(ii) $|\mathbf{a} \times \mathbf{c}|$,
(iii) $|\mathbf{a} \times (\mathbf{b} + \mathbf{c})|$.

(b) Is $|\mathbf{a} \times \mathbf{b}| + |\mathbf{a} \times \mathbf{c}| = |\mathbf{a} \times (\mathbf{b} + \mathbf{c})|$? Give reasons.

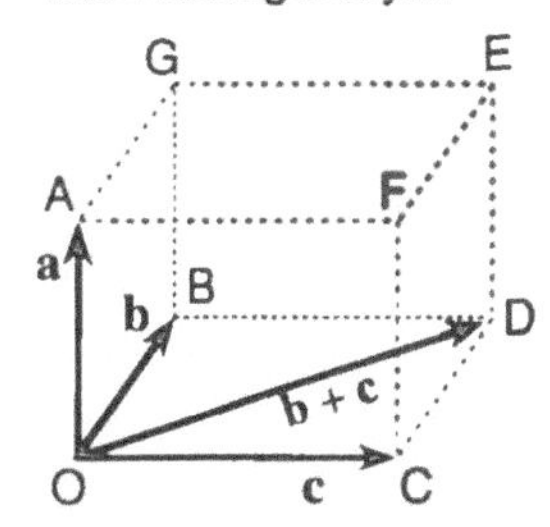

Fig 2.17
Vector **a** is normal to the plane of **b** and **c**.

(c) Make a sketch, in the plane of **b** and **c**, showing **b**, **c**, **b** + **c**, **a** × **b**, **a** × **c** and **a** × (**b** + **c**). Assume for convenience that $|\mathbf{a}| = 1/2$. Specify the parallelogram that illustrates the distributive rule (2.24).

Solution 4.3

(a) (i) $|\mathbf{a} \times \mathbf{b}| = ab\sin 90° = ab$, where $a = |\mathbf{a}|$ and $b = |\mathbf{b}|$, i.e. $|\mathbf{a} \times \mathbf{b}|$ is the area of the rectangle OAGB.

(ii) Similarly, $|\mathbf{a} \times \mathbf{c}| = ac$, the area of rectangle OAFC, and

(iii) $|\mathbf{a} \times (\mathbf{b} + \mathbf{c})| = a|\mathbf{b} + \mathbf{c}|$, the area of rectangle ODEA.

(b) No. The three rectangles involved all have the same height a but they stand on lines OB, OC and OD that have lengths corresponding to three sides of a triangle. The sum of the two smaller rectangular areas must therefore be greater than the larger rectangular area standing on OD, since CD + OC > OD and OB = CD.

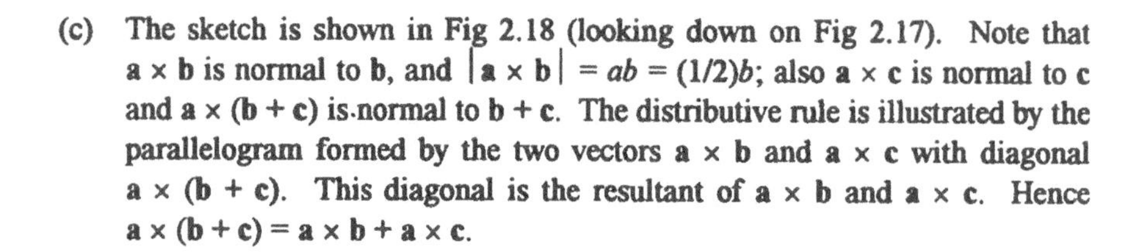

(c) The sketch is shown in Fig 2.18 (looking down on Fig 2.17). Note that **a** × **b** is normal to **b**, and $|\mathbf{a} \times \mathbf{b}| = ab = (1/2)b$; also **a** × **c** is normal to **c** and **a** × (**b** + **c**) is normal to **b** + **c**. The distributive rule is illustrated by the parallelogram formed by the two vectors **a** × **b** and **a** × **c** with diagonal **a** × (**b** + **c**). This diagonal is the resultant of **a** × **b** and **a** × **c**. Hence $\mathbf{a} \times (\mathbf{b} + \mathbf{c}) = \mathbf{a} \times \mathbf{b} + \mathbf{a} \times \mathbf{c}$.

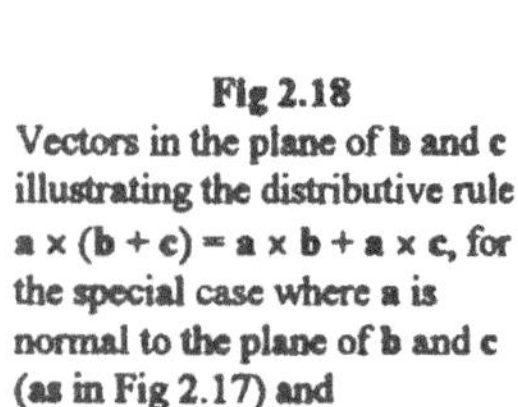

Fig 2.18
Vectors in the plane of **b** and **c** illustrating the distributive rule $\mathbf{a} \times (\mathbf{b} + \mathbf{c}) = \mathbf{a} \times \mathbf{b} + \mathbf{a} \times \mathbf{c}$, for the special case where **a** is normal to the plane of **b** and **c** (as in Fig 2.17) and $|\mathbf{a}| = a = 1/2$.

Example 4.4 (*Objectives 3,9*) Given the equation $\mathbf{a} = \beta\mathbf{b} + \gamma\mathbf{c}$, select the options that represent correct statements, giving your reasoning.

A $|\mathbf{a}|^2 = \beta\mathbf{a} \cdot \mathbf{b} + \gamma\mathbf{a} \cdot \mathbf{c}$

B $\mathbf{a} \times \mathbf{b} = \gamma\mathbf{b} \times \mathbf{c}$

C $\gamma = (\mathbf{c} \cdot \mathbf{a} - \beta\mathbf{b} \cdot \mathbf{c})/|\mathbf{c}|^2$

D $0 = \beta\mathbf{a} \times \mathbf{b} + \gamma\mathbf{a} \times \mathbf{c}$

E $\mathbf{a} \times \mathbf{b}/\gamma = -\mathbf{b} \times \mathbf{c}$

F $|\mathbf{a}|^2 = \beta^2 + \gamma^2$

Solution 4.4 A is correct. Take the scalar product of **a** with both sides of the given equation to give $\mathbf{a} \cdot \mathbf{a} = \mathbf{a} \cdot \beta\mathbf{b} + \mathbf{a} \cdot \gamma\mathbf{c}$. Then put $\mathbf{a} \cdot \mathbf{a} = |\mathbf{a}|^2$ and take β and γ outside the scalar products (Eq 2.7).

C is correct. Take the scalar product of **c** with both sides of the given equation and rearrange.

D is correct. Take the vector product of **a** with both sides of the given equation, put $\mathbf{a} \times \mathbf{a} = \mathbf{0}$ and take β and γ outside the vector products (Eq 2.25).

E is correct. Take the vector product of **b** with both sides, put $\mathbf{b} \times \beta\mathbf{b} = \mathbf{0}$, use $\mathbf{b} \times \mathbf{a} = -\mathbf{a} \times \mathbf{b}$ and rearrange.

(B is incorrect: $\mathbf{a} \times \mathbf{b} = -\gamma\mathbf{b} \times \mathbf{c}$. This is obtained by taking the vector product of both sides with **b**, putting $\mathbf{b} \times \gamma\mathbf{b} = \mathbf{0}$, $\mathbf{b} \times \mathbf{a} = -\mathbf{a} \times \mathbf{b}$ and rearranging.

F is incorrect. Taking the scalar product of each side of the given equation with itself yields $|\mathbf{a}|^2 = (\beta\mathbf{b} + \gamma\mathbf{c}) . (\beta\mathbf{b} + \gamma\mathbf{c}) = \beta^2|\mathbf{b}|^2 + 2\beta\gamma\mathbf{b} . \mathbf{c} + \gamma^2|\mathbf{c}|^2$ where the commutative property $\mathbf{c} . \mathbf{b} = \mathbf{b} . \mathbf{c}$ has been used.)

Problem 4.1 (*Objectives 7,8,9*) In triangle ABC of Fig 2.19, $|\mathbf{b}| = 3$ and $|\mathbf{c}| = 2$.

(a) Specify the vector product $\mathbf{b} \times \mathbf{c}$ and hence state the area of the triangle.
(b) Determine $\mathbf{b} \times (\mathbf{b} \times \mathbf{c})$, i.e. the vector product of **b** with the vector $\mathbf{b} \times \mathbf{c}$.
(c) State the direction of $\mathbf{b} - \mathbf{c}$. If $\mathbf{a} = \mathbf{BC}$, explain why $\mathbf{a} \times (\mathbf{b} - \mathbf{c}) = \mathbf{0}$.

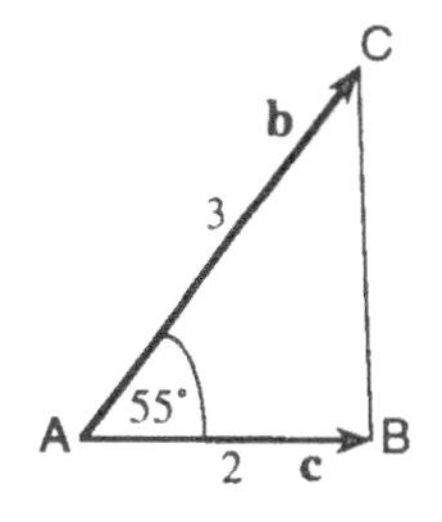

Fig 2.19
The triangle ABC.

Problem 4.2 (*Objectives 7,9*) $\mathbf{e}_1$ and $\mathbf{e}_2$ are orthogonal unit vectors and vector **F** is of length 5 and directed normal to the plane of $\mathbf{e}_1$ and $\mathbf{e}_2$. Determine $|\mathbf{F} \times \mathbf{e}_1|$ and $|\mathbf{F} \times (2\mathbf{e}_1 - \mathbf{e}_2)|$.

Problem 4.3 (*Objectives 3,9*) Some of the options below express possible relationships among certain vectors and scalars; others are meaningless. Select the options that are meaningless, giving your reasons.

A	$\mathbf{p} \times \mathbf{q} - \mathbf{p} . \mathbf{q} = \mathbf{r}$	E	$\mathbf{p} \times \mathbf{t} = |\mathbf{v}|/\mathbf{v}$
B	$\mathbf{a} . (\mathbf{p} \times \mathbf{q}) = \alpha$	F	$(\mathbf{a} . \mathbf{b})\mathbf{c} = \mathbf{r} \times \mathbf{s}$
C	$\mathbf{p} \times \mathbf{q} - \mathbf{s} = \mathbf{0}$	G	$\mathbf{a} + \mathbf{b} \times \mathbf{c} = \mathbf{a} . \mathbf{v}$
D	$|\mathbf{r} \times (\mathbf{t} - \mathbf{b})| = s$	H	$\mathbf{Q} \times (\mathbf{R} \times \mathbf{S}) = \mathbf{T}$

Problem 4.4 (*Objectives 8,9*) Refer to the three vectors forming the sides of the triangle shown in Fig 2.20. Use the relationship between the vector product and the area of a triangle to derive the **sine rule**:

$$\frac{a}{\sin\alpha} = \frac{b}{\sin\beta} = \frac{c}{\sin\gamma}$$

where $a = |\mathbf{a}|$, etc.

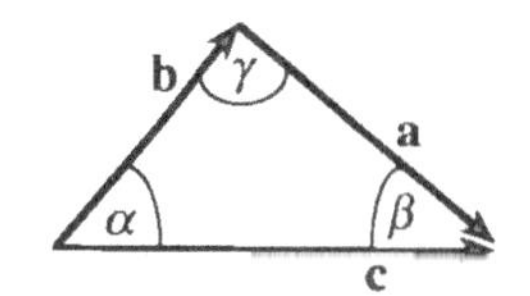

Fig 2.20
The triangle formed from **a**, **b** and **c**.

Problem 4.5 (*Objective 7*) Specify all possible vector products of the cartesian unit vectors.

2.5 CARTESIAN FORM OF THE VECTOR PRODUCT

You have seen that the scalar product of two cartesian vectors is given by a simple formula involving only the cartesian components, Eq (2.11). This formula was obtained by making use of the scalar products of the cartesian unit vectors, Eqs (2.10). In a similar way we now derive a formula for the vector product of two cartesian vectors. The first step is to determine the vector products of the cartesian unit vectors.

It follows from Eq (2.22) that $\mathbf{i} \times \mathbf{i} = \mathbf{0}$, $\mathbf{j} \times \mathbf{j} = \mathbf{0}$ and $\mathbf{k} \times \mathbf{k} = \mathbf{0}$. We also have $|\mathbf{i} \times \mathbf{j}| = |\mathbf{i}|\,|\mathbf{j}|\sin 90° = 1$. The direction of $\mathbf{i} \times \mathbf{j}$ is found from the right-hand rule to be that of $\mathbf{k}$. Thus $\mathbf{i} \times \mathbf{j} = \mathbf{k}$, and similarly for the vector products of the other unit vectors. Collecting these results together we have

$$\mathbf{i} \times \mathbf{i} = \mathbf{j} \times \mathbf{j} = \mathbf{k} \times \mathbf{k} = \mathbf{0} \qquad (2.26a)$$

$$\mathbf{i} \times \mathbf{j} = \mathbf{k}, \quad \mathbf{j} \times \mathbf{k} = \mathbf{i}, \quad \mathbf{k} \times \mathbf{i} = \mathbf{j} \qquad (2.26b)$$

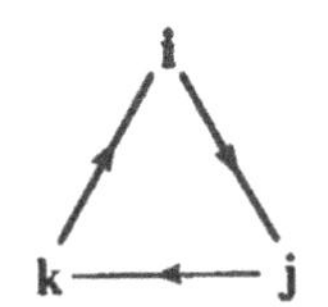

Fig 2.21
Cyclic orders of the three unit vectors are indicated by a clockwise rotation.

with $\mathbf{j} \times \mathbf{i} = -\mathbf{k}$, etc. Notice the cyclic order of the vectors in Eqs (2.26b). The vector product of any two cartesian unit vectors, taken in the clockwise cyclic order indicated in Fig 2.21, yields the third cartesian unit vector; the anticlockwise order introduces a minus sign.

Now consider the vector product of two cartesian vectors $\mathbf{a}$ and $\mathbf{b}$,

$$\mathbf{a} \times \mathbf{b} = (a_x\mathbf{i} + a_y\mathbf{j} + a_z\mathbf{k}) \times (b_x\mathbf{i} + b_y\mathbf{j} + b_z\mathbf{k}) \qquad (2.27)$$

The rules of algebra for the vector product (Eqs (2.23) to (2.25)) allow us to multiply the brackets out and simplify, in the same way as we multiply brackets and simplify in ordinary algebra, except that here it is important to respect the order in a vector product. Thus, for example, the product of the first two terms in each bracket gives $a_x\mathbf{i} \times b_x\mathbf{i}$ which can be written as $a_xb_x\mathbf{i} \times \mathbf{i} = \mathbf{0}$ by Eq (2.26a). The product of the first term in the first bracket and the second term in the second bracket gives $a_x\mathbf{i} \times b_y\mathbf{j}$ which simplifies to $a_xb_y\mathbf{i} \times \mathbf{j} = a_xb_y\mathbf{k}$, where Eq (2.26b) has been used. Proceeding in this way we obtain from Eq (2.27) nine terms of which three are zero. The remaining six terms can be grouped to give

$$\mathbf{a} \times \mathbf{b} = (a_yb_z - a_zb_y)\mathbf{i} + (a_zb_x - a_xb_z)\mathbf{j} + (a_xb_y - a_yb_x)\mathbf{k} \qquad (2.28a)$$

This formula gives the vector product of two cartesian vectors in terms of the components of the two vectors. Despite its apparent complexity, the right-hand side of Eq (2.28a) exhibits various cyclic symmetries that make it easy to use and remember. For example, notice that the x-component of the vector product

$$(\mathbf{a} \times \mathbf{b})_x = a_yb_z - a_zb_y \qquad (2.29)$$

involves only the y- and z-components of the two vectors. Similarly, the y-component of the vector product depends only on the z- and x- components of the two vectors, and is obtained from Eq (2.29) by the cyclic replacement of the subscripts (Fig 2.21): $x \to y$, $y \to z$, $z \to x$, and similarly for the z-component of the vector product. This results in a cyclic order of corresponding terms in each bracket. For example, the symbols appearing first in each of the three brackets in Eq (2.28a) are, respectively, a_y, a_z, a_x. Other symbols in corresponding positions in the three brackets also exhibit a cyclic order.

The important result, Eq (2.28a), can be expressed as an ordered triple

$$\mathbf{a} \times \mathbf{b} = (a_y b_z - a_z b_y, a_z b_x - a_x b_z, a_x b_y - a_y b_x) \tag{2.28b}$$

or as a determinant

$$\mathbf{a} \times \mathbf{b} = \begin{vmatrix} \mathbf{i} & \mathbf{j} & \mathbf{k} \\ a_x & a_y & a_z \\ b_x & b_y & b_z \end{vmatrix} \tag{2.28c}$$

Recall that the vector product $\mathbf{a} \times \mathbf{b}$ is orthogonal to both $\mathbf{a}$ and $\mathbf{b}$. We can express this by using the orthogonality condition, Eq (2.2). For example, the orthogonality of the two vectors $\mathbf{a}$ and $\mathbf{a} \times \mathbf{b}$ is expressed by

$$\mathbf{a} \,.\, (\mathbf{a} \times \mathbf{b}) = 0 \tag{2.30}$$

and similarly $\mathbf{b} \,.\, (\mathbf{a} \times \mathbf{b}) = 0$. This property is used in the Examples below to check calculations of vector products.

Summary of section 2.5

- The vector products of the cartesian unit vectors are

$$\mathbf{i} \times \mathbf{i} = \mathbf{j} \times \mathbf{j} = \mathbf{k} \times \mathbf{k} = 0 \tag{2.26a}$$

$$\mathbf{i} \times \mathbf{j} = \mathbf{k}, \quad \mathbf{j} \times \mathbf{k} = \mathbf{i}, \quad \mathbf{k} \times \mathbf{i} = \mathbf{j} \qquad (\mathbf{j} \times \mathbf{i} = -\mathbf{k}, \text{ etc.}) \tag{2.26b}$$

- The vector product of any two cartesian vectors is

$$\mathbf{a} \times \mathbf{b} = (a_y b_z - a_z b_y)\mathbf{i} + (a_z b_x - a_x b_z)\mathbf{j} + (a_x b_y - a_y b_x)\mathbf{k} \tag{2.28a}$$

Example 5.1 (*Objective 10*) Consider the cartesian vectors

$$\mathbf{p} = \mathbf{i} - 3\mathbf{j} + 2\mathbf{k}, \quad \mathbf{q} = 3\mathbf{i} + 2\mathbf{j} - \mathbf{k} \quad \text{and} \quad \mathbf{r} = \mathbf{j} + \mathbf{k}$$

Determine $\mathbf{p} \times \mathbf{q}$, $\mathbf{q} \times \mathbf{r}$, $\mathbf{k} \times \mathbf{p}$ and $(\mathbf{r} \times \mathbf{p})_y$.

Solution 5.1 Using Eq (2.28a), we have

$$\mathbf{p} \times \mathbf{q} = ((-3)(-1) - 2\times2)\mathbf{i} + (2\times3 - 1\times(-1))\mathbf{j} + (1\times2 - (-3)\times3)\mathbf{k}$$

$$= -\mathbf{i} + 7\mathbf{j} + 11\mathbf{k}$$

$$\mathbf{q} \times \mathbf{r} = (2\times1 - (-1)\times1)\mathbf{i} + (-1\times0 - 3\times1)\mathbf{j} + (3\times1 - 2\times0)\mathbf{k}$$

$$= 3\mathbf{i} - 3\mathbf{j} + 3\mathbf{k}$$

$$\mathbf{k} \times \mathbf{p} = (0\times2 - 1\times(-3))\mathbf{i} + (1\times1 - 0\times2)\mathbf{j} + (0\times(-3) - 0\times1)\mathbf{k}$$

$$= 3\mathbf{i} + \mathbf{j}$$

(Check: a vector product is orthogonal to each of the two vectors (Eq (2.30)). We can use this property to check our answer for $\mathbf{p} \times \mathbf{q}$ by calculating, say, $\mathbf{p}\,.\,(\mathbf{p} \times \mathbf{q}) = (\mathbf{i} - 3\mathbf{j} + 2\mathbf{k})\,.\,(-\mathbf{i} + 7\mathbf{j} + 11\mathbf{k}) = 1\times(-1) + (-3)\times7 + 2\times11 = 0$. We could instead have confirmed that $\mathbf{q}\,.\,(\mathbf{p} \times \mathbf{q}) = 0$. Similarly we check $\mathbf{q}\,.\,(\mathbf{q} \times \mathbf{r}) = 3\times3 + 2\times(-3) + (-1)\times3 = 0$, etc.)

Finally we have to determine $(\mathbf{r} \times \mathbf{p})_y$, the y-component of $\mathbf{r} \times \mathbf{p}$, which is $r_z p_x - r_x p_z = 1\times1 - 0\times2 = 1$.

Example 5.2 (*Objective 10*) Determine the vector product $\mathbf{u} \times \mathbf{v}$ where $\mathbf{u} = 3\mathbf{i} - 2\mathbf{j}$ and $\mathbf{v} = 5\mathbf{i} + 4\mathbf{j}$,

(a) by multiplying out the brackets $(3\mathbf{i} - 2\mathbf{j}) \times (5\mathbf{i} + 4\mathbf{j})$ and using the orthonormal properties of the unit vectors, and

(b) by using Eq (2.28a).

Solution 5.2

(a) $(3\mathbf{i} - 2\mathbf{j}) \times (5\mathbf{i} + 4\mathbf{j})$

$$= 3\mathbf{i} \times 5\mathbf{i} + 3\mathbf{i} \times 4\mathbf{j} + (-2\mathbf{j} \times 5\mathbf{i}) + (-2\mathbf{j} \times 4\mathbf{j})$$

$$= 15\mathbf{i} \times \mathbf{i} + 12\mathbf{i} \times \mathbf{j} - 10\mathbf{j} \times \mathbf{i} - 8\mathbf{j} \times \mathbf{j}$$

$$= 0 + 12\mathbf{k} - 10(-\mathbf{k}) - 8 \times 0 = 22\mathbf{k}$$

(b) $\mathbf{u} \times \mathbf{v} = ((-2)\times0 - 0\times4)\mathbf{i} + (0\times5 - 3\times0)\mathbf{j} + (3\times4 - (-2)\times5)\mathbf{k} = 22\mathbf{k}$

(Check: We note that $\mathbf{u}$ and $\mathbf{v}$ are both in the x-y plane and our answer, $22\mathbf{k}$, is parallel to the z-axis and is therefore perpendicular to both $\mathbf{u}$ and $\mathbf{v}$ as required.)

Example 5.3 (*Objectives 8,10*) Find the area of the triangle defined by the three points A(1,0,2), B(3,2,1) and C(–1,–5,0) and specify a unit vector normal to the plane of the triangle.

Solution 5.3 The area of any triangle is $|\mathbf{a} \times \mathbf{b}|/2$ where the vectors **a** and **b** make two sides. A sketch in the plane of the given triangle is shown in Fig 2.22. Vectors representing two sides of the triangle are in ordered triple notation,

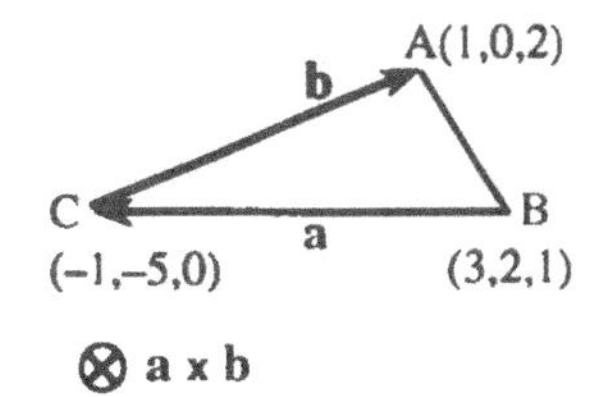

Fig 2.22
The direction of **a** × **b** is normal to the plane of **a** and **b** into the page pointing away from you. This direction is indicated in the figure by the cross in a circle.

$$\mathbf{a} = \mathbf{BC} = (-1-3, -5-2, 0-1) = (-4,-7,-1)$$

and

$$\mathbf{b} = \mathbf{CA} = (1-(-1), 0-(-5), 2-0) = (2,5,2)$$

The vector product is, using Eq (2.28b),

$$\mathbf{a} \times \mathbf{b} = (-7\times2-(-1)\times5, -1\times2-(-4)\times2, -4\times5-(-7)\times2) = (-9,6,-6).$$

The required area is therefore

$$A = \frac{1}{2}|(-9,6,-6)| = \frac{1}{2}(81+36+36)^{1/2} = 6.185$$

The vector product **a** × **b** is a vector directed normal to the plane of the triangle and into the plane of the page, as indicated by the symbol ⊗ in Fig 2.22. A unit vector normal to the plane is therefore

$$\frac{\mathbf{a}\times\mathbf{b}}{|\mathbf{a}\times\mathbf{b}|} = \frac{(-9,6,-6)}{2\times6.185} = 0.243(-3,2,-2)$$

(Another acceptable answer is –0.243(–3,2,–2), the unit vector in the opposite direction. Note that the same answers to all parts of this question would be obtained from any pair of vectors representing any two sides of the triangle.)

Problem 5.1 (*Objectives 9,10*) Calculate

(a) $\mathbf{r} \times \mathbf{p}$ where $\mathbf{r} = (2,3,5)$ and $\mathbf{p} = (1,0,-1)$
(b) $\mathbf{i} \times \mathbf{A}$ where $\mathbf{A} = \mathbf{j} + \mathbf{k}$
(c) $(\mathbf{i}+\mathbf{j}) \times (\mathbf{i}-\mathbf{j})$

Problem 5.2 (*Objective 8*) Find the area of the triangle formed from the three vectors $\mathbf{i} + 3\mathbf{j}$, $\mathbf{j} - \mathbf{k}$ and $\mathbf{i} + 4\mathbf{j} - \mathbf{k}$.

Problem 5.3 (*Objective 10*)

(a) Let $\mathbf{q}$ be any non-zero cartesian vector not collinear with $\mathbf{k}$. Specify the plane in which $\mathbf{k} \times \mathbf{q}$ lies.

(b) Find a vector in the x-y plane normal to $\mathbf{a} = 3\mathbf{i} - \mathbf{j} - 2\mathbf{k}$.

Problem 5.4 (*Objective 10,11*) Use the vector product to determine the conditions that the scalars α, β and γ must satisfy if the vector (α,β,γ) is parallel to the vector $(2,0,-3)$.

Problem 5.5 (*Objective 7*) A student calculates the vector product $\mathbf{a} \times \mathbf{b}$ of two given vectors $\mathbf{a}$ and $\mathbf{b}$. He checks that his answer satisfies $\mathbf{a} \,.\, (\mathbf{a} \times \mathbf{b}) = 0$ and $\mathbf{b} \,.\, (\mathbf{a} \times \mathbf{b}) = 0$. Do these checks guarantee that his answer is correct? Give reasons for your answer.

2.6 TRIPLE PRODUCTS OF VECTORS

There are no new principles introduced in this section. It deals with the scalar and vector products of two vectors when one of them is itself a vector product. Such expressions do occur in physical applications and so it is useful to become familiar with them and to know how to manipulate them efficiently. You will need to have developed a good facility with scalar and vector products from previous sections before attempting this section.

2.6.1 The scalar triple product

The vector product $\mathbf{a} \times \mathbf{b}$ is itself a vector. We can therefore form a scalar product of the vector $\mathbf{a} \times \mathbf{b}$ with a third vector $\mathbf{c}$ to give the scalar quantity $(\mathbf{a} \times \mathbf{b}) \,.\, \mathbf{c}$. Actually the brackets can be removed without ambiguity, $(\mathbf{a} \times \mathbf{b}) \,.\, \mathbf{c} = \mathbf{a} \times \mathbf{b} \,.\, \mathbf{c}$, because $\mathbf{b} \,.\, \mathbf{c}$ is a scalar and so $\mathbf{a} \times (\mathbf{b} \,.\, \mathbf{c})$ is meaningless. This way of forming a product of three vectors to give a scalar is called a **scalar triple product**,

$$\mathbf{a} \times \mathbf{b} \,.\, \mathbf{c} \qquad \text{(scalar triple product)} \tag{2.31}$$

The order of the two vectors in the scalar product is of course unimportant and so

$$\mathbf{a} \times \mathbf{b} \,.\, \mathbf{c} = \mathbf{c} \,.\, \mathbf{a} \times \mathbf{b} \tag{2.32}$$

We have already seen examples of scalar triple products such as $\mathbf{a} \,.\, \mathbf{a} \times \mathbf{b}$ in the context of checking answers to vector products (see Eq (2.30) and its application in the Examples of the previous section).

The scalar triple product has a simple geometrical interpretation: the magnitude of the scalar triple product, $|\mathbf{a} \times \mathbf{b} \,.\, \mathbf{c}|$, is the volume of the parallelepiped formed from the three vectors. This is illustrated in Fig 2.23a in the case where the angle γ between **a** × **b** and **c** is an acute angle and the scalar triple product is a positive number. When γ is an obtuse angle the scalar triple product is negative. It follows from this geometrical interpretation that the magnitude of the scalar triple product is independent of any permutation (i.e. change of order) of the three vectors or interchange of the dot and cross symbols (see for example Fig 2.23b).

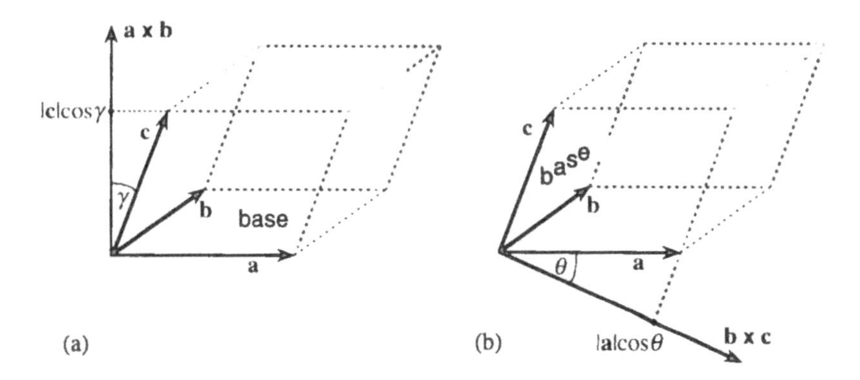

Fig 2.23
Vectors **a**, **b** and **c** define a parallelepiped. The volume of the parallelepiped is V = area of base × vertical height.
(a) The base is the parallelogram defined by **a** and **b** of area $|\mathbf{a} \times \mathbf{b}|$ and the height is $|\mathbf{c}|\cos\gamma$. Hence the volume is $V = |\mathbf{a} \times \mathbf{b}|\,|\mathbf{c}|\cos\gamma = |\mathbf{a} \times \mathbf{b} \,.\, \mathbf{c}|$ or $|\mathbf{c} \,.\, \mathbf{a} \times \mathbf{b}|$.
(b) Regard the base of the parallelepiped to be the parallelogram defined by **b** and **c** of area $|\mathbf{b} \times \mathbf{c}|$. The height is then $|\mathbf{a}|\cos\theta$ and the volume is $V = |\mathbf{b} \times \mathbf{c}|\,|\mathbf{a}|\cos\theta = |\mathbf{b} \times \mathbf{c} \,.\, \mathbf{a}|$ or $|\mathbf{a} \,.\, \mathbf{b} \times \mathbf{c}|$.

Furthermore, any cyclic permutation of the three vectors leaves the sign (as well as the magnitude or volume) of the scalar triple product unchanged, whereas changing the cyclic order changes the sign. These properties of the scalar triple product can be summed up by the identities

$$\mathbf{a} \times \mathbf{b} \,.\, \mathbf{c} = \mathbf{c} \times \mathbf{a} \,.\, \mathbf{b} = \mathbf{b} \times \mathbf{c} \,.\, \mathbf{a} \qquad \text{(cyclic permutations)} \qquad (2.33)$$

$$\mathbf{a} \times \mathbf{b} \,.\, \mathbf{c} = -\mathbf{a} \times \mathbf{c} \,.\, \mathbf{b} \qquad \text{(an anticyclic permutation)} \qquad (2.34)$$

Eqs (2.33) and (2.34) can be derived algebraically from the rules of algebra for scalar and vector products obtained in Sections 2.1.2 and 2.4.2.

We also have, from Eq (2.33) and knowing that $\mathbf{b} \times \mathbf{c} \,.\, \mathbf{a} = \mathbf{a} \,.\, \mathbf{b} \times \mathbf{c}$ (from Eq 2.22)

$$\mathbf{a} \times \mathbf{b} \,.\, \mathbf{c} = \mathbf{a} \,.\, \mathbf{b} \times \mathbf{c} \qquad (2.35)$$

This shows that the dot and cross symbols can be interchanged without changing the value of a scalar triple product.

An application of the scalar triple product is to test for coplanar vectors. For suppose **a**, **b** and **c** are non-zero coplanar vectors. Then the vector **a** × **b**, being normal to the plane of **a** and **b**, is also normal to **c**, and so **a** × **b** . **c** = 0. The "parallelepiped" is in this case a flat figure of zero volume. Thus **the test for three coplanar vectors** is to determine the scalar triple product; if it is zero then the vectors are coplanar. A special case is when two of the three vectors are equal or collinear, $\mathbf{a} = \alpha\mathbf{b}$ for example. We then have $\alpha\mathbf{b} \times \mathbf{b} \,.\, \mathbf{c} = 0$. Thus a scalar triple product is zero whenever one vector appears more than once.

2.6.2 The vector triple product

We can take the vector product of the vector $\mathbf{a} \times \mathbf{b}$ with a vector $\mathbf{c}$ to give the vector $(\mathbf{a} \times \mathbf{b}) \times \mathbf{c}$. This triple product is called a **vector triple product**

$$(\mathbf{a} \times \mathbf{b}) \times \mathbf{c} \qquad \text{(vector triple product)} \qquad (2.36)$$

The brackets are necessary here because $(\mathbf{a} \times \mathbf{b}) \times \mathbf{c}$ is not equal to $\mathbf{a} \times (\mathbf{b} \times \mathbf{c})$. Note that the vector $(\mathbf{a} \times \mathbf{b}) \times \mathbf{c}$ is normal to $\mathbf{a} \times \mathbf{b}$, and $\mathbf{a} \times \mathbf{b}$ is itself normal to $\mathbf{a}$ and $\mathbf{b}$. It follows that $(\mathbf{a} \times \mathbf{b}) \times \mathbf{c}$ is a vector in the plane of $\mathbf{a}$ and $\mathbf{b}$ and so it can be expressed as a linear combination, $\alpha\mathbf{a} + \beta\mathbf{b}$. It can be shown that $\alpha = -\mathbf{b} \,.\, \mathbf{c}$ and $\beta = \mathbf{a} \,.\, \mathbf{c}$, but we shall not show the details here. Thus we have the identity

$$(\mathbf{a} \times \mathbf{b}) \times \mathbf{c} = (\mathbf{a} \,.\, \mathbf{c})\mathbf{b} - (\mathbf{b} \,.\, \mathbf{c})\mathbf{a} \qquad (2.37)$$

This is a useful result because the right-hand side contains scalar products only and it is often easier to work out the scalar products than the vector products on the left-hand side.

Summary of section 2.6

- The triple product $\mathbf{a} \times \mathbf{b} \,.\, \mathbf{c}$ is a scalar called a **scalar triple product**. Its magnitude represents the volume of the parallelepiped formed by the three vectors. It is zero when the three vectors are coplanar thus providing a **test for coplanar vectors**.

- The scalar triple product is unchanged by any cyclic permutation of the three vectors or by interchange of the dot and cross symbols

$$\mathbf{a} \times \mathbf{b} \,.\, \mathbf{c} = \mathbf{c} \times \mathbf{a} \,.\, \mathbf{b} = \mathbf{b} \times \mathbf{c} \,.\, \mathbf{a} \qquad (2.33)$$

$$\mathbf{a} \times \mathbf{b} \,.\, \mathbf{c} = \mathbf{a} \,.\, \mathbf{b} \times \mathbf{c} \qquad (2.35)$$

- The triple product $(\mathbf{a} \times \mathbf{b}) \times \mathbf{c}$ is a vector called a **vector triple product**; it lies in the plane of $\mathbf{a}$ and $\mathbf{b}$ and can be expanded to give the identity

$$(\mathbf{a} \times \mathbf{b}) \times \mathbf{c} = (\mathbf{a} \,.\, \mathbf{c})\mathbf{b} - (\mathbf{b} \,.\, \mathbf{c})\mathbf{a} \qquad (2.37)$$

Example 6.1 (*Objective 12*) Describe where possible each of the following expressions in terms of scaled vectors, scalar products, vector products, and triple products, and state whether the expression represents a scalar or a vector or is meaningless:

(i) $(\mathbf{p}\,.\,\mathbf{q})\mathbf{r}$	(iv) $(\mathbf{p}\,.\,\mathbf{q}\times\mathbf{r})\times\mathbf{s}$	(vii) $\mathbf{C}\times\mathbf{A}\,.\,\mathbf{B}\,.\,\mathbf{C}$
(ii) $\mathbf{A}\,.\,(\mathbf{B}\times\mathbf{C})\times\mathbf{C}$	(v) $(\mathbf{s}\times\mathbf{t})\times(\mathbf{u}\times\mathbf{v})$	
(iii) $(\mathbf{A}\,.\,\mathbf{B})(\mathbf{C}\,.\,\mathbf{E}\times\mathbf{F})$	(vi) $(\mathbf{a}\,.\,\mathbf{b}\times\mathbf{c})\mathbf{u}\times\mathbf{v}$	

Solution 6.1 (i) is a vector $\mathbf{r}$ scaled by the scalar product $\mathbf{p}\,.\,\mathbf{q}$; it is a vector. (ii) is the scalar product of $\mathbf{A}$ with a vector triple product; it is a scalar. (Note that the first three vectors cannot be interpreted as a scalar triple product, i.e. as a scalar, because they are immediately followed by a cross symbol and another vector.) (iii) is a scalar product times a scalar triple product, i.e. it is a scalar. (iv) is meaningless. The bracketed quantity is a scalar which cannot be immediately followed by a cross symbol and another vector. (v) is a vector product of two vector products, i.e. it is a vector. (vi) is a vector product, $\mathbf{u}\times\mathbf{v}$, scaled by a scalar triple product, i.e. it is a vector. (vii) is meaningless whichever operation you try to perform first.

Example 6.2 (*Objective 12*) Verify the vector identity, Eq (2.37), in the case of the three vectors: $\mathbf{a} = (1,0,2)$, $\mathbf{b} = (3,2,1)$ and $\mathbf{c} = (0,0,4)$.

Solution 6.2 We find $\mathbf{a}\times\mathbf{b} = (1,0,2)\times(3,2,1) = (-4,5,2)$ and $(\mathbf{a}\times\mathbf{b})\times\mathbf{c} = (-4,5,2)\times(0,0,4) = (20,16,0)$.

We now calculate $\mathbf{a}\,.\,\mathbf{c} = (1,0,2)\,.\,(0,0,4) = 8$ and $\mathbf{b}\,.\,\mathbf{c} = 4$, and so the right-hand side of Eq (2.37) is

$$(\mathbf{a}\,.\,\mathbf{c})\mathbf{b} - (\mathbf{b}\,.\,\mathbf{c})\mathbf{a} = 8(3,2,1) - 4(1,0,2) = (20,16,0)$$

in agreement with $(\mathbf{a}\times\mathbf{b})\times\mathbf{c}$.

Example 6.3 (*Objective 13*) Derive the identity

$$(\mathbf{p}\times\mathbf{q})\,.\,(\mathbf{r}\times\mathbf{s}) = (\mathbf{p}\,.\,\mathbf{r})(\mathbf{q}\,.\,\mathbf{s}) - (\mathbf{p}\,.\,\mathbf{s})(\mathbf{q}\,.\,\mathbf{r})$$

Solution 6.3 Recognise the left-hand side as the scalar triple product of vector $\mathbf{p}\times\mathbf{q}$ with the vectors $\mathbf{r}$ and $\mathbf{s}$. Using Eqs (2.33) and (2.35) we can write $(\mathbf{p}\times\mathbf{q})\,.\,(\mathbf{r}\times\mathbf{s}) = \mathbf{r}\,.\,\mathbf{s}\times(\mathbf{p}\times\mathbf{q})$. Now recognise this as the scalar product of $\mathbf{r}$ with the vector triple product $\mathbf{s}\times(\mathbf{p}\times\mathbf{q}) = -(\mathbf{p}\times\mathbf{q})\times\mathbf{s}$ and use Eq (2.37) to expand the triple product. Thus we can express $(\mathbf{p}\times\mathbf{q})\,.\,(\mathbf{r}\times\mathbf{s})$ as

$$\mathbf{r}\,.\,(-(\mathbf{p}\times\mathbf{q})\times\mathbf{s}) = \mathbf{r}\,.\,[(\mathbf{s}\,.\,\mathbf{q})\mathbf{p} - (\mathbf{s}\,.\,\mathbf{p})\mathbf{q}] = (\mathbf{s}\,.\,\mathbf{q})(\mathbf{r}\,.\,\mathbf{p}) - (\mathbf{s}\,.\,\mathbf{p})(\mathbf{r}\,.\,\mathbf{q})$$

which is equal to the right-hand side of the given identity.

Problem 6.1 (*Objective 12*) Given $\mathbf{a} = (1,2,3)$, $\mathbf{b} = (0,4,1)$ and $\mathbf{c} = (5,-1,0)$, calculate

(a) $\mathbf{a} \cdot \mathbf{b} \times \mathbf{c}$,
(b) $\mathbf{a} \times (\mathbf{b} \times \mathbf{c})$
(c) $(\mathbf{a} \times \mathbf{b}) \times \mathbf{c}$

Problem 6.2 (*Objectives 12,13*) Derive the identity

$$(\mathbf{c} \times \mathbf{a}) \times (\mathbf{b} \times \mathbf{a}) = (\mathbf{b} \cdot \mathbf{a} \times \mathbf{c})\mathbf{a}$$

and confirm this result for the special case where **a**, **b** and **c** are the vectors given in Problem 6.1.

Problem 6.3 (*Objectives 12,13*)

(a) Derive the identity

$$|\mathbf{a} \times \mathbf{b}|^2 = |\mathbf{a}|^2|\mathbf{b}|^2 - (\mathbf{a} \cdot \mathbf{b})^2$$

(b) Calculate $|(5,-3,1) \times (-2,7,9)|^2$.

Problem 6.4 (*Objective 12*) Calculate $\mathbf{u} \times (\mathbf{v} \times \mathbf{w})$ for $\mathbf{u} = (1,0,-1)$, $\mathbf{v} = (3,2,1)$ and $\mathbf{w} = (0,1,0)$.

2.7 SCALAR AND VECTOR PRODUCTS IN SCIENCE AND ENGINEERING

No new vector principles are taught in this section. The aim is to illustrate scalar products and vector products in a scientific context by worked examples and problems. Topics are drawn from the areas of mechanics and electromagnetism. The text is restricted to brief background summaries of the scientific terms and definitions that are referred to in the Examples and Problems. These summaries are not intended to be studied in the usual sense. You should start straightaway on the Examples and refer to the summary when cued to do so.

2.7.1 Background summary: Forces, torques and equilibrium

The Examples and Problems in this section relate to the following:

1. When a force **F** is applied at a point P, the **torque** of **F** about a point O is defined to be the vector

$$\boldsymbol{\Gamma} = \mathbf{r} \times \mathbf{F} \qquad \text{(m N or N m)} \qquad (2.38)$$

where $\mathbf{r} = \mathbf{OP}$ (Fig 2.24). Note that the torque depends on the position of the point O as well as the force **F** and the point P.

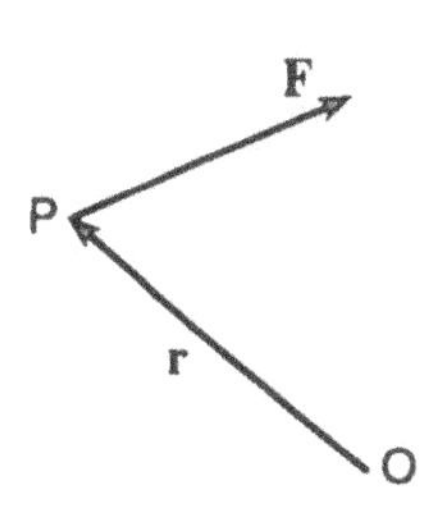

Fig 2.24
The torque of **F** about O is $\Gamma = \mathbf{r} \times \mathbf{F}$.

2. A pair of forces of the same magnitude but opposite directions is called a **couple** (Fig 2.25).

3. A body is in **translational equilibrium** when the resultant force acting on it is zero.

4. A body that is in translational equilibrium is also in **rotational equilibrium** when the resultant torque about *any* point is zero.

5. A system of forces is said to be **equivalent** to another system of forces if both systems have the same resultant force and the same resultant torque about any point.

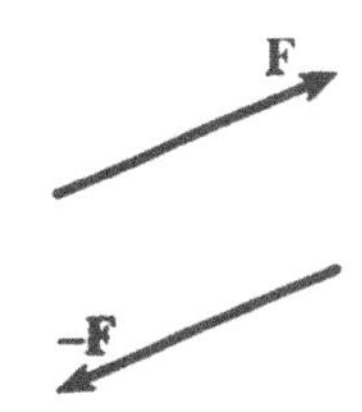

Fig 2.25
The two antiparallel forces **F** and **–F** constitute a couple.

Example 7.1 (*Objectives 4,10,14*) Fig 2.26 shows three forces acting at points O, R and S in the *y-z* plane.

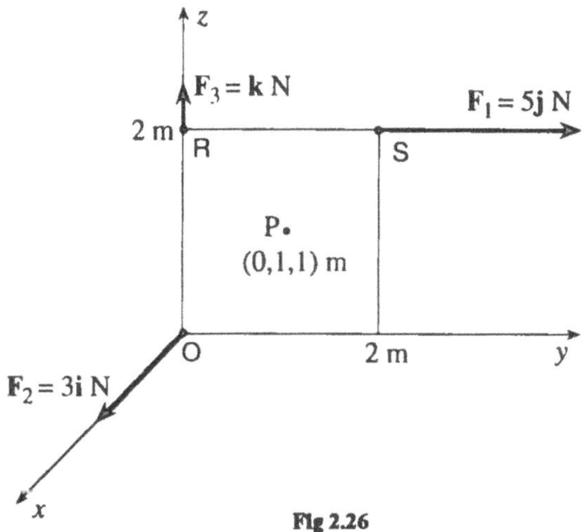

Fig 2.26
Three forces acting at O, R and S.

(a) Use the definition of **torque** (Eq (2.38)) to determine the torques of each of the three forces about (i) the point O(0,0,0) and (ii) the point P(0,1,1).

(b) Determine the resultant force **F** and the resultant torque **G** about P.

(c) Are **F** and **G** perpendicular to one another? Are they collinear?

Solution 7.1

(a) (i) The force $\mathbf{F}_1$ acting at point S has a torque about O of

$$\Gamma_1 = \mathbf{OS} \times \mathbf{F}_1 = (2\mathbf{j} + 2\mathbf{k}) \times 5\mathbf{j}$$

To work out this vector product we could use Eqs (2.28), but it is quicker simply to multiply out the brackets and recall the vector products of the cartesian unit vectors (Eqs (2.26)). Thus

$$\Gamma_1 = 10\mathbf{j} \times \mathbf{j} + 10\mathbf{k} \times \mathbf{j} = 0 + (-10\mathbf{i}) = -10\mathbf{i}$$

Putting in the units, we have the answer $\Gamma_1 = -10\mathbf{i}$ Nm.

We now have to find the torque Γ_2 of $\mathbf{F}_2$ about O and the torque Γ_3 of $\mathbf{F}_3$ about O. Proceeding in a similar way we find

$$\Gamma_2 = \mathbf{0} \times 3\mathbf{i} = \mathbf{0} \quad \text{and} \quad \Gamma_3 = \mathbf{OR} \times \mathbf{F}_3 = 2\mathbf{k} \times \mathbf{k} = \mathbf{0}$$

(This illustrates the fact that the torque of a force about any point lying on the line of action of the force is always zero.)

(ii) We now find the torques about the point P(0,1,1).

$$\Gamma_1 = \mathbf{PS} \times \mathbf{F}_1 = ((2-1)\mathbf{j} + (2-1)\mathbf{k}) \times 5\mathbf{j} \text{ Nm}$$

$$= (\mathbf{j} \times 5\mathbf{j} + \mathbf{k} \times 5\mathbf{j}) \text{ Nm} = -5\mathbf{i} \text{ Nm}$$

$$\Gamma_2 = \mathbf{PO} \times \mathbf{F}_2 = (-\mathbf{j} - \mathbf{k}) \times 3\mathbf{i} \text{ Nm} = (3\mathbf{k} - 3\mathbf{j}) \text{ Nm}$$

$$\Gamma_3 = \mathbf{PR} \times \mathbf{F}_3 = (-\mathbf{j} + \mathbf{k}) \times \mathbf{k} \text{ Nm} = -\mathbf{i} \text{ Nm}$$

(b) The resultant force is $\mathbf{F} = \mathbf{F}_1 + \mathbf{F}_2 + \mathbf{F}_3 = (3\mathbf{i} + 5\mathbf{j} + \mathbf{k})$ N.
The resultant torque about P is, from part (a) (ii), $\mathbf{G} = (-5\mathbf{i} + (3\mathbf{k} - 3\mathbf{j}) + (-\mathbf{i}))$ Nm $= (-6\mathbf{i} - 3\mathbf{j} + 3\mathbf{k})$ Nm.

(c) No, because $\mathbf{F} \cdot \mathbf{G} = (-18 - 15 + 3)\ \text{N}^2\text{m} \neq 0$. No, because $\mathbf{F} \times \mathbf{G} = (18\mathbf{i} + -15\mathbf{j}) + 21\mathbf{k})\ \text{N}^2\text{m} \neq \mathbf{0}$. (Here we have used the scalar product test for orthogonality and the vector product test for collinear vectors.)

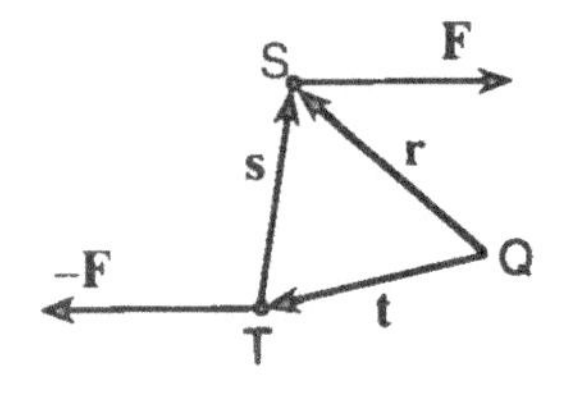

Fig 2.27
The couple consisting of **F** acting at S and **-F** acting at T.

Example 7.2 (*Objectives 9, 14*) Fig 2.27 shows a **couple** (see Background Summary) consisting of two forces **F** and **-F** acting at points S and T with **s** = **TS**. Find expressions for the resultant of the torques of the two forces about (a) the point T, (b) the point S, and (c) an arbitrary point Q not necessarily in the plane of **F** and **s**.

Solution 7.2

(a) Resultant torque about T is $\mathbf{s} \times \mathbf{F} + \mathbf{0} \times (-\mathbf{F}) = \mathbf{s} \times \mathbf{F}$.

(b) Resultant torque about S is $\mathbf{0} \times \mathbf{F} + (-\mathbf{s}) \times (-\mathbf{F}) = \mathbf{s} \times \mathbf{F}$.

(c) Resultant torque about Q is $\mathbf{r} \times \mathbf{F} + \mathbf{t} \times (-\mathbf{F}) = (\mathbf{r} - \mathbf{t}) \times \mathbf{F}$. But $\mathbf{r} - \mathbf{t} = \mathbf{s}$ and so the resultant torque is $\mathbf{s} \times \mathbf{F}$, the same as in parts (a) and (b).

(This example demonstrates an important property of a couple: the torque of a couple is the same about any point.)

Example 7.3 (*Objectives 7, 14*) Fig 2.28 shows a uniform rod of length L and weight $\mathbf{W}$ smoothly hinged at one end S with the other end O held by a horizontal force $\mathbf{F}$. The rod is translational and rotational equilibrium.

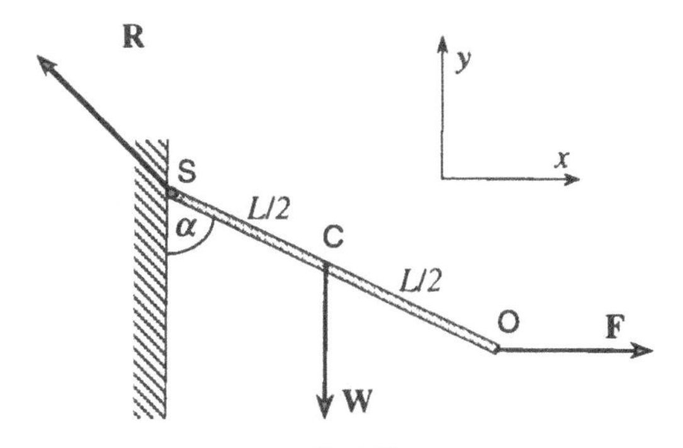

Fig 2.28
A hinged rod held in equilibrium by a horizontal force **F**.

(a) Use the condition for **rotational equilibrium** (see Background Summary) to determine the magnitude F of $\mathbf{F}$ in terms of the magnitude W of the weight and the angle α, and calculate F when $W = 10$ N and $\alpha = 60°$.

(b) Determine the magnitude and direction of the force $\mathbf{R}$ of the hinge on the rod.

Solution 7.3

(a) The force $\mathbf{R}$ at the hinge is of unknown magnitude and direction. We therefore choose the hinged end S of the rod as the point about which to calculate torques when applying the condition for rotational equilibrium. The weight $\mathbf{W}$ acts at the centre-of-mass assumed to be the centre point C of the uniform rod and so the torque of $\mathbf{W}$ about the hinge point S is $\mathbf{SC} \times \mathbf{W}$. The torque of the force $\mathbf{F}$ about S is $\mathbf{SO} \times \mathbf{F}$ and of course the torque of the hinge force $\mathbf{R}$ about S is zero. Thus equating the resultant torque about S to zero we have

$$\mathbf{SC} \times \mathbf{W} + \mathbf{SO} \times \mathbf{F} = 0 \tag{2.39}$$

When the forces all lie in a plane, as in this problem, the torques of the forces about any point in the plane are directed normal to the plane. The projections of the torques onto a unit normal vector **n** are called the **moments of the forces.**

Both torques are directed at right-angles to the plane of the page. Let **n** be a unit vector directed *into* the page. Then $\mathbf{SC} \times \mathbf{W} = (\mathrm{SC} \times W \sin\alpha)\mathbf{n}$ and $\mathbf{SO} \times \mathbf{F} = (\mathrm{SO} \times \mathrm{F} \sin(90° - \alpha))(-\mathbf{n})$. Thus projecting both sides of Eq (2.39) onto **n** and using SO = 2SC, we have

$$W\sin\alpha - 2F\cos\alpha = 0$$

This gives the answer $F = (W/2)\tan\alpha$. For $W = 10$ N and $\alpha = 60°$, $F = 8.66$ N.

(b) We can find the force **R** by using the condition for **translational equilibrium** (see Background Summary). This gives $\mathbf{R} + \mathbf{W} + \mathbf{F} = \mathbf{0}$, and so

$$\mathbf{R} = -\mathbf{W} - \mathbf{F}$$

W and **F** are directed vertically downwards (the negative y-direction) and horizontally (the x-direction) respectively. Thus projecting both sides of the above equation onto the cartesian unit vectors **i** and **j** we obtain the two scalar component equations

$$R_x = -0 - 8.66 \text{ N} = -8.66 \text{ N} \quad \text{and} \quad R_y = -(-10 \text{ N}) + 0 = 10 \text{ N}$$

Finally, $|\mathbf{R}| = ((-8.66)^2 + 10^2)^{1/2}\text{N} = 13.2$ N. The direction cosines of **R** are given by $\cos\theta_x = -8.66/13.2$ and $\cos\theta_y = 10/13.2$. Hence $\theta_x = 131°$ and $\theta_y = 41°$.

Problem 7.1 (*Objectives 7, 14*)

(a) Let a force **F** be applied at a point P. Show that the torque of the force **F** about a point O is of magnitude Fl where l is the shortest distance of the point O from the line of action of **F**, and state the direction of the torque.

(b) Consider a couple consisting of the force **F** at P and a force − **F** applied at some other point. Show that the magnitude of the torque of the couple is Fl where l is the shortest distance between the lines of action of **F** and −**F**.

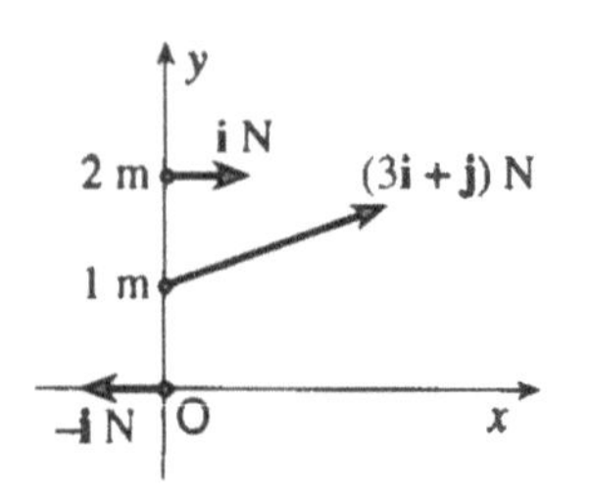

Fig 2.29
A force and a couple.

Problem 7.2 (*Objectives 10, 14*) Fig 2.29 shows a force of (3**i** + **j**) N acting at the point (0,1 m) and a couple consisting of forces −**i** N acting at the origin point O and **i** N acting at a point (0,2)m.

(a) State the resultant force, determine the torque of the couple and the resultant torque about O.

(b) The force and the couple are **equivalent** (see Background Summary) to a single force acting at a point $(0,y)$. Specify the single force and determine y.

Problem 7.3 (*Objective 7, 14*) A ladder of mass 25 kg rests against a smooth (i.e. frictionless) vertical wall at an angle of 70° with the horizontal. The forces holding the ladder in equilibrium are shown in Fig 2.30. S is the normal force of the wall on the ladder; the force **R** and the friction force **F** are the vertical and horizontal resolutes of the force of the ground on the ladder and **W** is the weight of the ladder. Write down two vector equations expressing the condition for translational equilibrium and the condition for rotational equilibrium, and hence find the magnitude F of the friction force.

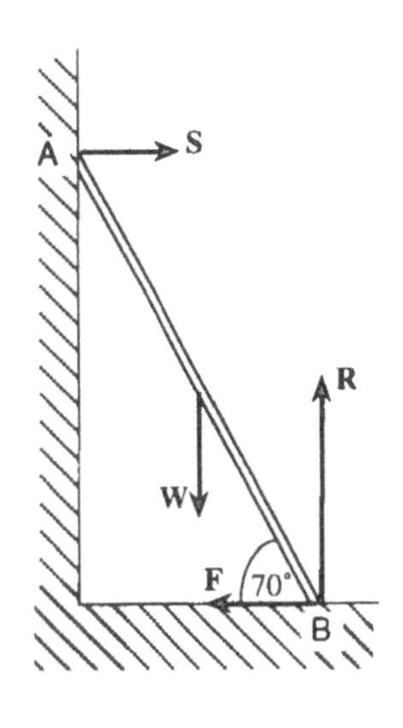

Fig 2.30
The ladder AB.

Problem 7.4 (*Objectives 10, 14*) Consider a uniform square plate of side length 1m (Fig 2.31) lying in the x-y plane. Forces of magnitudes 10 N, 15 N and 25 N are applied to the plate as shown.

(a) Find the resultant of the three forces and the resultant torque about the origin.

(b) The plate is to be held in equilibrium by a fourth force **f** applied at the centre of the plate together with a couple of torque Γ. Specify **f** and Γ.

Problem 7.5 (*Objectives 7,14*) Consider any three non-collinear forces $\mathbf{F}_1$, $\mathbf{F}_2$ and $\mathbf{F}_3$ acting at arbitrary points A, B and C on a rigid body. Suppose the forces hold the body is equilibrium, and let P be the point where the lines of action of $\mathbf{F}_1$ and $\mathbf{F}_2$ intersect. Show that the line of action of $\mathbf{F}_3$ must also pass through P.

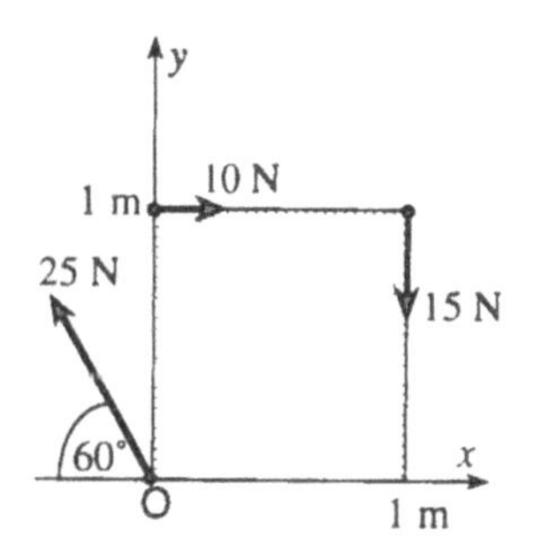

Fig 2.31
Three forces acting on a square plate. The forces are all in the plane of the plate and are labelled by their magnitudes.

Problem 7.6 (*Objectives 2,5,10,14*) The two forces shown in Fig 2.32 are equivalent to a single force **R** acting at the origin and a couple of torque **G**.

(a) Find **R** and **G**.

(b) Determine $(\mathbf{R}\,.\,\mathbf{G})\mathbf{R}/|\mathbf{R}|^2$ and $\mathbf{G} - (\mathbf{R}\,.\,\mathbf{G})\mathbf{R}/|\mathbf{R}|^2$, and interpret these quantities in terms of projections of **G**.

2.7.2 Background summary: Work and energy

The Examples and Problems in this section relate to the following:

1. (a) The **work** done by a constant force **F** acting on a particle that undergoes a displacement **s** is the scalar product

$$W = \mathbf{F}\,.\,\mathbf{s} \qquad \text{(J)} \qquad (2.40)$$

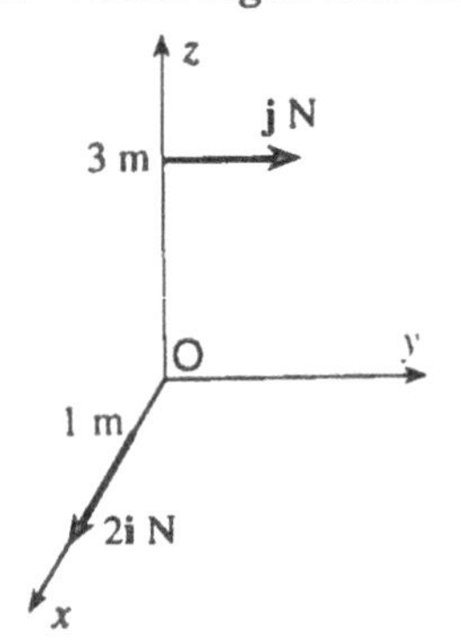

Fig 2.32
The two forces are equivalent to a single force acting at O and a couple.

(b) The rate at which work is done by a force **F** that acts on a particle moving with velocity **v** is

$$\mathbf{R} = \mathbf{F} \cdot \mathbf{v} \qquad (\mathrm{J\,s^{-1}} \text{ or W}) \qquad (2.41)$$

2. The **momentum** of a particle of mass m moving with velocity **v** is the vector

$$\mathbf{p} = m\mathbf{v} \qquad (\mathrm{kg\,m\,s^{-1}}) \qquad (2.42)$$

3. The **kinetic energy** of a particle is the scalar

$$K = (\mathbf{p} \cdot \mathbf{p})/2m = (1/2)mv^2 \qquad (\mathrm{J}) \qquad (2.43)$$

The symbol Δ in ΔK is not a physical quantity but a shorthand for "the change in". In general, if X is any physical quantity then ΔX means: final value of X – initial value of X.

4. Let $\mathbf{F}_{res}$ be the resultant force acting on a particle. The work done by $\mathbf{F}_{res}$ when the particle undergoes a displacement **s** is equal to the change of kinetic energy of the particle, ΔK. Thus

$$\mathbf{F}_{res} \cdot \mathbf{s} = \Delta K \qquad (\mathrm{J}) \qquad (2.44)$$

This result is known as the **work kinetic energy theorem**.

5. When a *conservative force* $\mathbf{F}_{cons}$ acts on a particle which undergoes a displacement **s**, the work done by the conservative force is equal to the loss of *potential energy*. Thus, using Eq (2.40), the change ΔU of potential energy U is

$$\Delta U = -\mathbf{F}_{cons} \cdot \mathbf{s} \qquad (\mathrm{J}) \qquad (2.45)$$

Only changes of potential energy are defined. (Conservative forces and potential energy are discussed in detail in Chapters 5 and 6.)

6. It follows from 4) and 5) that when the resultant force acting on a particle is a conservative force,

$$\Delta K = -\Delta U \qquad (\mathrm{J}) \qquad (2.46)$$

i.e. the gain in kinetic energy is equal to the loss of potential energy. This is the **law of conservation of mechanical energy**.

7. When a particle carrying an electric charge q moves with velocity **v** in a region where there is an electric field **E** and a magnetic field **B**, the electromagnetic force on the particle is given by the **Lorentz force law**

$$\mathbf{F} = q\mathbf{E} + q\mathbf{v} \times \mathbf{B} \qquad (\mathrm{N}) \qquad (2.47)$$

Example 7.4 (*Objectives 13,14*) Use Eqs (2.41) and (2.47) to show that the magnetic force acting on a moving charged particle does no work.

Solution 7.4 The magnetic force $\mathbf{F}_{mag}$ is the second term in the Lorenz force law (Eq (2.47)), $\mathbf{F}_{mag} = q\mathbf{v} \times \mathbf{B}$. The rate at which this force does work is found from Eq (2.41) to be $\mathbf{F}_{mag} \cdot \mathbf{v} = (q\mathbf{v} \times \mathbf{B}) \cdot \mathbf{v}$. This is a scalar triple product in which two of the vectors, $q\mathbf{v}$ and $\mathbf{v}$, are collinear, and so the scalar triple product is zero. Thus the magnetic force does no work.

Example 7.5 (*Objectives 1,14*) A simple pendulum consists of a small body of mass 250 grams fixed at the end of a string of constant length 75 cm and of negligible mass. The other end of the string is attached to a rigid support at S (Fig 2.33). The body is held to one side at A so that the string makes an angle of 35° with the vertical and then released. Determine the kinetic energy and the speed of the body as it passes through its lowest point O. Take $g = 10 \text{ m s}^{-2}$.

1 gram = 10^{-3} kg.

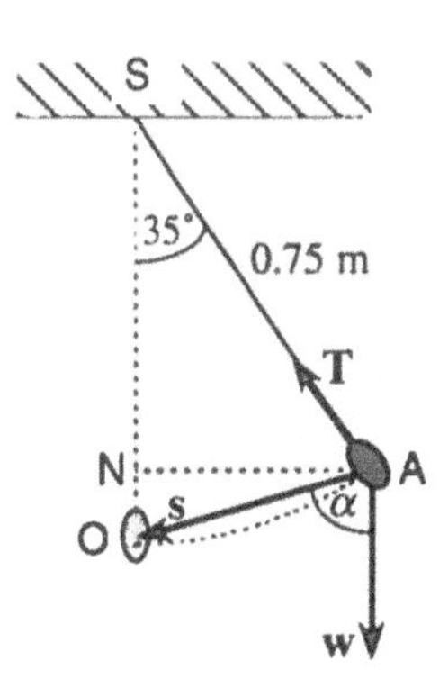

Fig 2.33
The tension force **T** is always directed normal to the direction of the velocity vector and therefore does no work as the particle swings.

Solution 7.5 We can find the kinetic energy by using Eq (2.44). Two forces act on the body: its weight **w** and the tension force **T** in the string. The tension force **T** does no work as the body swings because at each instant it is directed at right-angles to the velocity **v**, and so the work rate (Eq (2.41)) is $\mathbf{T} \cdot \mathbf{v} = 0$. The weight **w** is a constant force of magnitude $0.250 \text{ kg} \times 10 \text{ m s}^{-2} = 2.50 \text{ N}$, acting vertically downwards. The displacement of the body from its initial position to its lowest point is $\mathbf{s} = \mathbf{AO}$ (Fig 2.33), and the angle between **w** and **s** is α. Thus the work done by the weight force is, from Eq (2.40),

$$W_{weight} = \mathbf{w} \cdot \mathbf{s} = (2.50 \text{ N})\,|\mathbf{s}|\,\cos\alpha$$

Now $|\mathbf{s}|\cos$ is equal to the length NO = SO − SN = SO − SAcos35° where SO = SA = 0.75 m, the length of the string. Thus

$$|\mathbf{s}|\cos\alpha = (0.75 \text{ m})(1 - \cos 35°) = 0.136 \text{ m}$$

and so the work done is

$$W_{weight} = (2.50 \text{ N}) \times (0.136 \text{ m}) = 0.340 \text{ J}$$

Since the force **T** does no work, W_{weight} is the work done by the resultant force $\mathbf{F}_{res} = \mathbf{w} + \mathbf{T}$. We now use Eq (2.44) to give

$$\Delta T = W_{weight} = 0.340 \text{ J}$$

This is the kinetic energy at O since the initial kinetic energy at A is zero. Finally, using Eq (2.43), we find the speed is $v = (2 \times 0.340/0.250)^{1/2} \text{ m s}^{-1} = 1.65 \text{ m s}^{-1}$.

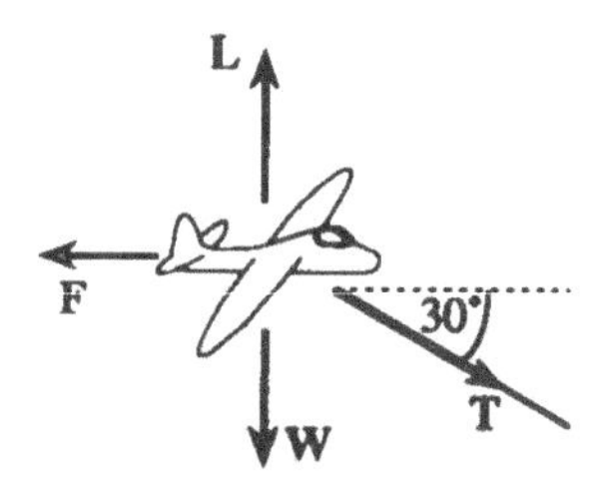

Fig 2.34
Four forces acting on a glider.

Example 7.6 (*Objectives 1,14*) A glider is towed at constant velocity in a horizontal direction over a distance of 1.2 km. The forces acting on it (Fig 2.34) are: its weight **W**, the lift force **L**, a horizontal air friction force **F** and the towing force **T**. The towing cable makes an angle of 30° with the horizontal and exerts a force directed along its length of magnitude $|\mathbf{T}| = 1800$ N.

(a) Determine the magnitude of the air friction force **F**. (Use the fact that a body moving with constant velocity is in translational equilibrium.)

(b) Determine the work done by each of the four forces.

Solution 7.6

$\cos(150°) = -\cos(30°)$

(a) Since the glider is in translational equilibrium, the resultant force acting on it must be zero. Hence $\mathbf{W} + \mathbf{L} + \mathbf{F} + \mathbf{T} = \mathbf{0}$. Projecting both sides of this vector equation onto a horizontal unit vector in the direction of **F**, gives the scalar equation for the horizontal components, $|\mathbf{F}| + |\mathbf{T}|\cos(150°) = 0$. Thus $|\mathbf{F}| = (1800\text{ N})\cos(30°) = 1560$ N.

(b) The forces **W** and **L** do no work because they are directed at right-angles to the horizontal direction of motion.

The work done by the cable force is $\mathbf{T}\cdot\mathbf{s}$ where **s** is the horizontal displacement of magnitude 1.2 km pointing in the direction of motion. Thus $\mathbf{T}\cdot\mathbf{s} = Ts\cos(30°) = (1800\text{ N})\times 1.2\times 10^3\text{ m}\times\cos(30°) = 1.87\times 10^6$ J.

The work done by friction is $\mathbf{F}\cdot\mathbf{s} = Fs\cos(180°) = -Fs = (-1560\text{ N})\times(1.2\times 10^3\text{ m}) = -1.87\times 10^6$ J.

Example 7.7 (*Objectives 10,14*) An electron moves in a region where there is a uniform electrostatic field of magnitude $1.50\times 10^4\text{ Vm}^{-1}$ directed in the positive *y*-direction and a uniform magnetic field of magnitude 1.35×10^{-2} T directed in the positive *z*-direction. Find the magnitude and direction of the resultant force acting on the electron at an instant when it is moving with speed $5.50\times 10^6\text{ ms}^{-1}$ (a) in the *x*-direction, (b) in the *y*-direction and (c) in the *z*-direction.

Solution 7.7 Use the Lorentz force law (Eq (2.47)) with $q = -1.60\times 10^{-19}$ C. For convenience we suppress the units when doing the calculations.

(a) $\mathbf{F} = q(\mathbf{E} + \mathbf{v}\times\mathbf{B})$

$$= (-1.60\times 10^{-19})[1.50\times 10^4\mathbf{j} + (5.50\times 10^6\mathbf{i})\times(1.35\times 10^{-2})\mathbf{k}]$$

Now use $\mathbf{i}\times\mathbf{k} = -\mathbf{j}$ and we have

$$\mathbf{F} = -1.60 \times 10^{-19}[1.50 \times 10^4 - 7.43 \times 10^4]\mathbf{j}$$

$$= 9.5 \times 10^{-15}\mathbf{j}$$

Hence the force is of magnitude 9.5×10^{-15} N and acts in the positive y direction.

(b) Note that the electrostatic force is independent of the velocity of the electron. Thus case (b) differs from case (a) only in the direction of the magnetic force $q\mathbf{v} \times \mathbf{B}$. Now $\mathbf{v}$ is in the y direction while $\mathbf{B}$ remains in the z direction. Hence $\mathbf{v} \times \mathbf{B}$ is in the direction of $\mathbf{j} \times \mathbf{k} = \mathbf{i}$. Remembering that q is negative, we now find the magnetic force to be in the direction of $-\mathbf{i}$. Thus,

$$\mathbf{F} = -1.60 \times 10^{-19}[1.50 \times 10^4\mathbf{j} + 7.43 \times 10^4\mathbf{i}]$$

$$= -(2.40\mathbf{j} + 11.9\mathbf{i}) \times 10^{-15}$$

Hence the force is of magnitude $(2.40^2 + 11.9^2)^{1/2} \times 10^{-15}\,\text{N} = 12.1 \times 10^{-15}\,\text{N}$. The direction cosines are $\cos\theta_x = -11.9/12.1$ and $\cos\theta_y = -2.40/12.1$. Hence $\theta_x = 170°$ and $\theta_y = 101°$.

(c) In this case the magnetic force term is zero because $\mathbf{v}$ and $\mathbf{B}$ are in the same direction giving $\mathbf{v} \times \mathbf{B} = \mathbf{0}$, and so we are left with the electrostatic force of magnitude 2.40×10^{-15} N in the negative y direction.

Problem 7.7 (*Objectives 1,14*) A cannon on a cliff 180 m high fires a 5 kg ball horizontally out to sea with an initial speed of 70 m s^{-1} (Fig 2.35). Determine the work done on the ball during its flight by the force of gravity (use Eq (2.40)). Determine also the change in the gravitational potential energy of the ball and the speed with which the ball hits the sea (use Eqs (2.43), (2.44) and (2.45)). Neglect all effects of friction and take g to be 10 m s^{-2}.

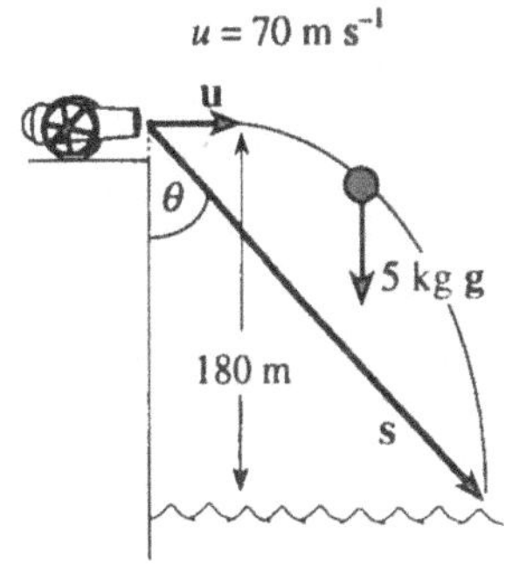

Fig 2.35
A cannon ball fired out to sea.

Problem 7.8 (*Objectives 1,14*) The handbrake fails to hold a car of mass 1750 kg initially parked on a slope inclined at 20° to the horizontal. Assume that the car moves under the following forces (Fig 2.36): the friction force $\mathbf{f}$ supplied by the handbrake of magnitude 4800 N directed up the slope, the normal force $\mathbf{F}$ of the road on the car and the weight $\mathbf{w}$ of magnitude (1750 kg) $\times$ g where g is the acceleration of gravity (assume g = 10 m s^{-2}). Determine

(a) The work done by each of the three forces as the car rolls 100 m down the slope (use Eq (2.40)),

(b) the change in potential energy of the car after it has rolled down 100 m (use Eq (2.45)),

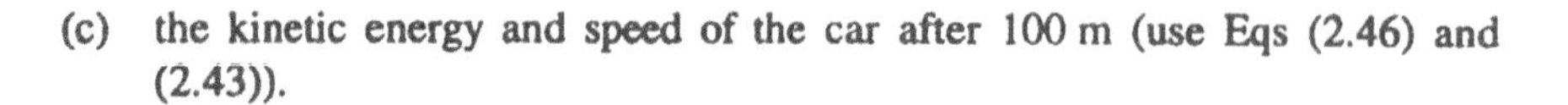

(c) the kinetic energy and speed of the car after 100 m (use Eqs (2.46) and (2.43)).

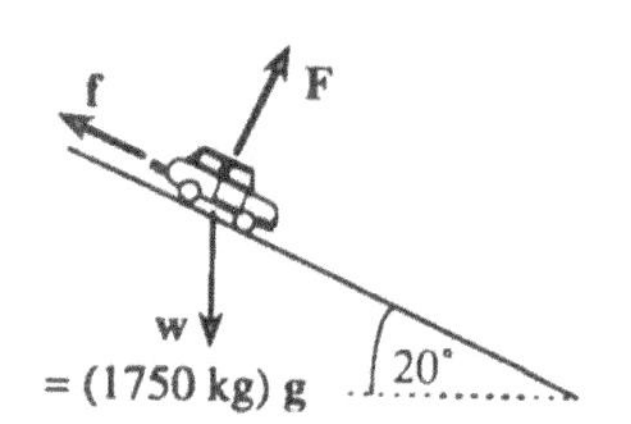

Fig 2.36
The car rolling down a slope.

Problem 7.9 (Objectives 10,14) A *velocity selector* for charged particles consists of an evacuated region with uniform electrostatic and magnetic fields directed at right-angles to one another. A beam of charged particles is injected into the region in a direction at right-angles to both fields. Each particle in the beam experiences a Lorentz force (Eq (2.47)) which depends on the charge q and velocity $\mathbf{v}$ of the particle. For particular values of q and $\mathbf{v}$ however, the electric and magnetic forces cancel and so the Lorentz force is zero and the particles pass through the region undeflected. Thus particles with these particular values of q and $\mathbf{v}$ can be selected from other particles which experience a non-zero Lorentz force and are deflected away. Specify the magnetic field required to select protons of speed 2.54×10^{6} m s^{-1} from a beam of protons initially moving along the z-axis, when the electrostatic field is of magnitude 3.00×10^{4} V m^{-1} directed parallel to the positive y-axis.

Electrostatic forces are conservative.

Problem 7.10 (*Objectives 5,14*) A proton undergoes a displacement $\mathbf{r} = (0,1,1)10^{-3}$ m in a uniform electrostatic field $\mathbf{E} = (5,10,20)$ V m^{-1}. Determine the work done on the proton, the change in the potential energy of the proton and the final speed of the proton if its initial velocity is $\mathbf{u} = (1,0,1)10^{3}$ m s^{-1}. (Take the mass and electric charge of a proton to be $m_p = 1.7 \times 10^{-27}$ kg and $e = 1.6 \times 10^{-19}$ C. Use equations given in the Background Summary.)

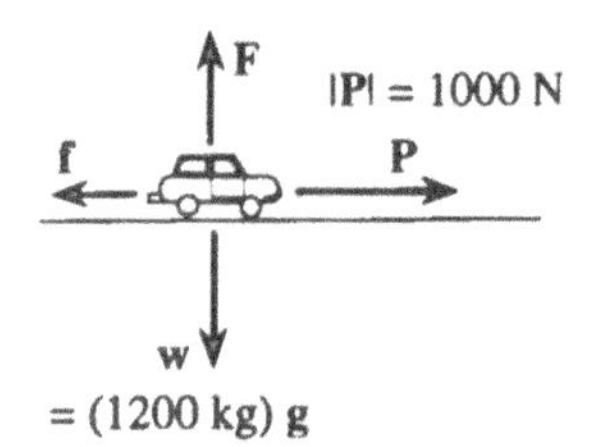

Fig 2.37
A car driven at constant speed along a straight horizontal road.

Problem 7.11 (*Objectives 1,14*) A car of mass 1200 kg moves at a constant speed of 40 km per hour due north on a horizontal road. The engine produces a constant forward force on the car of magnitude 1000 N. Fig 2.37 shows the forces acting on the car. Specify the magnitudes of the friction force $\mathbf{f}$, the weight $\mathbf{w}$ and the upward normal force $\mathbf{F}$ of the road, and determine the work that each force does over a horizontal distance of 1 km. (Take $g = 10$ m s^{-2}).

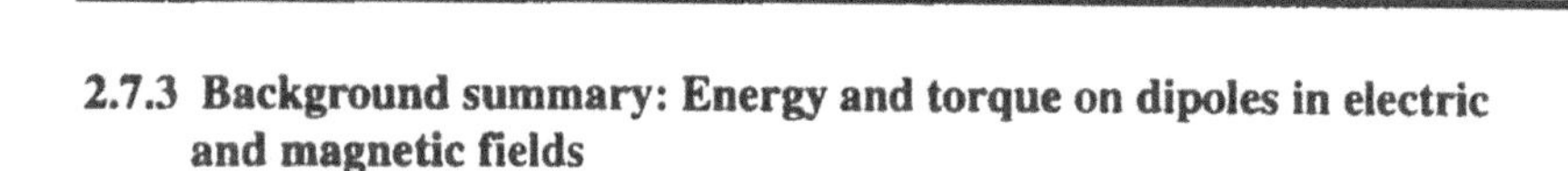

2.7.3 Background summary: Energy and torque on dipoles in electric and magnetic fields

The Examples and Problems in this section relate to the following:

1. An **electric dipole** consists of a pair of electric charges of the same magnitude and opposite sign at a fixed distance l apart (Fig 2.38). The **electric dipole moment** of an electric dipole is defined to be the vector

$$\mathbf{d} = |q|\mathbf{l} \qquad \text{(C m)} \qquad (2.48)$$

where $|q|$ is the magnitude of each charge and the vector **l** is the displacement of the positive charge from the negative charge.

Fig 2.38
An electric dipole.

2. When an electric dipole is in a uniform electrostatic field **E** there is no resultant electrostatic force acting on it, but it experiences a couple of torque

$$\Gamma = \mathbf{d} \times \mathbf{E} \qquad \text{(Nm)} \qquad (2.49)$$

3. The **potential energy of an electric dipole** of dipole moment **d** in a uniform electrostatic field **E** is defined to be

$$U_e = -\mathbf{d} \cdot \mathbf{E} \qquad \text{(J)} \qquad (2.50)$$

4. **Magnetic dipoles** are associated with current loops and bar magnets (Fig 2.39), and some elementary particles such as the proton. The **magnetic dipole moment** of a magnetic dipole is a vector $\boldsymbol{\mu}$. In the case of the plane current loop of area A and carrying an electric current I, the magnitude of the magnetic dipole moment is

$$\mu = |\boldsymbol{\mu}| = IA \qquad (\text{Am}^2) \qquad (2.51)$$

The direction of $\boldsymbol{\mu}$ is determined by the direction of the current and the right-hand rule.

5. When a magnetic dipole is in a uniform magnetic field **B** there is no net force acting on it, but it experiences a torque

$$\Gamma = \boldsymbol{\mu} \times \mathbf{B} \qquad \text{(Nm)} \qquad (2.52)$$

6. The **potential energy of a magnetic dipole** in a uniform magnetic field **B** is defined to be

$$U_m = -\boldsymbol{\mu} \cdot \mathbf{B} \qquad \text{(J)} \qquad (2.53)$$

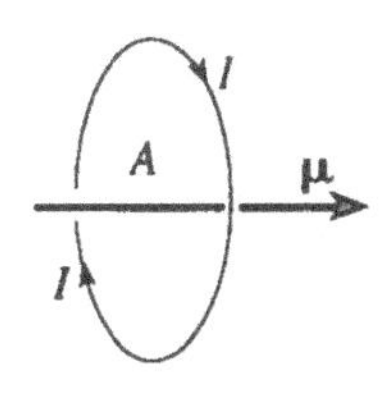

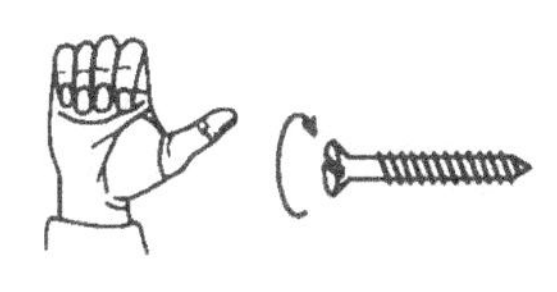

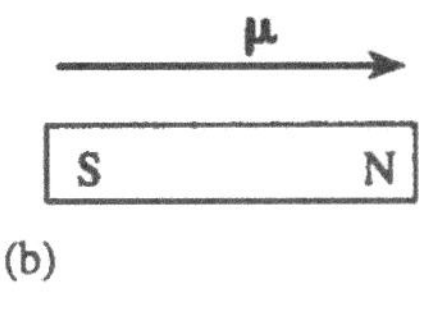

Fig 2.39
Magnetic dipoles. (a) The plane current loop has a magnetic dipole moment $\boldsymbol{\mu}$ of magnitude IA where I is the magnitude of the current and A is the area of the loop. The direction of $\boldsymbol{\mu}$ is determined from the direction of the current using the right-hand rule or the screw rule. (b) The magnetic dipole moment $\boldsymbol{\mu}$ of a bar magnet is directed from the south pole to the north pole of the magnet.

Example 7.8 (*Objectives 1,7,14*) A water molecule has an electric dipole moment of magnitude 6.2×10^{-30} Cm. Suppose the molecule is oriented with its dipole moment at an angle of 25° with the direction of a uniform electric field of magnitude 1.0×10^5 V m^{-1}. Determine

(a) the magnitude of the torque on the molecule, and
(b) the potential energy of the molecule.

Solution 7.8

(a) Use Eq (2.49).

$$|\boldsymbol{\Gamma}| = |\mathbf{d} \times \mathbf{E}| = dE\sin(25°)$$

$$= (6.2 \times 10^{-30}\,\mathrm{Cm}) \times (1.0 \times 10^{5}\,\mathrm{Vm}^{-1})\sin(25°)$$

$$= 2.6 \times 10^{-25}\,\mathrm{Nm}.$$

(The units come out as coulomb volt (CV) which is equivalent to Nm.)

(b) Use Eq (2.50).

$$U_e = -\mathbf{d} \,.\, \mathbf{E}$$

$$= -(6.2 \times 10^{-30}\,\mathrm{Cm}) \times (1.0 \times 10^{5}\,\mathrm{Vm}^{-1}) \times \cos(25°)$$

$$= -5.6 \times 10^{-25}\,\mathrm{J}.$$

(Again the units come out as CV or Nm, but here, because the answer is work rather than torque, we can use the energy unit joule (J).)

Example 7.9 (*Objectives 5,10,14*) A coil carrying a constant current is initially oriented such that its magnetic moment is $\boldsymbol{\mu} = 1.2\mathbf{i}\,\mathrm{Am}^2$ in a region where there is a uniform magnetic field $\mathbf{B} = (1.1\mathbf{j} - 3.1\mathbf{k}) \times 10^{-2}\,\mathrm{T}$.

(a) Determine the magnitude and direction of the torque on the coil.

(b) Suppose the coil is turned through 90° so that its dipole moment points in the *y* direction. (Assume that the magnitude of the dipole moment is unchanged.) Determine the change of potential energy.

(c) Suppose the coil is turned through a further 90°, reversing the direction of the dipole moment relative to its initial orientation. Determine the total change of potential energy.

Solution 7.9

(a) The torque (Eq 2.52) is

$$\boldsymbol{\Gamma} = \boldsymbol{\mu} \times \mathbf{B} = 1.2\mathbf{i} \times (1.1\mathbf{j} - 3.1\mathbf{k})10^{-2}\,\mathrm{Am}^2\mathrm{T}$$

$$= (1.2\,(1.1)\mathbf{k} + 1.2\,(3.1)\mathbf{j})10^{-2}\,\mathrm{Am}^2\mathrm{T} = (3.72\mathbf{j} + 1.32\mathbf{k})10^{-2}\,\mathrm{Nm}$$

(The units $\mathrm{A\,m^2\,T}$ are equivalent to Nm). The magnitude of the torque is $|\boldsymbol{\Gamma}| = (3.72^2 + 1.32^2)^{1/2} \times 10^{-2}\,\mathrm{Nm} = 3.95 \times 10^{-2}\,\mathrm{Nm}$.

The direction cosines are $\cos\theta_x = 0$, $\cos\theta_y = 3.72/3.95$ and $\cos\theta_z = 1.32/3.95$. Hence $\theta_x = 90$, $\theta_y = 19.5°$, $\theta_z = 70.5°$.

(b) The potential energy when the coil is in its initial position is (Eq 2.53)

$$U_m = -\boldsymbol{\mu} \cdot \mathbf{B} = -(1.2\mathbf{i}\,\mathrm{A\,m^2}) \cdot (1.1\mathbf{j} - 3.1\mathbf{k}) \times 10^{-2}\,\mathrm{T} = 0$$

The magnetic moment was initially pointing in the x direction ($\boldsymbol{\mu} = 1.2\mathbf{i}\,\mathrm{A\,m^2}$). After the 90° anticlockwise rotation it points in the y direction and so the new magnetic moment of the coil is $1.2\mathbf{j}\,\mathrm{A\,m^2}$. The potential energy in this new orientation is

$$U_m = -(1.2\mathbf{j}\,\mathrm{A\,m^2}) \cdot (1.1\mathbf{j} - 3.1\mathbf{k}) \times 10^{-2}\,\mathrm{T} = -1.2 \times 1.1 \times 10^{-2}\,\mathrm{A\,m^2\,T}$$

$$= -1.32 \times 10^{-2}\,\mathrm{J}$$

(Here the units $\mathrm{A\,m^2\,T}$ are equivalent to the joule.) Thus the change of potential energy is $(-1.32 \times 10^{-2} - 0)\,\mathrm{J} = -1.32 \times 10^{-2}\,\mathrm{J}$, i.e. the potential energy decreases by $1.32 \times 10^{-2}\,\mathrm{J}$.

(c) The potential energy when the orientation of the coil is reversed relative to its initial orientation is found by putting $\boldsymbol{\mu} \to -\boldsymbol{\mu} = -1.2\mathbf{i}\,\mathrm{A\,m^2}$ in part (a) giving $U_m = 0$ again, so there is no net change of potential energy.

Problem 7.12 (*Objectives 1,14*) The proton (nucleus) of a hydrogen atom experiences a magnetic field of magnitude 34 T due to the presence of the orbiting electron. According to the laws of quantum mechanics, the magnetic moment of the proton cannot have an arbitrary orientation with respect to this magnetic field. Only two orientations are allowed: those in which the projection of the magnetic dipole moment onto the field direction is $1.41 \times 10^{-26}\,\mathrm{A\,m^2}$ and $-1.41 \times 10^{-26}\,\mathrm{A\,m^2}$. Determine the magnitude of the change in potential energy when the proton "flips" from one orientation to the other (use Eq 2.53).

Problem 7.13 (*Objectives 5,10,14*) An electric dipole consists of electric charges of 1.6×10^{-3} C at position (2.1, 3.6, 8.1) mm and -1.6×10^{-3} C at (0.0, 1.9, 5.5) mm. There is a uniform electric field of magnitude $5.2\,\mathrm{kV\,m^{-1}}$ directed parallel to the z-axis. Determine the torque on the dipole and the potential energy (use Eqs (2.49) and (2.50). ($1\,\mathrm{mm} = 10^{-3}$ m and $1\,\mathrm{kV} = 10^3$ V.)

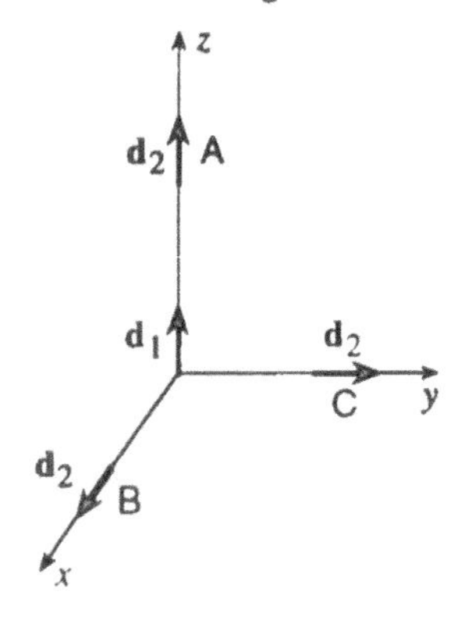

Fig 2.40
The three positions of the electric dipole $\mathbf{d}_2$.

Problem 7.14 (*Objectives 5,14*) An electric dipole of dipole moment $\mathbf{d}_1$ produces an electrostatic field. This field is not uniform and will exert a non-zero resultant force as well as a torque on another electric dipole $\mathbf{d}_2$ a distance R away. The two electric dipoles have a potential energy $U = -W$ where W is the work done by the electromagnetic forces and torques in bringing the dipoles to their positions from an infinite separation. For R much larger than the size l of each dipole, this potential energy is given by

$$U = \frac{1}{R^3}\,[\mathbf{d}_1 \cdot \mathbf{d}_2 - 3(\mathbf{n} \cdot \mathbf{d}_1)(\mathbf{n} \cdot \mathbf{d}_2)] \qquad (R >> l)$$

where $\mathbf{n}$ is a unit vector directed from $\mathbf{d}_1$ to $\mathbf{d}_2$. Fig 2.40 shows $\mathbf{d}_1$ at the origin of coordinates. Consider $\mathbf{d}_2$ to be a distance R from the origin at one of the three positions A, B, and C with the orientations shown. For $|\mathbf{d}_1| = |\mathbf{d}_2| = d$, determine the potential energy in each of the three cases.

3

Time-dependent vectors

Objectives

After you have studied this chapter you should be able to

- Determine vector function values for given values of the independent scalar variable (*Objective 1*).
- Obtain the three scalar functions for the cartesian components of a given vector function and use these to derive an equation for the space curve of the function (*Objective 2*).
- Recall and use the definitions of: average velocity, instantaneous velocity, average acceleration and instantaneous acceleration (*Objective 3*).
- Outline the steps involved in obtaining the derivative of a vector function from first principles (*Objective 4*).
- Use rules of differentiation to find the derivatives of vector functions and in particular to determine the velocity and acceleration of a particle from a given displacement function (*Objective 5*).
- Specify the angular speed and angular velocity of a rotating body and determine the velocity and acceleration of a point in the body (*Objective 6*).
- Determine the derivative of a rotating vector of constant magnitude (*Objective 7*).
- Determine the inertial forces observed in accelerating and rotating frames of reference (*Objective 8*).

Chapters 1 and 2 were concerned with vectors that remained constant in time. We now consider vectors that vary in time, i.e. time-dependent vectors. An important application is the study of moving bodies. The position vector of a moving body generally varies in magnitude and direction. The velocity and acceleration vectors may also vary. The acceleration vector is of particular interest because acceleration is produced by forces, as described by Newton's second law of motion. The velocity and acceleration vectors are obtained from the position vector by the methods of differential calculus. The application of calculus to vectors is introduced in this chapter and is in fact the main theme of the remainder of this book.

The study of moving bodies, without regard to the causes of motion, is called **kinematics**; while the study of the effects of forces on moving bodies is called **dynamics**.

Sections 3.1 and 3.2 introduce the concept of a vector function of time and show how a vector function can be differentiated from first principles. Rules for differentiating sums and products of vector functions are stated (but not derived) in Section 3.3. Important examples of particle motion such as projectile motion and motion in a circle are considered. The angular velocity vector is introduced in Section 3.4. The final sections describe applications to relative motion, including the derivation of inertial forces in accelerating and rotating frames of reference.

3.1 INTRODUCING VECTOR FUNCTIONS

We begin with a brief review of ordinary scalar functions of a single scalar variable in order to establish some notation and definitions.

3.1.1 Scalar functions - a review

A scalar **function** f is defined by a *rule* and a *domain*. The **rule** specifies a unique scalar function value $f(x)$ for each value x of the independent scalar variable. The rule is usually in the form of an equation, such as $f(x) = x^2 + 2$. The set of values of the independent variable x over which the rule is to be used is called the **domain** of the function. Thus we have, for example, the scalar function $f(x)$ defined by

$$f(x) = x^2 + 2 \qquad (x \in \Re) \tag{3.1}$$

Some other terminology is in common use: A function f is an *operation* on an *input variable* x which results in the value $y = f(x)$, where y is the *output variable*. x is also called the *argument* and y the *dependent variable* or the *image of* x *under* f.

The domain of this function is all the real numbers, as specified by the domain statement $(x \in \Re)$ where the symbol $\Re$ denotes the set of all real numbers and the Greek symbol $\in$ is a shorthand for "is a member of" or "belongs to". Although both rule and domain are necessary for the complete specification of a function, it is common to omit explicit reference to the domain and to refer to the function simply by giving the rule $f(x)$. We can use Eq (3.1) to find particular function values. For example, the function value at $x = 0$ is found by replacing x by 0 in the rule, giving $f(0) = 0^2 + 2 = 2$. Similarly we find $f(1) = 1^2 + 2 = 3$, $f(-5) = (-5)^2 + 2 = 27$, and so on. Another example of a scalar function is the square root function

$$h(x) = x^{1/2} \quad \text{or} \quad \sqrt{x} \qquad (x \in \Re, x \geq 0) \tag{3.2}$$

Here it is necessary to restrict the domain to non-negative numbers since negative numbers do not have real square roots. Thus we have for example, $h(9) = 9^{1/2} = 3$, but $h(-9)$ does not exist.

The use of the symbol x as the independent variable is just a convention. Any other scalar symbol such as y, λ, t, etc. will do. Thus the function f of Eq (3.1) could equally well be written as $f(t) = t^2 + 2$ $(t \in \Re)$.

A function specified by a symbol f or $f(x)$ consists of the set of all possible function values $f(x)$ and the corresponding set of all numbers x in the domain, i.e. it is the set of all number pairs $(x, f(x))$. The function can be represented pictorially by drawing the **graph** in which the x-axis represents the set of x values in the domain and the ordinate (y-axis) represents the set of function

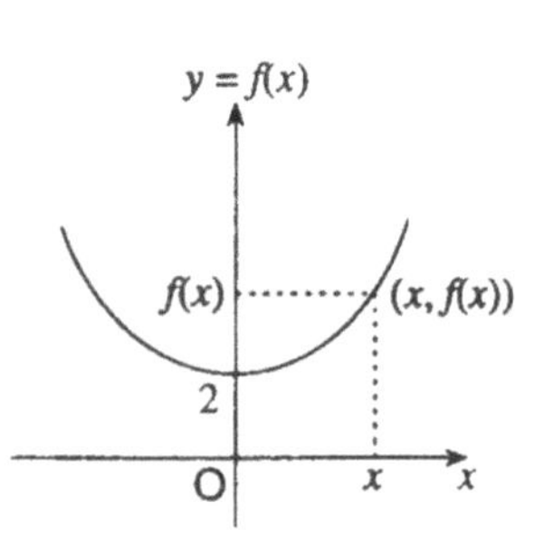

Fig 3.1
The graph of the function $f(x) = x^2 + 2$ $(x \in \Re)$ is the curve made up of the set of points $(x, f(x))$.

values $f(x)$. Each number pair $(x, f(x))$ gives a point on the graph. The graph of f is the curve made up of the set of all points with coordinates $(x, f(x))$ (Fig 3.1).

The symbol $f(x)$ has a dual use: $f(x)$ is used to denote the function value (i.e. a number) at a particular but unspecified value of x; it is also often used to denote the function f itself, i.e. the set of number pairs $(x, f(x))$. These two usages of the symbol $f(x)$ do not normally cause confusion since the meaning should be clear from the context.

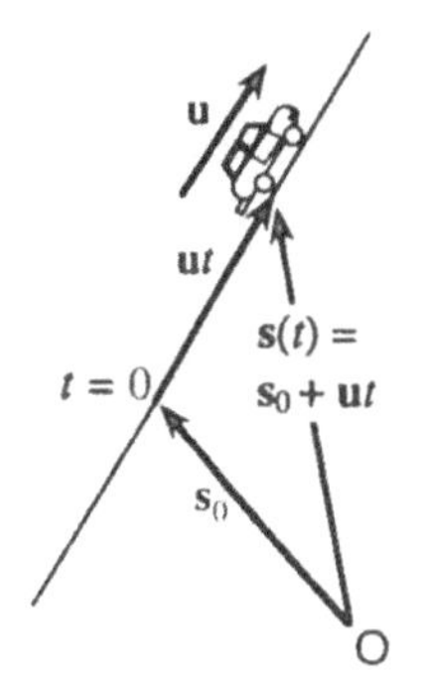

Fig 3.2
A car travelling along a straight road.

3.1.2 Vector functions of time

A **vector function f** of a scalar variable such as time t is defined by a rule and a domain. The rule assigns a unique vector $\mathbf{f}(t)$ to each value of t in the domain of the function. Vector functions of time are important in the study of moving bodies, i.e. kinematics. Consider a car moving along a straight road at a constant velocity $\mathbf{u}$. The displacement of the car measured from a fixed reference point O has a unique value $\mathbf{s}(t)$ at any instant t. The displacement at time $t = 0$ is shown in Fig 3.2 as the vector $\mathbf{s}_0$. Thus $\mathbf{s}(0) = \mathbf{s}_0$. The displacement $\mathbf{s}(t)$ at some later time t is given simply by the vector sum of $\mathbf{s}_0$ and the additional displacement $\mathbf{u}t$ occurring in the time t, i.e. by the vector equation $\mathbf{s}(t) = \mathbf{s}_0 + \mathbf{u}t$. Suppose the car moves with the constant velocity $\mathbf{u}$ during the interval from $t = -3$ s to $t = 5$ s. Then the displacement of the car in this interval is described by the vector function of time

We neglect the size and structure of the car and idealise it (i.e. model it) as a single point or particle.

$$\mathbf{s}(t) = \mathbf{s}_0 + \mathbf{u}t \qquad (-3 \le t \le 5) \qquad (3.3)$$

The rule of the function is the vector equation $\mathbf{s}(t) = \mathbf{s}_0 + \mathbf{u}t$, and the domain is the interval $(-3 \le t \le 5)$. Very often we omit reference to the domain and refer to the function simply by giving the rule $\mathbf{s}(t)$.

It is useful to have a pictorial representation of a vector function **f**. Suppose all vectors $\mathbf{f}(t)$ are drawn from the same beginning point O. Then the end points lie on a curve in space called a **space curve**. When the function **f** is the displacement **s** of a particle, the space curve is simply the path of the particle. In the example of Eq (3.3), the space curve is the straight line AB shown in Fig 3.3. Particular points on the space curve can be labelled by the values of the time t, such as the initial point $t = -3$, the final point $t = 5$ and the point $t = 0$. An arrowhead on the space curve shows the direction of increasing t.

We can introduce a cartesian coordinate system and project a vector function onto the cartesian unit vectors to obtain three scalar functions. Consider the function **s** defined by Eq (3.3) and let the x and y axes lie in the plane of the vectors $\mathbf{s}(t)$. We project both sides of the vector equation $\mathbf{s} = \mathbf{s}_0 + \mathbf{u}t$ onto the cartesian unit vectors to give scalar equations for the components

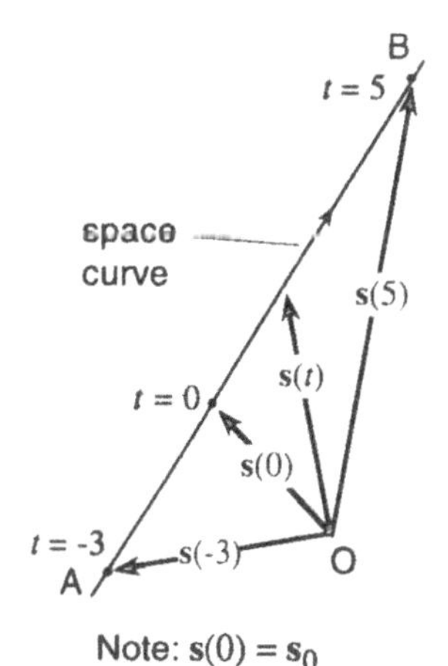

Fig 3.3
The space curve of $\mathbf{s}(t)$ is the path AB of the car.

$$\left.\begin{aligned} s_x(t) &= s_{0x} + u_x t \qquad (-3 \le t \le 5) \\ s_y(t) &= s_{0y} + u_y t \qquad (-3 \le t \le 5) \\ s_z(t) &= 0 \qquad (-3 \le t \le 5) \end{aligned}\right\} \tag{3.4}$$

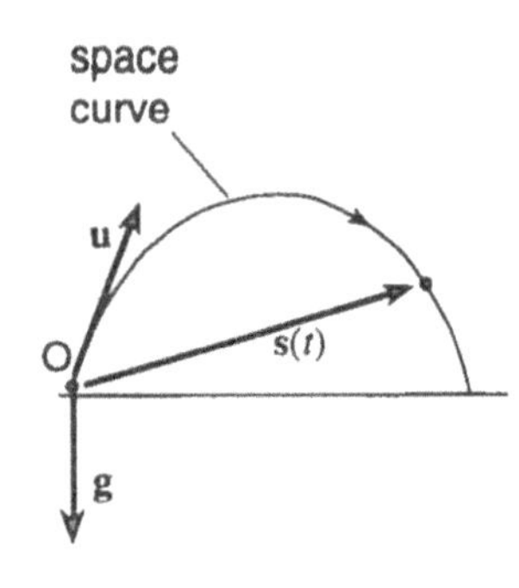

Fig 3.4
The parabolic path of a projectile.

where $s_x(t) = \mathbf{i} \cdot \mathbf{s}(t)$, the x-component of the displacement at time t, and $s_{0x} = \mathbf{i} \cdot \mathbf{s}_0$, the x-component of $\mathbf{s}_0$, etc. This triple of scalar component functions (Eqs (3.4)) is equivalent to the vector function Eq (3.3) which can be written as

$$\mathbf{s}(t) = (s_x(t), s_y(t), s_z(t)) \tag{3.5}$$

It is common to use the symbol **r** for the position vector – hence the change from **s** to **r** here. Of course, any vector symbol is acceptable.

If the vectors $\mathbf{s}(t)$ have their common beginning points at the origin of coordinates, then $\mathbf{s}(t)$ is the position vector $\mathbf{r}(t)$ with components $x(t)$, $y(t)$ and $z(t)$ which are the time-dependent coordinates of the particle. Thus we can express Eq (3.5) as

$$\mathbf{r}(t) = (x(t), y(t), z(t)) \tag{3.6}$$

In the example of Eq (3.3), the space curve of the function, i.e. the path of the car, is a straight line. In more interesting cases bodies travel along curved paths. For example, the displacement of a projectile from its point of projection, ignoring the effects of air resistance, is described by the vector function

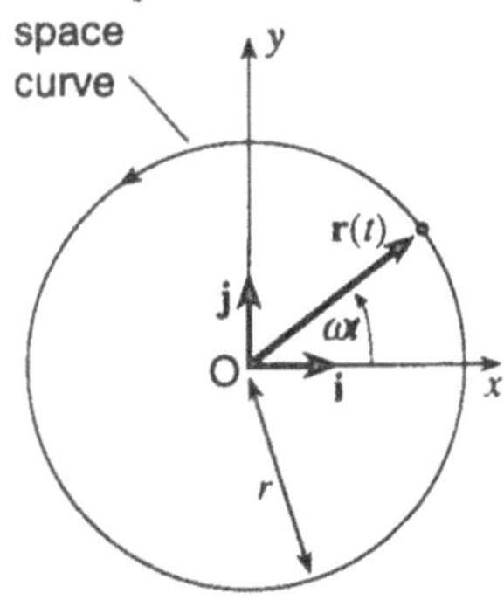

Fig 3.5
Circular motion.

$$\mathbf{s}(t) = \mathbf{u}t + \frac{1}{2}\mathbf{g}t^2 \qquad (0 \le t \le T) \tag{3.7}$$

where **u** is the velocity of projection at $t = 0$, T is the time of flight and **g** is the acceleration vector of gravity, directed vertically downwards. The space curve is a segment of a parabola (Fig 3.4).

Another important example is a body moving along a circular path, such as a particle on the circumference of a spinning wheel, or a satellite in orbit around the Earth. The space curve is the circular path. Of particular interest is the case of **uniform circular motion**, i.e. motion along a circular path at constant speed v. When the motion is in the x-y plane with the centre of the circle at the origin (Fig 3.5), the displacement (i.e. position vector) can be represented by the cartesian vector function of time

Satellites move in elliptical orbits that are often very nearly circular.

$$\mathbf{r}(t) = r(\mathbf{i} \cos \omega t + \mathbf{j} \sin \omega t) \tag{3.8}$$

Angular speed is defined in Section 3.4.

where r is the radius and $\omega = v/r$ is the constant angular speed in radians per second.

A curve in the x-y plane is described by an equation relating the cartesian variables x and y. The equation of a plane space curve can be found by eliminating the time t from the scalar component equations. For example, projecting Eq (3.8) onto **i** and **j** and writing x for $x(t)$ and y for $y(t)$, gives the scalar component equations

$$x = r\cos\omega t, \qquad y = r\sin\omega t \tag{3.9}$$

These two equations are referred to technically as the **parametric equations** of the space curve, the *parameter* being the variable t. There is a third equation ($z = 0$) expressing the fact that the motion is in the x-y plane. We can eliminate the parameter t by squaring and adding the two equations (3.9). This yields $x^2 + y^2 = r^2(\cos^2\omega t + \sin^2\omega t)$. We now use the well known identity $\cos^2\theta + \sin^2\theta = 1$, and so obtain the cartesian equation of the space curve,

$$x^2 + y^2 = r^2 \tag{3.10}$$

This is the equation of a circle in the x-y plane of radius r centred at the origin.

We have considered only vector functions that have plane space curves. The specification of a three-dimensional curve, such as the spiral path of a helter skelter (a helix), requires three parametric equations $x(t)$, $y(t)$ and $z(t)$ from which the parameter t can be eliminated to obtain two equations relating the cartesian variables. These two equations describe two surfaces which intersect in the space curve. We shall consider only plane space curves in this chapter.

In three dimensions we can regard Eq (3.10) as the equation of the curved surface of a cylinder of radius r with its axis lying on the z-axis. The circular path in the x-y plane is the intersection of this cylindrical surface with the plane surface described by the equation $z = 0$.

Summary of section 3.1

- A **vector function f** of time t is specified by a **rule** (a vector equation) for assigning a unique vector $\mathbf{f}(t)$ to each value of t in the **domain**.

- A vector function **f** can be represented pictorially by a curve in space, called the **space curve**, on which the end points of all the vectors $\mathbf{f}(t)$ lie when the beginning points are all at a common point. When $\mathbf{f}(t)$ is the position vector $\mathbf{r}(t)$ of a particle, the space curve is the path of the particle.

- A vector function is equivalent to the three scalar component functions which are the **parametric equations** of the space curve. The cartesian form of the space curve is found by eliminating the parameter t.

Example 1.1 (*Objective 1*) Given the vector function $\mathbf{q}(t) = \mathbf{i} - \mathbf{j}t + \mathbf{k}t^2$ ($t \geq 0$), specify where possible $\mathbf{q}(0)$, $\mathbf{q}(1/2)$, $\mathbf{q}(\pi)$ and $\mathbf{q}(-1)$.

Solution 1.1 $\mathbf{q}(0) = \mathbf{i} - \mathbf{j}0 + \mathbf{k}0^2 = \mathbf{i}$; $\mathbf{q}(1/2) = \mathbf{i} - \mathbf{j}/2 + \mathbf{k}/4$; $\mathbf{q}(\pi) = \mathbf{i} - \pi\mathbf{j} + \pi^2\mathbf{k}$. $\mathbf{q}(-1)$ does not exist because -1 is not in the domain ($t \geq 0$) of the function.

Example 1.2 (*Objective 1*) Refer to Eq (3.8) giving the position vector of a particle in uniform circular motion. Specify the location of the particle at $t = 0$, $t = \pi/2\omega$ and $t = 2\pi/\omega$.

Solution 1.2 The location of the particle at any time t is specified by the position vector $\mathbf{r}(t)$. At $t = 0$,

$$\mathbf{r}(0) = r(\mathbf{i}\cos(\omega\times 0) + \mathbf{j}\sin(\omega\times 0)) = r\mathbf{i}$$

Thus the particle is on the positive x-axis a distance r from O.

At $t = \pi/2\omega$, we have

$$\mathbf{r}(\pi/2\omega) = r(\mathbf{i}\cos(\omega\pi/2\omega) + \mathbf{j}\sin(\omega\pi/2\omega)) = r\mathbf{j}$$

Thus at $t = \pi/2\omega$ the particle is on the positive y-axis a distance r from O. Similarly we find that at $t = 2\pi/\omega$ the particle has completed one revolution and returned to position $r\mathbf{i}$.

Example 1.3 (*Objective 2*) Eq (3.3) describes a straight line path (space curve) in the horizontal x-y plane. Derive the equation of this straight line by eliminating the variable t from the parametric equations, Eqs (3.4).

Solution 1.3 Assume the displacement vectors have their common beginning point at the origin of coordinates. Then the parametric equations (3.4) can be written as

$$x = x_0 + u_x t, \quad y = y_0 + u_y t, \quad z = 0$$

where $(x, y, 0)$ are the time-dependent coordinates of the end point of $\mathbf{s}(t)$ and $(x_0, y_0, 0)$ are the coordinates of the end point of $\mathbf{s}_0$. The first equation can be rearranged to give $t = (x - x_0)/u_x$. Substituting this expression for t into the second equation gives

$$y = y_0 + (x - x_0)\frac{u_y}{u_x} \tag{3.11}$$

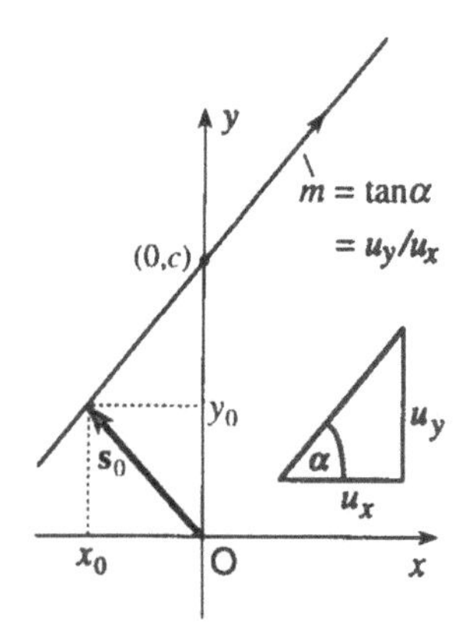

Fig 3.6
The straight-line path of the car showing the intercept c on the y-axis and the gradient m.

We have now eliminated t and obtained the equation for the space curve in the x-y plane. Remembering that x_0, y_0, u_x and u_y are all constants, you can see that Eq (3.11) is the equation of a straight line in the x-y plane, i.e. it is of the form

$$y = mx + c \qquad \text{(straight line: gradient } m\text{; } y\text{-intercept } c) \tag{3.12}$$

with gradient $m = (u_y/u_x)$ and intercept $c = (y_0 - x_0 u_y/u_x)$. See Fig 3.6.

Problem 1.1 (*Objective 1*) Given the vector function

$$\mathbf{f}(t) = \mathbf{i}\cos(\pi t) - \mathbf{j}\sin(\pi t) \qquad (0 \le t \le 1)$$

specify where possible $\mathbf{f}(0)$, $\mathbf{f}(0.25)$, $\mathbf{f}(0.5)$, $\mathbf{f}(0.75)$, $\mathbf{f}(1)$ and $\mathbf{f}(2)$.

Problem 1.2 (*Objective 2*) The displacement of a projectile is given by $\mathbf{s}(t) = \mathbf{u}t + \mathbf{g}t^2/2$ where **u** and **g** are constant vectors. Referring to the coordinate system of Fig 3.7, show that the projectile moves along a parabola described by $y = Ax + Bx^2$ where $A = u_y/u_x$ and $B = -g/2u_x^2$.

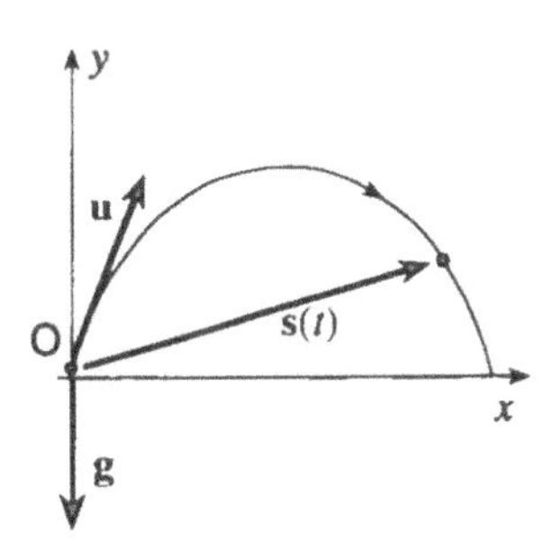

Fig 3.7
Projectile motion in the *x-y* plane.

Problem 1.3 (*Objectives 1,2*) The position of the fluorescent spot on an oscilloscope screen is given by $\mathbf{s}(t) = \mathbf{i}t + \mathbf{j}\sin(\pi t)$ $(0 \le t \le 5)$, where distances are in cm and time in seconds. The cartesian unit vectors **i** and **j** are directed horizontally and vertically respectively in the plane of the screen.

(a) What are the horizontal and vertical ranges of the spot?

(b) Write down the two scalar equations for the components of $\mathbf{s}(t)$ and hence derive the equation for the path traced out by the spot on the screen. Make a rough sketch of the path.

Problem 1.4 (*Objectives 1,2*) The position vector of a particle is given by

$$\mathbf{r}(t) = \mathbf{i}\cos(\pi t) - \mathbf{j}\sin(\pi t) + 2\mathbf{k} \qquad (t \ge 0)$$

(a) Where is the particle at $t = 0$ and at $t = 2$?
(b) Describe the path of the particle.

3.2 DIFFERENTIATING VECTOR FUNCTIONS - DEFINITIONS OF VELOCITY AND ACCELERATION

We now turn our attention to the velocity and acceleration of a particle. When the velocity varies with time we have to consider the *instantaneous velocity*, the velocity at an instant. The instantaneous velocity, like the displacement, is a vector function of time. It can be derived from the displacement function by *differentiation* with respect to time. Differentiation of a vector function from first principles involves calculating a limit by methods that are similar to those used to find the derivative df/dx of a scalar function $f(x)$. In this section we outline the method of differentiating vector functions by calculating a limit, i.e. finding the derivatives of vector functions from first principles. These calculations give insight into the vector nature of velocity and acceleration. In practice however it is usually easier to obtain derivatives of vector functions by applying *rules of differentiation*. Section 3.3 gives statements of some of these rules and practice in using them.

We begin with a brief review of the differentiation of ordinary scalar functions.

3.2.1 Differentiation of a scalar function - a review

A scalar function $f(x)$ whose graph varies smoothly, i.e. without discontinuities or sharp corners, can be differentiated. The result of differentiating $f(x)$ is another scalar function called the *derivative* of $f(x)$ denoted by the symbol df/dx, or by $f'(x)$. The derivative describes how the function $f(x)$ changes when x changes by a small amount, and is defined as a limit

$$\frac{df}{dx} = \lim_{\Delta x \to 0}\left[\frac{f(x+\Delta x) - f(x)}{\Delta x}\right] \tag{3.13}$$

Here the symbol Δx stands for a change in the value of x, and the limit symbol $\lim_{\Delta x \to 0}[\ ]$ means the value of the quantity in square brackets as Δx becomes smaller and smaller and is eventually zero. To see how the limit is calculated in a simple case, consider the function $f(x) = x^2 - 1$. We first find

$$f(x + \Delta x) = ((x + \Delta x)^2 - 1) = x^2 + 2x\Delta x + (\Delta x)^2 - 1$$

and then obtain

$$f(x + \Delta x) - f(x) = (x^2 + 2x\Delta x + (\Delta x)^2 - 1) - (x^2 - 1)$$

$$= 2x\Delta x + (\Delta x)^2$$

Now substitute into Eq (3.13) and evaluate the limit,

$$\frac{df}{dx} = \lim_{\Delta x \to 0}\left[\frac{2x\Delta x + (\Delta x)^2}{\Delta x}\right] = \lim_{\Delta x \to 0}[2x + \Delta x]$$

The limit of $2x + \Delta x$ as Δx goes to zero is simply $2x$, and so we obtain the derivative $df/dx = 2x$.

We don't normally need to go through the process of evaluating derivatives as limits because the derivatives of elementary functions are tabulated, and there are rules for differentiating sums and products of functions and composite functions. See Appendix B.

3.2.2 Differentiation of a vector function

A vector function $\mathbf{f}(t)$ that varies sufficiently smoothly with t can be differentiated to give another vector function called the derivative of $\mathbf{f}$ and denoted by the symbol $d\mathbf{f}/dt$ or by $\mathbf{f}'(t)$. The derivative of $\mathbf{f}$ describes how the magnitude and direction of $\mathbf{f}$ change when t changes by a small amount and is defined as a limit analogous to Eq (3.13). Thus the **derivative of a vector function f** is defined to be

$$\frac{d\mathbf{f}}{dt} = \lim_{\Delta t \to 0}\left[\frac{\mathbf{f}(t+\Delta t) - \mathbf{f}(t)}{\Delta t}\right] \tag{3.14}$$

The vector difference $\mathbf{f}(t + \Delta t) - \mathbf{f}(t)$ is the change in the vector **f** occurring in the interval Δt and is often denoted by the symbol $\Delta\mathbf{f}$. Thus Eq (3.14) can be written as

$$\frac{d\mathbf{f}}{dt} = \lim_{\Delta t \to 0}\left[\frac{\Delta\mathbf{f}}{\Delta t}\right]$$

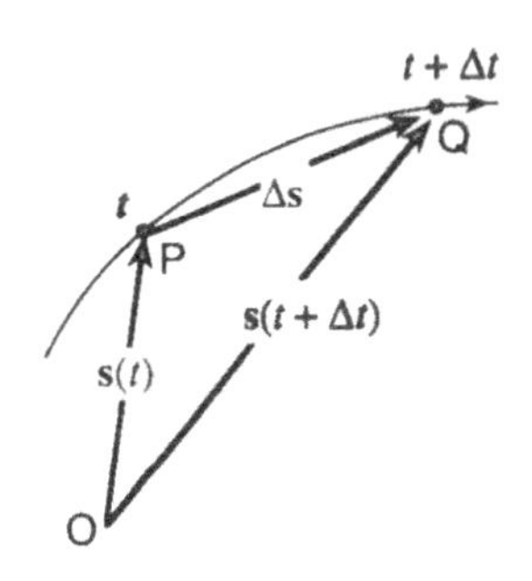

Fig 3.8
The displacements of a particle at t and $t + \Delta t$.

We now illustrate an important property of the derivative by applying Eq (3.14) to the displacement **s** of a particle. Fig 3.8 shows the path of a particle which is at P at time t and at Q at time $t + \Delta t$. The displacements of the particle at these two instants, measured from a fixed reference point O, are shown as $\mathbf{s}(t) = \mathbf{OP}$ and $\mathbf{s}(t + \Delta t) = \mathbf{OQ}$. Shown also in Fig 3.8 is the displacement $\Delta\mathbf{s} = \mathbf{PQ}$ that occurs in the interval between times t and $t + \Delta t$, where Δt is positive. Thus we have

$$\Delta\mathbf{s} = \mathbf{s}(t + \Delta t) - \mathbf{s}(t) = \mathbf{PQ}$$

Applying the definition Eq (3.14), we consider the limit as Δt goes to zero ($\Delta t \to 0$) of the vector

$$\frac{\Delta\mathbf{s}}{\Delta t} = \frac{\mathbf{s}(t+\Delta t) - \mathbf{s}(t)}{\Delta t} \tag{3.15}$$

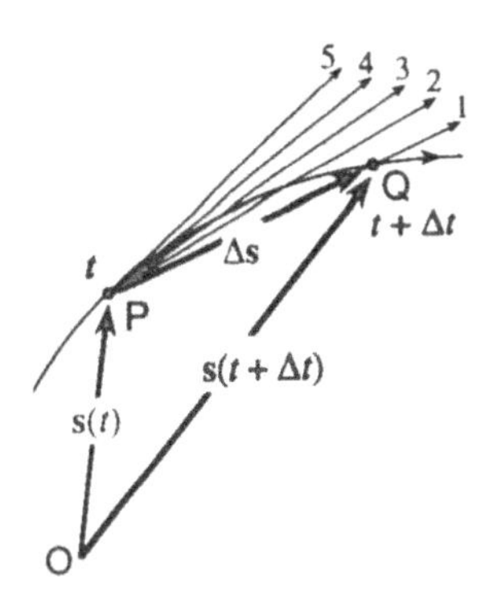

Fig 3.9
The directed straight lines labelled 1 to 5 show the directions of $\Delta\mathbf{s}$ for successively smaller values of Δt. Line 5 is the tangent to the curve at time t and gives the direction of $\Delta\mathbf{s}$ in the limit $\Delta t \to 0$.

As the limit ($\Delta t \to 0$) is approached, the numerator of Eq (3.15) goes closer and closer to the zero vector, but the ratio of magnitudes $|\Delta\mathbf{s}|/\Delta t$ tends towards a definite scalar limit which is the magnitude of the derivative. The direction of the derivative is the direction of the vector $\Delta\mathbf{s}$ in the numerator of Eq (3.15) as it approaches the zero vector. This is illustrated in Fig 3.9. In the limit, the direction of $\Delta\mathbf{s}$ is along the tangent to the path at P. We conclude that the direction of the derivative $d\mathbf{s}/dt$ at any instant t is along the tangent at the point t on the path of the particle. This illustrates an important general result:

> the direction of the derivative $d\mathbf{f}/dt$ at any instant is tangential to the space curve of **f** and points in the direction of increasing t.

3.2.3 Definitions of velocity and acceleration

The vector $\Delta\mathbf{s}/\Delta t$ of Eq (3.15) is called the **average velocity** of the particle in the interval Δt between times t and $t + \Delta t$. The derivative $d\mathbf{s}/dt$ is the instantaneous **velocity** $\mathbf{v}(t)$ at time t. Thus we have the definition

$$\mathbf{v}(t) = d\mathbf{s}/dt \qquad \text{(definition of velocity)} \tag{3.16}$$

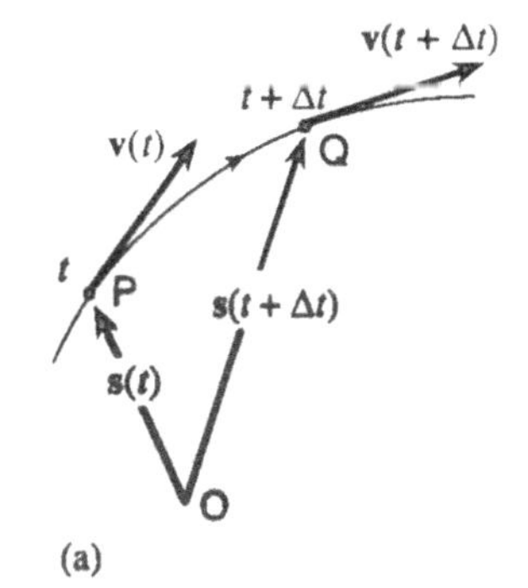

Fig 3.10a
The vectors $\mathbf{v}(t)$ and $\mathbf{v}(t + \Delta t)$ are tangential to the path of the particle. The beginning points are drawn from the positions of the particle at times t and $t + \Delta t$.

The velocity $\mathbf{v}(t)$ is itself a vector function of time and can be differentiated in the same way. Fig 3.10(a) shows the vectors involved in defining the derivative of $\mathbf{v}(t)$ according to Eq (3.14). Vectors $\mathbf{v}(t)$ and $\mathbf{v}(t + \Delta t)$ are the velocities of the particle as it passes through points P and Q at times t and $t + \Delta t$ respectively. The two velocity vectors are shown in Fig 3.10a with their beginning points at P and Q; they are directed along the tangents to the path at these points, as established at the end of Section 3.2.2. The **average acceleration** in the interval between t and $t + \Delta t$ is defined to be

$$\frac{\Delta \mathbf{v}}{\Delta t} = \frac{\mathbf{v}(t+\Delta t) - \mathbf{v}(t)}{\Delta t} \tag{3.17}$$

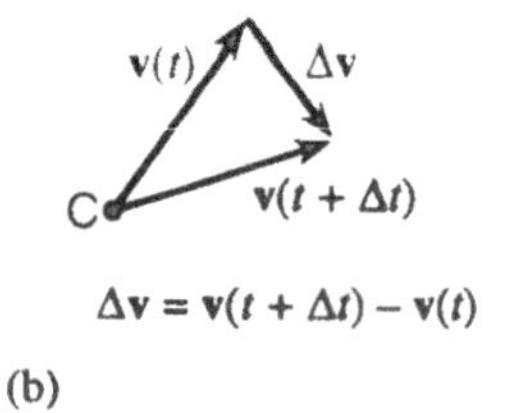

$\Delta \mathbf{v} = \mathbf{v}(t + \Delta t) - \mathbf{v}(t)$

(b)

Fig 3.10b
The two velocity vectors are redrawn with a common beginning point C so that the change in velocity $\Delta \mathbf{v}$ is easily drawn using the triangle rule.

The subtraction of the velocity vectors in the numerator of Eq (3.17) is illustrated in Fig 3.10b where the two velocity vectors are redrawn with their beginning points at a common point C. The derivative $\mathrm{d}\mathbf{v}/\mathrm{d}t$ is the limit ($\Delta t \to 0$) of Eq (3.17). This derivative is the instantaneous **acceleration** $\mathbf{a}(t)$ of the particle

$$\mathbf{a}(t) = \frac{\mathrm{d}\mathbf{v}}{\mathrm{d}t} = \frac{\mathrm{d}^2\mathbf{s}}{\mathrm{d}t^2} \qquad \text{(definition of acceleration)} \tag{3.18}$$

The symbol $\mathrm{d}^2\mathbf{s}/\mathrm{d}t^2$ denotes the *second derivative* of **s**, i.e. the derivative of $\mathrm{d}\mathbf{s}/\mathrm{d}t$. Thus $\mathrm{d}^2\mathbf{s}/\mathrm{d}t^2 = \mathrm{d}(\mathrm{d}\mathbf{s}/\mathrm{d}t)/\mathrm{d}t$.

Just as the velocity is always tangential to the path of the particle, so the acceleration is always tangential to the space curve of the velocity function $\mathbf{v}(t)$. Note that the acceleration is not tangential to the path of the particle (unless the path is a straight line). In projectile motion for example the acceleration is $\mathbf{a}(t) = \mathbf{g}$, a constant vector directed vertically downwards (Fig 3.11).

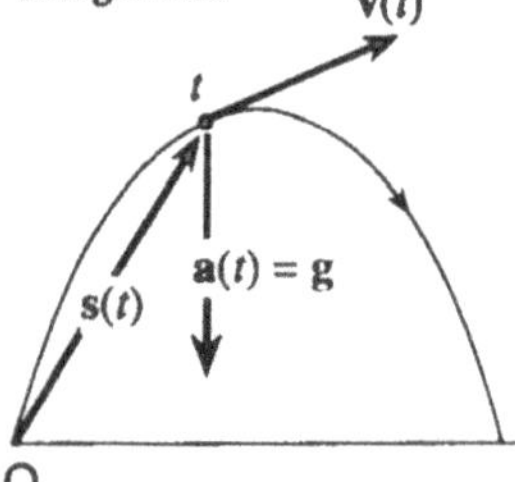

Fig 3.11
The parabolic path of a projectile. The displacement, velocity and acceleration vectors are shown at an instant t.

Summary of section 3.2

- The **derivative of a vector function f**(t) is a vector function denoted by $\mathrm{d}\mathbf{f}/\mathrm{d}t$ or $\mathbf{f}'(t)$ and defined by Eq (3.14). The derivative is always tangential to the space curve of **f** and points in the direction of increasing t.

- The **velocity** is defined to be the derivative of the displacement, and the **acceleration** is defined to be the derivative of the velocity (Eqs (3.16) and (3.18).) The **average velocity** and **average acceleration** are given by Eqs (3.15) and (3.17).

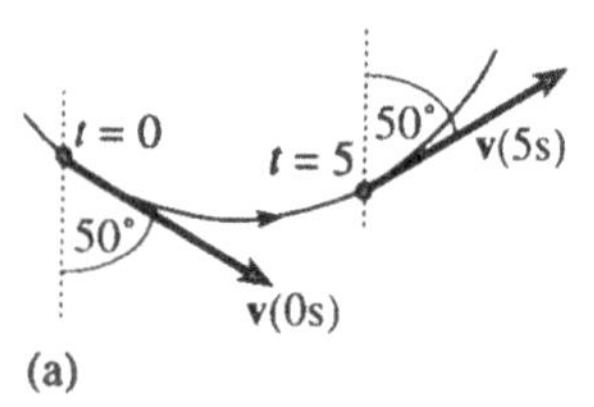

(a)

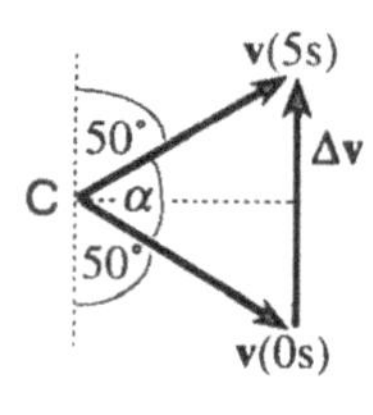

(b)

Fig 3.12
(a) The path of a diving aeroplane showing the velocity vectors at $t = 0$ s and $t = 5$ s. (b) The velocity vectors redrawn with a common beginning point. $\Delta \mathbf{v}$ is the change in velocity in the 5 s interval

Example 2.1 (*Objective 3*) Fig 3.12a shows the curved path of an aircraft executing a dive in a vertical plane. The directions of the velocity vectors at $t = 0$ s and $t = 5$ s are also shown. Given that $|\mathbf{v}(0\ \mathrm{s})| = |\mathbf{v}(5\ \mathrm{s})| = 150\ \mathrm{m\,s^{-1}}$, determine the average acceleration of the aircraft during the 5 s interval.

Solution 2.1 The average acceleration is

$$\frac{\Delta \mathbf{v}}{\Delta t} = \frac{\mathbf{v}(5\,\text{s}) - \mathbf{v}(0\text{s})}{5\,\text{s}}$$

The velocity vectors are drawn in Fig 3.12b from a common beginning point C. The angle α is $180° - 2 \times 50° = 80°$ and so the magnitude of $\Delta \mathbf{v}$ is

$$|\Delta \mathbf{v}| = 2|\mathbf{v}(0\text{s})|\sin(\alpha/2) = 2(150\text{ m s}^{-1})\sin(40°) = 192.8\text{ ms}^{-1}$$

The magnitude of the average acceleration is therefore $(192.8\text{ ms}^{-1})/5\text{ s} = 38.6\text{ ms}^{-2}$. The direction of the average acceleration is the direction of $\Delta \mathbf{v}$ which is seen from the figure to be directed vertically upwards.

Example 2.2 (*Objectives 1,3,4*) The velocity of a projectile moving under the Earth's gravity without air friction is given by the vector function

$$\mathbf{v}(t) = \mathbf{u} + \mathbf{g}t \qquad (0 \le t \le T)$$

where $\mathbf{u}$ is the initial velocity of projection at $t = 0$, $\mathbf{g}$ is the acceleration of gravity directed vertically downwards and of magnitude 10 ms^{-2} and T is the duration of flight. In a particular case $\mathbf{u}$ is directed at 30° above the horizontal with $|\mathbf{u}| = 20\text{ ms}^{-1}$ and $T = 2$ s. The time to reach the highest point can be shown to be $t = 1$ s.

(a) Use the equation $\mathbf{v}(t) = \mathbf{u} + \mathbf{g}t$ to determine $|\mathbf{v}(0\text{s})|$, $|\mathbf{v}(1\text{s})|$ and $|\mathbf{v}(2\text{s})|$.

(b) Make a rough sketch of the parabolic path of the projectile showing the vectors $\mathbf{v}(0\text{s})$, $\mathbf{v}(1\text{s})$ and $\mathbf{v}(2\text{s})$ drawn from the corresponding points on the path of the particle.

(c) Sketch a triangle addition diagram showing $\mathbf{v}(1\text{s})$ as the resultant of $\mathbf{u}$ and $\mathbf{g}(1\text{s})$, and another diagram showing $\mathbf{v}(2\text{s})$ as the resultant of $\mathbf{u}$ and $\mathbf{g}(2\text{s})$. Sketch the space curve of the velocity function $\mathbf{v}(t)$.

Be aware of the two usages of brackets here: $\mathbf{g}(1\text{s})$ denotes the vector g scaled by the scalar 1 s, i.e. it is a velocity. The symbol $\mathbf{v}(1\text{s})$ denotes the velocity at $t = 1\text{s}$.

(d) Determine the average acceleration $\Delta\mathbf{v}/\Delta t$ and deduce the limit of this for $\Delta t \to 0$.

Solution 2.2

(a) We use the given equation to find $\mathbf{v}(t)$ and then evaluate the magnitude. Thus putting $t = 0$ s we have $\mathbf{v}(0\text{s}) = \mathbf{u}$ and

$$|\mathbf{v}(0\text{ s})| = |\mathbf{u}| = 20\text{ m s}^{-1}$$

Putting $t = 1$ s, and omitting the physical units (which would clutter up the algebra), we have

Note that by omitting units and writing $\mathbf{u} + \mathbf{g}$ we appear to be adding an acceleration to a velocity which would be nonsense. In fact we are adding the velocity $\mathbf{g}(1\text{ s})$ to the velocity $\mathbf{u}$.

$$|\mathbf{v}(1)|^2 = \mathbf{v}(1) \cdot \mathbf{v}(1)$$

$$= (\mathbf{u} + \mathbf{g}) \cdot (\mathbf{u} + \mathbf{g}) = \mathbf{u} \cdot \mathbf{u} + 2\mathbf{u} \cdot \mathbf{g} + \mathbf{g} \cdot \mathbf{g}$$

$$= 20^2 + 2 \times 20 \times 10 \cos(90° + 30°) + 10^2 = 300$$

Thus $|\mathbf{v}(1\text{s})| = \sqrt{300}\text{ m s}^{-1} = 17.3\text{ m s}^{-1}$. Similarly, putting $t = 2$ s we have

$$|\mathbf{v}(2\text{s})|^2 = (\mathbf{u} + 2\mathbf{g}) \cdot (\mathbf{u} + 2\mathbf{g}) = \mathbf{u} \cdot \mathbf{u} + 4\mathbf{u} \cdot \mathbf{g} + 4\mathbf{g} \cdot \mathbf{g} = 400$$

and so $|\mathbf{v}(2\text{s})| = 20\text{ m s}^{-1}$.

(b) The path and the velocity vectors are shown in Fig 3.13a. Note that each velocity vector is tangential to the path.

(c) The triangle addition diagrams are shown in Fig 3.13b. The space curve of $\mathbf{v}(t)$ is the downwards-directed line AB traced out by the endpoints of the velocity vectors when they are all drawn from the same point C (Fig 3.13c).

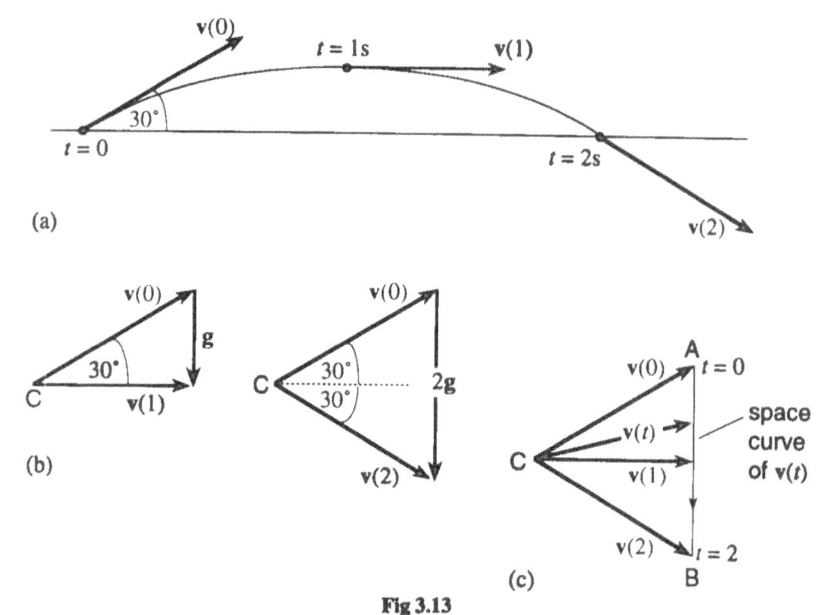

Fig 3.13
The projectile in Example 2.2. (a) The path and velocity vectors. (b) Velocity triangles. (c) The space curve of the velocity function is the vertical line AB.

(d) The average acceleration in the interval between t and $t + \Delta t$ is

$$\frac{\Delta \mathbf{v}}{\Delta t} = \frac{\mathbf{v}(t+\Delta t)-\mathbf{v}(t)}{\Delta t}$$

$$= \frac{\mathbf{u}+\mathbf{g}(t+\Delta t)-(\mathbf{u}+\mathbf{g}t)}{\Delta t} = \frac{\mathbf{g}\Delta t}{\Delta t} = \mathbf{g}$$

This is independent of Δt and so the instantaneous acceleration and the average acceleration are both equal to the constant **g**, as expected for projectile motion.

Problem 2.1 (*Objective 3*) A radar station observes a ship to be at a distance of 1.0 km in a direction due north. Three minutes later the ship is at a bearing of 5° east of north and at a distance of 800 m. Determine the average velocity of the ship in the three minute interval.

Problem 2.2 (*Objectives 1,3*) A particle moves with constant speed $v = 3\ \mathrm{ms}^{-1}$ along a circular path of radius $r = 5$ m.

(a) Determine the period T of the motion (i.e. the time it takes for the particle to make one complete revolution).

(b) Let $\mathbf{r}(t)$ and $\mathbf{v}(t)$ be the position vector and the velocity of the particle respectively with $\mathbf{r}(t)$ measured from the centre of the path, and suppose the particle passes through a point P at time $t = 0$ s. Make a rough sketch showing the path of the particle, the position vectors $\mathbf{r}(0\mathrm{s})$, $\mathbf{r}(5\mathrm{s})$, $\mathbf{r}(7\mathrm{s})$, and the velocity vectors $\mathbf{v}(0\mathrm{s})$, $\mathbf{v}(5\mathrm{s})$ and $\mathbf{v}(7\mathrm{s})$. Show the velocity vectors with their beginning points at the end points of the corresponding position vectors.

(c) Make another sketch showing the three velocity vectors drawn from a common point C. Describe in words the space curve of $\mathbf{v}(t)$.

Problem 2.3 (*Objective 4. This long problem leads you through the calculation of the acceleration of a particle in uniform circular motion from first principles using the definition of a derivative, Eq (3.14).*)

Consider a particle moving at constant speed v along a circular path of radius r.

(a) Sketch the path and show two displacements $\mathbf{s}(t)$ and $\mathbf{s}(t + \Delta t)$ measured from the centre O; show also the velocity vectors $\mathbf{v}(t)$ and $\mathbf{v}(t + \Delta t)$ with their beginning points at the corresponding points on the path.

(b) Express the time T for one complete revolution in terms of r and v.

(c) Sketch the two vectors $\mathbf{v}(t)$ and $\mathbf{v}(t + \Delta t)$ from a common beginning point C and sketch the vector $\Delta\mathbf{v} = \mathbf{v}(t + \Delta t) - \mathbf{v}(t)$. Express the angle α between $\mathbf{v}(t)$ and $\mathbf{v}(t + \Delta t)$ in terms of T and Δt.

(d) Obtain an expression for $|\Delta\mathbf{v}|$ in terms of α and v.

(e) Use the results of parts (b), (c) and (d) to show that the average acceleration of the particle is of magnitude $(v^2/r)(\sin(\alpha/2))/(\alpha/2)$.

(f) Determine the instantaneous acceleration $\mathbf{a} = \mathrm{d}\mathbf{v}/\mathrm{d}t$, given the limit $\sin(\theta)/\theta \to 1$ as $\theta \to 0$.

3.3 RULES OF DIFFERENTIATION OF VECTOR FUNCTIONS

The rules of differentiation enable us to determine derivatives of vector functions without having to explicitly calculate limits. This section gives statements of some of the most important rules, and applies them to finding the velocities and accelerations of particles.

Let $\mathbf{F}(t)$ and $\mathbf{G}(t)$ be any two vector functions and let $f(t)$ be a scalar function. For convenience we denote these functions simply by $\mathbf{F}$, $\mathbf{G}$ and f. The following rules give the derivatives of sums and products:

$$\frac{\mathrm{d}(\mathbf{F}+\mathbf{G})}{\mathrm{d}t} = \frac{\mathrm{d}\mathbf{F}}{\mathrm{d}t} + \frac{\mathrm{d}\mathbf{G}}{\mathrm{d}t} \tag{3.19}$$

$$\frac{\mathrm{d}(\mathbf{F}\cdot\mathbf{G})}{\mathrm{d}t} = \frac{\mathrm{d}\mathbf{F}}{\mathrm{d}t}\cdot\mathbf{G} + \mathbf{F}\cdot\frac{\mathrm{d}\mathbf{G}}{\mathrm{d}t} \tag{3.20}$$

$$\frac{\mathrm{d}(\mathbf{F}\times\mathbf{G})}{\mathrm{d}t} = \frac{\mathrm{d}\mathbf{F}}{\mathrm{d}t}\times\mathbf{G} + \mathbf{F}\times\frac{\mathrm{d}\mathbf{G}}{\mathrm{d}t} \tag{3.21}$$

$$\frac{\mathrm{d}(f\,\mathbf{G})}{\mathrm{d}t} = \frac{\mathrm{d}f}{\mathrm{d}t}\mathbf{G} + f\frac{\mathrm{d}\mathbf{G}}{\mathrm{d}t} \tag{3.22}$$

A useful special case of Eq (3.22) is where $\mathbf{G}$ is a constant function, i.e. $\mathbf{G}$ is a constant vector independent of t. Then

$$\frac{\mathrm{d}(f\,\mathbf{G})}{\mathrm{d}t} = \frac{\mathrm{d}f}{\mathrm{d}t}\mathbf{G} \qquad (\mathbf{G}\text{ a constant vector}) \tag{3.23}$$

The order of the symbols in a scaling expression such as $f\mathbf{G}$ or in a scalar product such as $\mathbf{F}\cdot\mathbf{G}$ is unimportant, but the order of vectors in the vector products in Eq (3.21) must be respected – a change of order in a vector product introduces a minus sign. The above rules are to be supplemented with the rules for differentiating elementary scalar functions (Appendix B).

As an application of the above rules, consider the case of a particle moving in uniform circular motion in the x-y plane of radius r and constant speed v. The displacement measured from the centre of the circular path (i.e. the position vector) is

$$\mathbf{r}(t) = r(\mathbf{i}\cos\omega t + \mathbf{j}\sin\omega t) \tag{3.24}$$

where **i** and **j** are cartesian unit vectors and the constant ω is the angular speed v/r. The velocity of the particle is found by differentiating $\mathbf{r}(t)$, i.e. $\mathbf{v}(t) = \mathrm{d}\mathbf{r}/\mathrm{d}t$. To calculate this derivative we first use the rather obvious rule 3.19 with $\mathbf{F}(t) = r\,\mathbf{i}\cos\omega t$ and $\mathbf{G}(t) = r\,\mathbf{j}\sin\omega t$, to give

$$\mathbf{v}(t) = \frac{\mathrm{d}\mathbf{r}}{\mathrm{d}t} = \frac{\mathrm{d}(r\,\mathbf{i}\cos\omega t)}{\mathrm{d}t} + \frac{\mathrm{d}(r\,\mathbf{j}\sin\omega t)}{\mathrm{d}t}$$

Next we note that each of the two vector functions to be differentiated in the above equation is the product of a scalar function and a constant unit vector. Thus we use rule 3.23 and the results for differentiating the scalar trigonometric functions (Appendix B) to obtain

$$\mathbf{v}(t) = (-r\,\omega\sin\omega t)\mathbf{i} + (r\,\omega\cos\omega t)\mathbf{j} = -r\,\omega(\mathbf{i}\sin\omega t - \mathbf{j}\cos\omega t) \tag{3.25}$$

We can go on to find the acceleration of the particle by a second differentiation, i.e. $\mathbf{a} = \mathrm{d}\mathbf{v}/\mathrm{d}t$. Proceeding as before we obtain

$$\mathbf{a}(t) = \omega(-r\omega)(\mathbf{i}\cos\omega t + \mathbf{j}\sin\omega t) = -\omega^2\mathbf{r}$$

where the last step involves recognising that the two successive differentiations have returned the original function $\mathbf{r}(t)$ with a scaling factor $(-\omega^2)$, a familiar property of a sine or cosine function. Thus we have

Problem 2.3 indicates how Eq (3.26) can be derived from first principles using the definition of a derivative, Eq (3.14).

$$\mathbf{a}(t) = -\omega^2\mathbf{r}(t) \qquad \text{(centripetal acceleration)} \tag{3.26}$$

The minus sign shows that the acceleration (Eq (3.26)) is directed radially towards the centre of the circle. The acceleration of uniform circular motion is often called **centripetal acceleration**.

We can use rules 3.19 and 3.23 to derive a general result from which derivatives such as **v** and **a** above (Eqs (3.25) and (3.26)) can be found efficiently. Given a cartesian vector function $\mathbf{F} = F_x\mathbf{i} + F_y\mathbf{j} + F_z\mathbf{k}$, we can differentiate it to obtain

$$\frac{\mathrm{d}\mathbf{F}}{\mathrm{d}t} = \frac{\mathrm{d}(F_x\mathbf{i} + F_y\mathbf{j} + F_z\mathbf{k})}{\mathrm{d}t}$$

$$= \frac{\mathrm{d}(F_x\mathbf{i})}{\mathrm{d}t} + \frac{\mathrm{d}(F_y\mathbf{j})}{\mathrm{d}t} + \frac{\mathrm{d}(F_z\mathbf{k})}{\mathrm{d}t}$$

$$= \frac{dF_x}{dt}\mathbf{i} + \frac{dF_y}{dt}\mathbf{j} + \frac{dF_z}{dt}\mathbf{k} \tag{3.27a}$$

This shows that the x-component of the derivative is equal to the derivative of the x-component, i.e. $(d\mathbf{F}/dt)_x = dF_x/dt$, and similarly for the other components. Thus the derivative $d\mathbf{F}/dt$ can be written neatly as an ordered triple of scalar functions

$$\frac{d\mathbf{F}}{dt} = \frac{d(F_x, F_y, F_z)}{dt} = \left(\frac{dF_x}{dt}, \frac{dF_y}{dt}, \frac{dF_z}{dt}\right) \tag{3.27b}$$

You should confirm that Eqs (3.25) and (3.26) can be derived from Eqs (3.24) and (3.25) respectively, by using this rule.

Summary of section 3.3

- Rules for differentiating vector functions were stated: Eqs (3.19)–(3.23). These rules are similar to the rules for differentiating sums and products of scalar functions but the order of vectors in a vector product must be respected.

- The acceleration of uniform circular motion is called **centripetal acceleration** and is given by $\mathbf{a}(t) = -\omega^2\mathbf{r}(t)$ where $\omega = v/r$ and $\mathbf{r}(t)$ is the position vector measured from the centre of the path.

- The components of the derivative of a vector function are equal to the derivatives of the components ((3.27a and (b).

Example 3.1 (*Objective 5*) Given the time-dependent vectors

$$\mathbf{A}(t) = 2\mathbf{i} - \mathbf{k}t^2 \quad \text{and} \quad \mathbf{B}(t) = 3t\mathbf{i} + t^3\mathbf{j}$$

determine $\frac{d\mathbf{A}}{dt}, \frac{d\mathbf{B}}{dt}, \frac{dA_z}{dt}, \frac{d(\mathbf{A}-\mathbf{B})_y}{dt}, \frac{d(\mathbf{A}\,.\,\mathbf{B})}{dt}, \frac{d(\mathbf{A}\times\mathbf{B})}{dt}$.

Solution 3.1 Using the rules of differentiation, we find

$$\frac{d\mathbf{A}}{dt} = -\mathbf{k}2t$$

$$\frac{d\mathbf{B}}{dt} = 3\mathbf{i} + 3t^2\mathbf{j}$$

$$\frac{dA_z}{dt} = \frac{d(-t^2)}{dt} = -2t$$

$$\frac{d(\mathbf{A}-\mathbf{B})_y}{dt} = \frac{d(-t^3)}{dt} = -3t^2$$

$$\frac{d\mathbf{A}\,.\,\mathbf{B}}{dt} = (-\mathbf{k}2t)\,.\,(3t\mathbf{i} + t^3\mathbf{j}) + (2\mathbf{i} - \mathbf{k}t^2)\,.\,(3\mathbf{i} + 3t^2\mathbf{j}) = 6$$

$$\frac{d\mathbf{A} \times \mathbf{B}}{dt} = (-\mathbf{k}2t) \times (3t\mathbf{i} + t^3\mathbf{j}) + (2\mathbf{i} - \mathbf{k}t^2) \times (3\mathbf{i} + 3t^2\mathbf{j})$$

$$= 5t^4\mathbf{i} - 9t^2\mathbf{j} + 6t^2\mathbf{k}$$

(We could alternatively have evaluated the dot and cross products first and then differentiated.)

Example 3.2 (*Objective 5*) A particle moves at constant speed along a circular path in the x-y plane. Its displacement from the centre of the path is given by $\mathbf{s}(t) = 5(\mathbf{i}\cos 3\pi t + \mathbf{j}\sin 3\pi t)$.

(a) Write down three scalar equations for the cartesian components and express $\mathbf{s}(t)$ as an ordered triple of scalar functions.

(b) Obtain the velocity $\mathbf{v}(t)$ and the acceleration $\mathbf{a}(t)$ as ordered triples of scalar functions. Determine the magnitudes $|\mathbf{s}(t)|$, $|\mathbf{v}(t)|$ and $|\mathbf{a}(t)|$.

(c) Determine $\mathbf{s}(0)$, $\mathbf{v}(0)$, $\mathbf{a}(0)$ and $\mathbf{s}(1/3)$, $\mathbf{v}(1/3)$, $\mathbf{a}(1/3)$. Make a sketch of the circular path in the x-y plane and draw six arrows labelled $\mathbf{s}(0)$, $\mathbf{v}(0)$, etc. showing the directions of the vectors. How long does it take for the particle to make one complete revolution?

Solution 3.2

(a) Projecting $\mathbf{s}(t)$ onto cartesian unit vectors we have

$$s_x(t) = 5\cos(3\pi t), \quad s_y(t) = 5\sin(3\pi t), \quad s_z(t) = 0$$

Thus $\mathbf{s}(t) = (5\cos(3\pi t), 5\sin(3\pi t), 0)$.

(b) $\mathbf{v}(t)$ is the derivative $d\mathbf{s}/dt$. The components of a derivative are the derivatives of the components (Eq (3.27b)). Hence

See Appendix B for the derivatives of the sine and cosine functions.

$$\mathbf{v}(t) = (-3\pi\times5\sin(3\pi t), 3\pi\times5\cos(3\pi t), 0) = 3\pi\times5(-\sin(3\pi t), \cos(3\pi t), 0)$$

$$\mathbf{a}(t) = (3\pi)^2\times5(-\cos(3\pi t), -\sin(3\pi t), 0) = -(3\pi)^2\mathbf{s}(t)$$

The magnitude of $\mathbf{s}(t)$ is

$$|\mathbf{s}(t)| = (5^2\cos^2(3\pi t) + 5^2\sin^2(3\pi t) + 0^2)^{1/2} = 5$$

where we have used the identity $\sin^2\theta + \cos^2\theta = 1$. In the same way we find

$$|\mathbf{v}(t)| = 3\pi \times 5 = 15\pi \quad \text{and} \quad |\mathbf{a}(t)| = (3\pi)^2 \times 5 = 45\pi^2$$

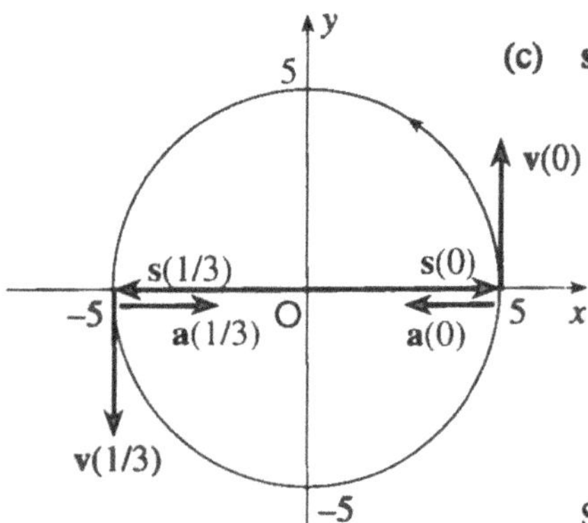

Fig 3.14
The position vectors, velocity vectors and acceleration vectors at $t = 0$ and $t = 1/3$, for Example 3.2.

(c) $\mathbf{s}(0) = (5\cos 0, 5\sin 0, 0) = (5,0,0)$. Similarly

$$\mathbf{v}(0) = (0,15\pi,0)$$

$$\mathbf{a}(0) = -(3\pi)^2(5,0,0) = (-45\pi^2,0,0)$$

$$\mathbf{s}(1/3) = (5\cos(3\pi\times 1/3), 5\sin(3\pi\times 1/3), 0)$$

$$= (5\cos\pi, 5\sin\pi, 0) = (-5,0,0)$$

Similarly $\mathbf{v}(1/3) = (0,-15\pi,0)$ and $\mathbf{a}(1/3) = (3\pi)^2(5,0,0)$. The path and the directions of the vectors are shown in Fig 3.14. It is seen from the figure that it takes a time 1/3 for one half of a complete revolution. The time for a complete revolution is therefore $T = 2/3$.

Example 3.3 (*Objective 5*) The position vector of a particle is given by

$$\mathbf{r}(t) = A\exp\left(-\frac{t}{5}\right)\mathbf{i} + Bt\,\mathbf{j} + C\,\mathbf{k}$$

where A, B and C are constants.

(a) Determine the velocity and acceleration at time $t = 10$.

(b) Sketch the path of the particle over the interval from $t = 0$ to $t = 5$ in the case where $A = 1$, $B = 0.2$ and $C = 0$.

Solution 3.3

(a) The velocity at time t is

$$\mathbf{v}(t) = \frac{d\mathbf{r}}{dt} = -\frac{1}{5}A\exp\left(-\frac{t}{5}\right)\mathbf{i} + B\mathbf{j}$$

The acceleration at time t is

$$\mathbf{a}(t) = \frac{d\mathbf{v}}{dt} = \left(-\frac{1}{5}\right)^2 A \exp\left(-\frac{t}{5}\right)\mathbf{i}$$

Evaluating these vector functions at $t = 10$, gives

$$\mathbf{v}(10) = -\frac{1}{5}A\exp(-2)\,\mathbf{i} + B\,\mathbf{j} = -0.0271A\mathbf{i} + B\mathbf{j}$$

$$\mathbf{a}(10) = \left(-\frac{1}{5}\right)^2 A\exp(-2)\,\mathbf{i} = 0.00541A\mathbf{i}$$

(b) The parametric equations of the path are the components of the position vector obtained by projecting $\mathbf{r}(t)$ onto the cartesian unit vectors. Thus

$$x = A\exp\left(-\frac{t}{5}\right), \quad y = Bt, \quad z = C$$

where $x = r_x(t)$, etc. Eliminating t gives

$$x = A\exp\left(-\frac{y}{5B}\right)$$

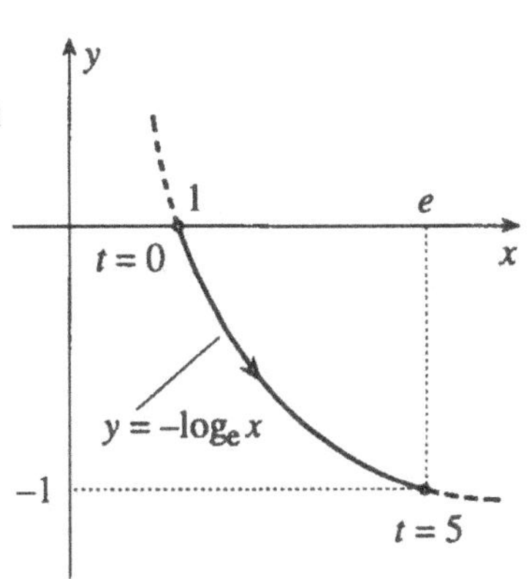

Fig 3.15 The path of the particle in Example 3.3 is a segment of the curve $y = -\log_e x$.

Putting in the given values of A and B we obtain $x = \exp(-y)$ or $y = -\log_e x$. The path is in the plane $z = C = 0$ and is sketched in Fig 3.15.

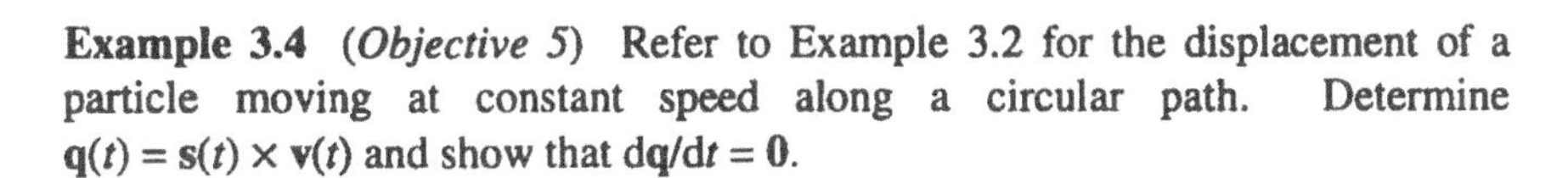

Example 3.4 (*Objective 5*) Refer to Example 3.2 for the displacement of a particle moving at constant speed along a circular path. Determine $\mathbf{q}(t) = \mathbf{s}(t) \times \mathbf{v}(t)$ and show that $d\mathbf{q}/dt = \mathbf{0}$.

Solution 3.4 We can work out the vector product first and then differentiate or we can use the rule for differentiating vector products (Eq (3.21)). Choosing the former, we have

$$\mathbf{q}(t) = \mathbf{s}(t) \times \mathbf{v}(t)$$

$$= 5(\mathbf{i}\cos 3\pi t + \mathbf{j}\sin 3\pi t) \times (-3\pi)5(\mathbf{i}\sin 3\pi t - \mathbf{j}\cos 3\pi t)$$

$$= 5(-3\pi)5[\mathbf{i} \times (-\mathbf{j})\cos^2 3\pi t + \mathbf{j} \times \mathbf{i}\sin^2 3\pi t]$$

where we have used $\mathbf{i} \times \mathbf{i} = \mathbf{j} \times \mathbf{j} = \mathbf{0}$. Now use $\mathbf{i} \times (-\mathbf{j}) = -\mathbf{i} \times \mathbf{j} = \mathbf{j} \times \mathbf{i} = -\mathbf{k}$ and the identity $\cos^2\theta + \sin^2\theta = 1$, and the expression for $\mathbf{q}(t)$ simplifies to

$$\mathbf{q}(t) = 75\pi\mathbf{k}$$

Thus $\mathbf{q}(t)$ is a constant vector and so $d\mathbf{q}/dt = \mathbf{0}$. (We could check the answer by using Eq (3.21); see Problem 3.7.)

Problem 3.1 (*Objective 5*) Given the position vector $\mathbf{r}(t) = 3\mathbf{i} - 5\mathbf{j}\exp(-2t)$, determine the velocity as a function of time and the speed at $t = 4$.

Problem 3.2 (*Objective 5*) The displacement of a projectile measured from the point of projection is given (in the absence of air resistance) by

$$\mathbf{s}(t) = \mathbf{u}t + \frac{1}{2}\mathbf{g}t^2$$

where $\mathbf{u}$ is the initial velocity of projection directed at an angle θ above the horizontal and $\mathbf{g}$ is the acceleration due to gravity directed downwards and of magnitude $g = 10$.

(a) Project $\mathbf{s}(t)$ onto horizontal and vertical axes in the plane of $\mathbf{u}$ and $\mathbf{g}$, and differentiate your projected equations. Given $u = 7$ and $\theta = 55°$ determine the horizontal and vertical components of the velocity of the particle at time $t = 0.2$. Use these components to determine the speed of the particle at this instant.

(b) Differentiate $\mathbf{s}(t)$ without introducing a coordinate system and hence specify $\mathbf{v}(0.2)$. Determine $\mathbf{v}(0.2) \cdot \mathbf{v}(0.2)$ and use the given specifications of $\mathbf{u}$ and $\mathbf{g}$ to determine the speed $v(0.2)$ thereby confirming your answer to part (a).

(c) Differentiate a second time to confirm that the acceleration is $\mathbf{g}$.

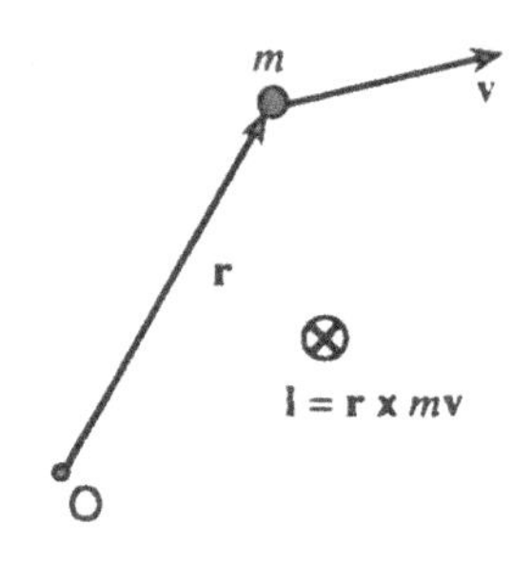

Fig 3.16
The angular momentum of a particle.

Problem 3.3 (*Objective 5*) A particle of mass m moving with velocity $\mathbf{v}$ has an **angular momentum** $\mathbf{l}$ about a point O defined by

$$\mathbf{l} = \mathbf{r} \times m\mathbf{v} \qquad \text{(angular momentum of a particle)}$$

where $\mathbf{r}$ is the displacement of the particle from O (Fig 3.16). Differentiate $\mathbf{l} = \mathbf{r} \times m\mathbf{v}$ and use Newton's second law of motion, $\mathbf{F} = m\mathbf{a}$, to show that

$$\frac{d\mathbf{l}}{dt} = \mathbf{\Gamma}$$

where $\mathbf{\Gamma} = \mathbf{r} \times \mathbf{F}$ is the torque of the force about O. Show that when the force $\mathbf{F}$ is a **central force**, i.e. when $\mathbf{F}$ is always directed towards or away from a fixed point O, the angular momentum of the particle remains constant in time. Give an example of such motion.

Problem 3.4 (*Objectives 2,5*) A particle moves in one-dimensional **simple harmonic motion** when it moves along a straight line with an acceleration directed towards a fixed point O on the line and of magnitude proportional to its distance from O. Show that the projection of uniform circular motion (Eq (3.24)) onto the x-axis is simple harmonic motion. What is the maximum speed of simple harmonic motion and the maximum magnitude of the acceleration, and when do these maximum values occur?

Problem 3.5 (*Objectives 2,5*) A particle moves at constant speed v along a circular path of radius r. What shape is the space curve of the velocity vector **v**?

The space curve of a velocity vector is sometimes called a *hodograph*.

Problem 3.6 (*orthogonal vectors*) Show (by calculating the scalar product) that the position and velocity vectors of a particle in uniform circular motion in the x-y plane (Eqs (3.24) and (3.25)) are orthogonal. Show that this orthogonality is maintained when the position vector is measured from any point on the z-axis.

Problem 3.7 (*Objective 5*) Confirm the result $d\mathbf{q}/dt = 0$ in Example 3.4 by using Eq (3.21).

3.4 ROTATIONAL MOTION – THE ANGULAR VELOCITY VECTOR

The study of rotating bodies is a very subtle and rich area with many fascinating applications. At centre stage is a vector quantity called *angular velocity*. We introduce the angular velocity vector in this section and show how it can be used to calculate the velocity of a point on a rotating rigid body; for example, your velocity through space due to the rotational motion of the Earth about its axis.

Consider a particle moving with speed v along a circular path of radius r in the x-y plane (Fig 3.17a). The angle $\phi(t)$ between the position vector and the positive x-axis is called the **angular displacement** of the particle. The average angular speed of the particle in the interval between t and $t + \Delta t$ (Fig 3.17b) is

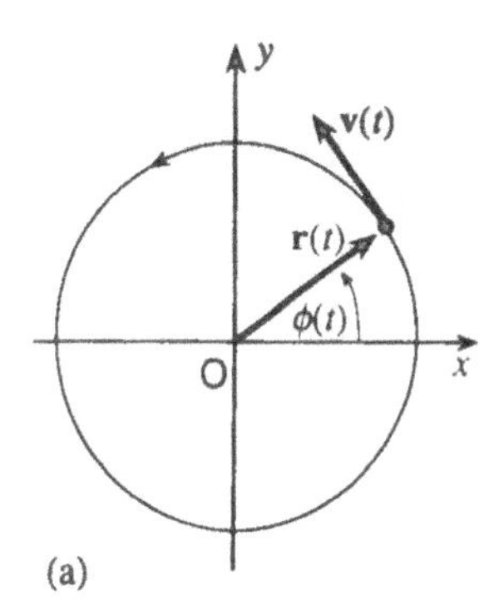

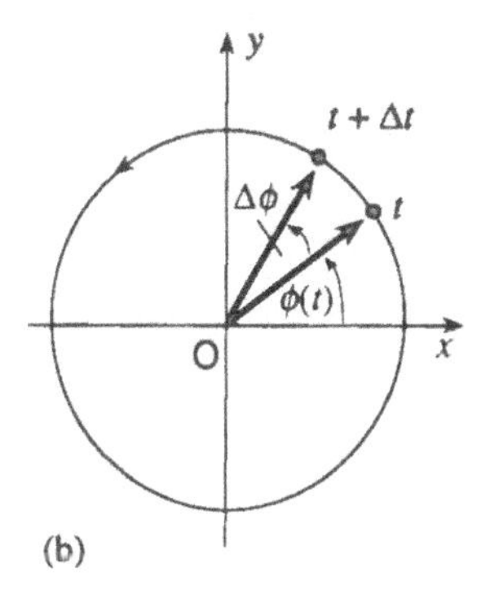

Fig 3.17
(a) The angular displacement $\phi(t)$ is measured anticlockwise from the positive x-axis.
(b) The angular displacements are ϕ and $\phi + \Delta\phi$ at times t and $t + \Delta t$.

$$\left|\frac{\Delta\phi}{\Delta t}\right| = \left|\frac{\phi(t+\Delta t)-\phi(t)}{\Delta t}\right| \tag{3.28}$$

The instantaneous **angular speed** ω of the particle at time t is defined to be the limit of Eq (3.28) as $\Delta t \to 0$

$$\omega = \left|\frac{d\phi}{dt}\right| \tag{3.29}$$

When the particle moves with constant speed, the instantaneous angular speed is constant and equal to the average angular speed in any interval. One complete revolution is an angular displacement of 2π radians and takes a time T called the **period**. Hence a constant angular speed ω is related to the period T by

$$\omega = \frac{2\pi}{T} \tag{3.30}$$

Since the particle travels a distance $2\pi r$ in the time T, its speed is $v = 2\pi r/T$. Using this result with Eq (3.30) we obtain the relationship between speed and angular speed

$$v = \omega r \quad \text{or} \quad \omega = v/r \tag{3.31}$$

The angular speed ω is also called the **angular frequency**. It is related to the **frequency** f by

$$\omega = 2\pi f \tag{3.32}$$

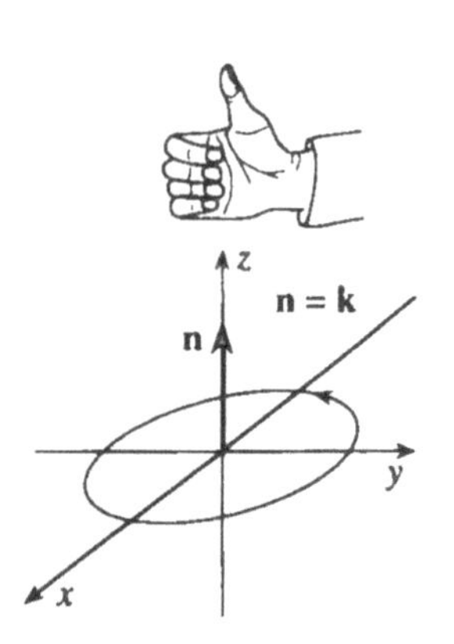

Fig 3.18
The right-hand rule.

where $f = 1/T$, the number of revolutions per second. The units of angular speed ω are those of angle divided by time, i.e. radians per second (rad s^{-1}), whereas the units of frequency f are hertz (Hz). 1 Hz = 2π rad s^{-1}.

Angular speed is a positive scalar. We obtain an angular velocity vector $\boldsymbol{\omega}$ by using the right-hand rule to specify a unit vector $\mathbf{n}$ in a direction normal to the plane of the circular path; then $\boldsymbol{\omega} = \omega\mathbf{n}$. For a particle moving anticlockwise in the x-y plane the right-hand rule specifies $\mathbf{n} = \mathbf{k}$, the unit vector in the positive z direction (Fig 3.18). Thus we define the **angular velocity** of a particle by Eq (3.29) combined with the right-hand rule:

$$\boldsymbol{\omega} = \left|\frac{d\phi}{dt}\right|\mathbf{n} \tag{3.33}$$

A **rigid body** is a body in which the distance between any two constituent particles is fixed.

A **fixed axis** is a line of points that remain at rest as the body rotates; if a rotating rigid body has two of its points held fixed in space then the straight line joining these points is a fixed axis.

We now turn our attention from a single particle to a **rigid body** rotating about a **fixed axis**. When a rigid body rotates about a fixed axis any constituent particle P is constrained to move along a circular path of radius ρ in a plane normal to the axis of rotation and with centre point C on the axis (Fig 3.19). Because the body is rigid, the angular displacement $\Delta\phi$ in an interval Δt is the same for all constituent particles of the body. It follows that the angular velocity is the same for all constituent particles and so it is a property of the rigid body as a whole. Thus the definition Eq (3.33) also gives the angular velocity of a rigid body. The angular velocity vector lies on the axis of rotation and points in the direction given by the right-hand rule (Fig 3.19).

The velocities of the constituent particles in a rotating rigid body are of course not all the same, since those near the axis of rotation move more slowly than the more distant ones, and so it is useful to know how to determine the velocity $\mathbf{v}$ of an arbitrary constituent particle. Let the position of a constituent particle at a point P be specified by the position vector $\mathbf{r}$ measured from an arbitrary fixed reference point O on the axis of rotation (Fig 3.20). The centre of the circular path of the particle is the point C on the axis, and the radius of the path is $\rho = |\mathbf{r}|\sin\alpha$. The direction of $\mathbf{v}$ is normal to the position vector $\mathbf{r}$ (Problem 3.6) and also normal to the axis of rotation; it is therefore normal to both $\mathbf{r}$ and $\boldsymbol{\omega}$, since $\boldsymbol{\omega}$ lies on the axis of rotation. Moreover the magnitude of $\mathbf{v}$

is $v = \omega\rho$; this follows from Eq (3.31) applied to circular motion of radius $\rho = CP$. Since $\rho = |\mathbf{r}|\sin\alpha$, we have $v = \omega|\mathbf{r}|\sin\alpha$. Putting direction and magnitude together it can be seen that the velocity **v** is given by the vector product

$$\mathbf{v} = \boldsymbol{\omega} \times \mathbf{r} \tag{3.34}$$

It should be noted that **r** is the position vector of the point (or particle) measured from an arbitrary origin point on the fixed axis of rotation, as indicated in Fig 3.20. When the origin point is chosen to be the centre C of the circular path of the particle, the vectors **v**, **ω** and **r** are mutually perpendicular and Eq (3.34) yields the magnitude $v = \omega r$ which is equivalent to Eq (3.31).

Eq (3.34) is an important equation in the study of rotational motion. It is not restricted to rotational motion about a fixed axis but applies to any arbitrary rotational motion of a rigid body about a fixed point O, with the point P constrained to move on a spherical surface of radius $r = |\mathbf{r}| = OP$. An example of rotational motion about a fixed point is a rod swinging or rotating in an arbitrary fashion from one end O held in a universal joint. There exists at each instant an instantaneous angular velocity **ω** lying on an instantaneous axis of rotation which always passes through the fixed point O but whose direction may vary in time. Eq (3.34) gives the instantaneous velocity of any point P whose position vector is $\mathbf{r} = \mathbf{OP}$. We shall however only consider cases of rotation about a fixed axis.

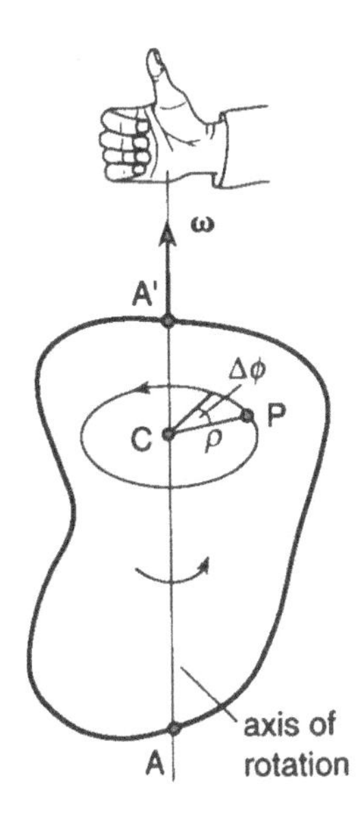

Fig 3.19
A rigid body rotating about a fixed axis AA′. A constituent particle at P moves along a circular path of radius ρ and centre point C on the axis of rotation.

Summary of section 3.4

- The average angular speed of a particle moving in a circular path in the *x-y* plane is $|\Delta\phi/\Delta t|$ where ϕ is the **angular displacement** measured anticlockwise from the positive *x*-axis. The instantaneous **angular speed (angular frequency)** is $\omega = |d\phi/dt|$ (rad s^{-1}).

- $T = 2\pi/\omega$ is the **period** and $f = 1/T$ (Hz) is the **frequency**.

- The speed v and the angular speed ω are related by $v = \omega r$ or $\omega = v/r$ where r is the radius.

- The **angular velocity** of a particle in circular motion is a vector **ω** of magnitude ω pointing at right angles to the plane of the path in the direction of the unit vector **n** given by the right-hand rule: $\boldsymbol{\omega} = \mathbf{n}|d\phi/dt|$.

- The angular velocity of a rotating rigid body is equal to the angular velocity of any one of its constituent particles.

- The velocity of any point P in a rigid body rotating with angular velocity **ω** is given by

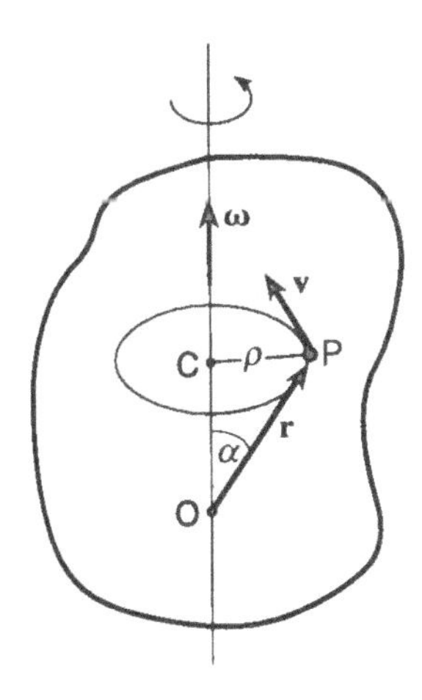

Fig 3.20
The velocity **v** of the point P is normal to **r** and **ω**.

$$\mathbf{v} = \boldsymbol{\omega} \times \mathbf{r} \qquad (3.34)$$

where **r** is the position vector of P measured from a fixed origin O on the axis of rotation.

Example 4.1 (*Objective 6*)

(a) Specify the angular velocity of the Earth as it rotates about its north-south polar axis.

(b) Determine the velocity of the citizens of Moscow due to the Earth's rotational motion.

(c) Determine the acceleration of the citizens of Moscow due to the Earth's rotation.

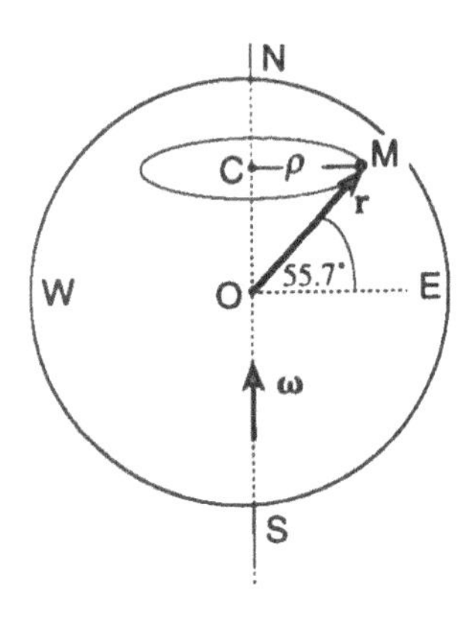

Fig 3.21
Moscow M is at a latitude 55.7° north.

Take the radius of the Earth to be 6370 km and the latitude of Moscow to be 55.7° north. Neglect the orbital motion of the Earth around the Sun.

Solution 4.1

(a) The Earth rotates through 2π radians in 24 hours. Hence

$$\omega = |\boldsymbol{\omega}| = \frac{2\pi}{(24\times 60\times 60)}\ \text{rad s}^{-1} = 7.27\times 10^{-5}\ \text{rad s}^{-1}$$

Observing that the Sun rises in the east and sets in the west, we can deduce that the Earth rotates from west to east, and so the direction of $\boldsymbol{\omega}$ (by the right-hand rule) is from the South Pole towards the North Pole (Fig 3.21).

(b) The velocity of Muscovites is given by $\mathbf{v} = \boldsymbol{\omega} \times \mathbf{r}$. Their speed is therefore $v = |\boldsymbol{\omega}|\,|\mathbf{r}| \sin\alpha$ where **r** is the position vector of Moscow as measured from any point on the axis of rotation. A convenient point is the centre O of the Earth so that $|\mathbf{r}|$ is equal to the Earth's radius and the angle between $\boldsymbol{\omega}$ and **r** is $\alpha = 90° - 55.7° = 34.3°$. Hence

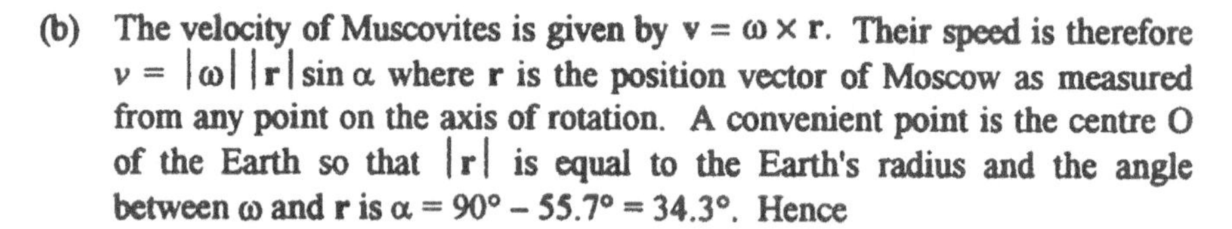

$$v = 7.27\times 10^{-5}\ \text{rad s}^{-1} \times 6.370\times 10^{6}\ \text{m} \times \sin 34.3° = 261\ \text{m s}^{-1}$$

The direction of **v** is towards the east.

(c) The acceleration is found from Eq (3.26): $\mathbf{a} = -\omega^2\mathbf{CM}$, where **CM** is the displacement of magnitude ρ measured from the centre C of the circular path (Fig 3.21). The direction of the acceleration is therefore parallel to $-\mathbf{CM} = \mathbf{MC}$, i.e. dipping by an angle $90° - 55.7° = 34.3°$ below the horizontal towards the north. The magnitude is

$$a = \omega^2\rho = (7.27 \times 10^{-5}\ \text{rad s}^{-1})^2 \times 6.370 \times 10^6\ \text{m} \times \cos 55.7° = 0.019\ \text{m s}^{-2}$$

Example 4.2 (*Objective 6*) The position vector of a particle moving anticlockwise along a circular path of radius ρ in the plane $z = c$ with angular speed ω is given by $\mathbf{r}(t) = (\rho\cos\omega t, \rho\sin\omega t, c)$. Determine the velocity $\mathbf{v}$ of the particle by

(a) using Eq (3.34), and

(b) by using the rules of differentiation.

(c) Evaluate the vector product $\mathbf{r} \times \mathbf{v}$. For what value of c is this vector product independent of time?

Solution 4.2

(a) Use $\mathbf{v} = \boldsymbol{\omega} \times \mathbf{r}$ with $\boldsymbol{\omega} = \omega\mathbf{k} = \omega(0,0,1)$. Then

$$\mathbf{v} = \omega(0,0,1) \times (\rho\cos\omega t, \rho\sin\omega t, c) = \omega\rho(-\sin\omega t, \cos\omega t, 0)$$

(b) Using Eq (3.27b) we differentiate each component of $\mathbf{r}(t)$:

$$\mathbf{v} = \frac{d\mathbf{r}}{dt} = \left(\frac{d(\rho\cos\omega t)}{dt}, \frac{d(\rho\sin\omega t)}{dt}, \frac{dc}{dt}\right)$$

$$= (-\omega\rho\sin\omega t, \omega\rho\cos\omega t, 0) = \omega\rho(-\sin\omega t, \cos\omega t, 0)$$

in agreement with part (a).

(c) $\mathbf{r} \times \mathbf{v} = (\rho\cos\omega t, \rho\sin\omega t, c) \times \omega\rho(-\sin\omega t, \cos\omega t, 0)$

$$= (-c\omega\rho\cos\omega t, -c\omega\rho\sin\omega t, \omega\rho^2\cos^2\omega t + \omega\rho^2\sin^2\omega t)$$

$$= (-c\omega\rho\cos\omega t, -c\omega\rho\sin\omega t, \omega\rho^2)$$

This is independent of time when $c = 0$, i.e. when the path is in the x-y plane.

Problem 4.1 (*Objective 6*) Determine the angular speed of the hour hand of a clock and the speed of a point on the hour hand 1cm from the centre of rotation, assuming the motion to be continuous.

Problem 4.2 (*Objective 6*) A rigid body rotates about the z-axis with angular velocity $\boldsymbol{\omega} = 5\mathbf{k}\ \text{rad s}^{-1}$. At time t_0 a point P fixed in the body has coordinates (3,0,1) cm. Determine the velocity of point P at t_0.

Problem 4.3 (*Objective 6*) Starting from $\mathbf{v} = \boldsymbol{\omega} \times \mathbf{r}$, make use of an identity for the vector triple product to show that

$$\boldsymbol{\omega} = \frac{1}{|\mathbf{r}|^2}(\mathbf{r} \times \mathbf{v} + (\mathbf{r} \cdot \boldsymbol{\omega})\mathbf{r})$$

Show that the vector product obtained in part (c) of Example 4.2 is consistent with this result.

Problem 4.4 (*Objective 6*) A flywheel of radius 12 cm spins at 250 Hz. Determine the speed of a point on its circumference.

3.5 ROTATING VECTORS OF CONSTANT MAGNITUDE

The study of rotating bodies often involves time-dependent vectors of constant magnitude. When a rigid body is rotating uniformly about a fixed axis (i.e. when $\boldsymbol{\omega}$ is constant in magnitude and direction) each constituent particle moves in uniform circular motion. The position vector, the velocity vector and the acceleration vector of a constituent particle have constant magnitudes, but the vectors change their directions continuously and uniformly as the body rotates. Other examples occur when rotating coordinate systems are used by observers on rotating platforms such as a passenger on a roundabout or the inhabitants of a spinning planet (you and me!). The cartesian unit vectors rotating with the coordinate system have a constant magnitude of unity but their directions are changing continuously. The derivatives of such vectors play an important role in many mechanical problems and can be found by using the rules of differentiation (Section 3.3). However it is often easier to obtain these derivatives by using a formula (obtained below) which does not explicitly involve differentiation and which can offer additional insight.

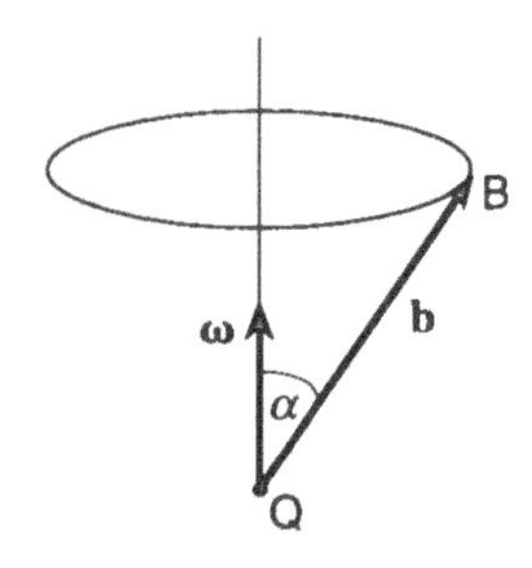

Fig 3.22
A time-varying vector **b** of constant magnitude. The figure shows the case where the rotation is about a fixed axis through Q and the endpoint B of **b** moves on a circle. More generally the direction of $\boldsymbol{\omega}$ may vary in time and B moves on a sphere of radius $|\mathbf{b}|$.

The time dependence of vectors of constant magnitude can be found by considering Eq (3.34) which gives the velocity of a point P in a rotating rigid body. We can express Eq (3.34) as

$$\frac{d\mathbf{r}}{dt} = \boldsymbol{\omega} \times \mathbf{r} \qquad (|\mathbf{r}| \text{ is constant}) \qquad (3.35)$$

since $\mathbf{v} = d\mathbf{r}/dt$. We can interpret Eq (3.35) as giving the derivative of a position vector of constant magnitude. Although this is just a change of notation and interpretation, it illustrates a very general result valid for any vector which like $\mathbf{r}$ is of constant magnitude. Consider any time-dependent vector $\mathbf{b}$ of constant magnitude. If the beginning point of $\mathbf{b}$ is fixed at a point Q, then the vector $\mathbf{b}$ rotates about Q like a rigid rod held by a universal joint. The special case where the angular velocity $\boldsymbol{\omega}$ of $\mathbf{b}$ is fixed in direction (i.e. rotation of $\mathbf{b}$ about a fixed axis) is illustrated in Fig 3.22; the end point B of $\mathbf{b}$ moves along a circular path in a plane normal to $\boldsymbol{\omega}$. The motion of the vector $\mathbf{b}$ is completely analogous to that of the position vector $\mathbf{r} = \mathbf{OP}$ of a point P in a rotating rigid body (compare Figs 3.21 and 3.22). The vector **r** obeys Eq (3.35) and **b** obeys

$$\frac{d\mathbf{b}}{dt} = \boldsymbol{\omega} \times \mathbf{b} \qquad (|\mathbf{b}| \text{ constant}) \qquad (3.36)$$

where ω is the angular velocity of rotation of the vector **b**.

To illustrate the application of Eq (3.36), we consider the velocity vector **v** of a point P in a rigid body rotating with constant angular velocity ω (as in Fig 3.20). The vector **v** is of constant magnitude and always lies in the plane of the circle in a direction tangential to the path and normal to the position vector **r**. It follows that the vector **v** rotates with the same angular velocity ω as the position vector **r**. Thus we can apply Eq (3.36) with $\mathbf{b} = \mathbf{v}$ to give

$$\frac{d\mathbf{v}}{dt} = \boldsymbol{\omega} \times \mathbf{v} \qquad (|\mathbf{v}| \text{ constant}) \qquad (3.37)$$

This is the acceleration of P. Now using $\mathbf{v} = \boldsymbol{\omega} \times \mathbf{r}$ we can write Eq (3.37) as

$$\frac{d\mathbf{v}}{dt} = \boldsymbol{\omega} \times (\boldsymbol{\omega} \times \mathbf{r})$$

$$= \boldsymbol{\omega}(\boldsymbol{\omega} \cdot \mathbf{r}) - \mathbf{r}(\boldsymbol{\omega} \cdot \boldsymbol{\omega}) = \boldsymbol{\omega}(\boldsymbol{\omega} \cdot \mathbf{r}) - \omega^2\mathbf{r} \qquad (3.38)$$

where we have expanded the vector triple product (Eq. 2.37) and put $\boldsymbol{\omega} \cdot \boldsymbol{\omega} = \omega^2$. This expression for the acceleration of P takes on a familiar form when we choose to measure **r** from the centre C of the circular path of P. The vectors ω and **r** are then orthogonal and so $\boldsymbol{\omega} \cdot \mathbf{r} = 0$ and Eq (3.38) gives

$$\frac{d\mathbf{v}}{dt} = -\omega^2\mathbf{r}$$

Thus we have derived the well-known expression (Eq (3.26)) for the centripetal acceleration of a particle moving in uniform circular motion.

Summary of section 3.5

If **b** is any time-varying vector of constant magnitude then

$$\frac{d\mathbf{b}}{dt} = \boldsymbol{\omega} \times \mathbf{b} \qquad (|\mathbf{b}| \text{ constant}) \qquad (3.36)$$

where ω is the angular velocity of rotation of the vector **b**.

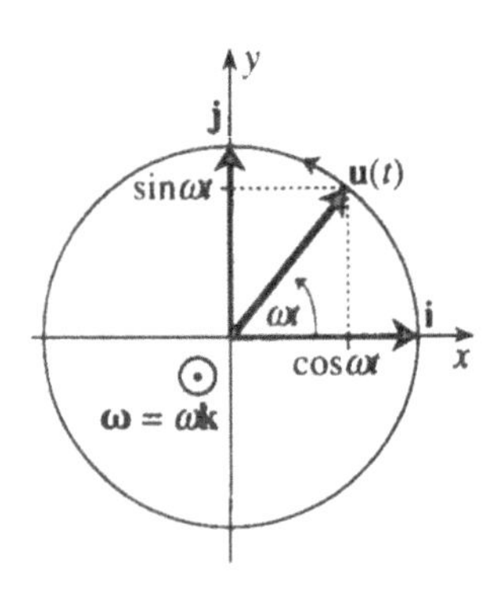

Fig 3.23
A unit vector rotating in the *x-y* plane.

Example 5.1 (*Objective 7*) Let $\mathbf{u}(t)$ be a unit vector rotating in the *x-y* plane with angular velocity $\boldsymbol{\omega} = \omega\mathbf{k}$ (Fig 3.23). The time dependence of $\mathbf{u}(t)$ is described by

$$\mathbf{u}(t) = \mathbf{i}\cos\omega t + \mathbf{j}\sin\omega t$$

Find $d\mathbf{u}/dt$ by

(a) direct differentiation of $\mathbf{u}(t)$, and
(b) by using Eq (3.36).

Solution 5.1

(a) Using Eq (3.27a) we have

$$\frac{d\mathbf{u}}{dt} = \mathbf{i}(-\omega\sin\omega t) + \mathbf{j}\omega\cos\omega t = \omega(-\mathbf{i}\sin\omega t + \mathbf{j}\cos\omega t)$$

(b) Using Eq (3.36) we have

$$\frac{d\mathbf{u}}{dt} = \boldsymbol{\omega}\times\mathbf{u}$$

$$= \omega\mathbf{k}\times\mathbf{u} = \omega(\mathbf{k}\times\mathbf{i}\cos\omega t + \mathbf{k}\times\mathbf{j}\sin\omega t)$$

$$= \omega(\mathbf{j}\cos\omega t + (-\mathbf{i})\sin\omega t) = \omega\,(-\mathbf{i}\sin\omega t + \mathbf{j}\cos\omega t)$$

where we have used Eq (2.26) for the vector products of the cartesian unit vectors. The derivative obtained agrees with that found in part (a).

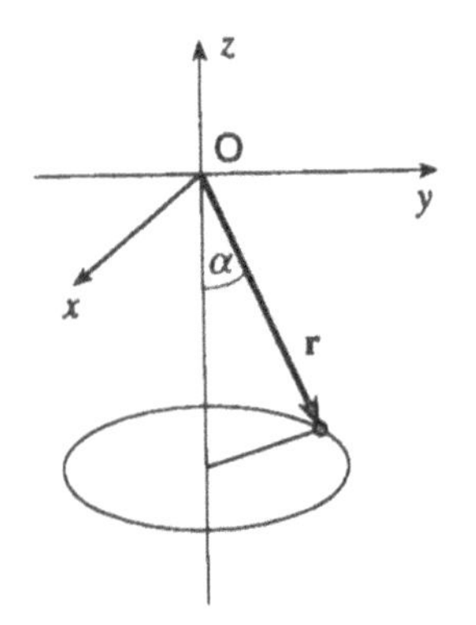

Fig 3.24
A conical pendulum.

Problem 5.1 (*Objective 7*) Show explicitly by using Eqs (3.20) and (3.36) that the derivative of $\mathbf{b}\,.\,\mathbf{b}$ is zero when $\mathbf{b}$ is a vector of constant magnitude rotating with angular velocity $\boldsymbol{\omega}$.

Problem 5.2 (*Objective 7*) A vector $\mathbf{b}$ of fixed magnitude rotates with constant angular velocity $\boldsymbol{\omega}$. What condition must be satisfied if $d^2\mathbf{b}/dt^2 = -\omega^2\mathbf{b}$?

Problem 5.3 (*Objectives 6,7*) A conical pendulum consists of a mass hanging on the end of a string and swinging around a horizontal circular path, with the string making a constant angle α with the vertical (Fig 3.24). The position vector of the mass, measured from the fixed point O from which the string is suspended, is given by

$$\mathbf{r}(t) = 0.1(\mathbf{i}\cos\pi t - \mathbf{j}\sin\pi t) - \mathbf{k}$$

(a) Specify the angular velocity vector ω and the period T. State the coordinates of the particle at $t = 0$, $t = T/4$ and $t = T/2$ where T is the period.

(b) Determine the angle α to the nearest degree.

(c) Determine the velocity and acceleration of the mass and evaluate the magnitude of the acceleration to two significant figures.

Problem 5.4 (*Objectives 6,7*) The position vector of a particle moving non-uniformly along a circular path is given by $\mathbf{r}(t) = A(\mathbf{i}\sin(\alpha t^3) - \mathbf{j}\cos(\alpha t^3))$. A and α are constants. Confirm that $|\mathbf{r}(t)|$ is constant and determine the velocity of the particle at $t = 2$ if $A = 0.1$ and $\alpha = 1$. Can Eq (3.36) be used to find the acceleration of the particle?

3.6 APPLICATION TO RELATIVE MOTION AND INERTIAL FORCES

There are many examples where two observers in relative motion make measurements of the velocity or acceleration of other moving bodies, and it is necessary for them to know how their respective measurements are related. Air traffic control and navigation in busy sea lanes are obvious examples. In this section we derive the relative velocity equations and the relative acceleration equations for *translational* and *rotational* motion.

Translational motion is motion without change in orientation, i.e. motion without rotation. Translational motion may occur along a straight line or along a curve. The chairs and passengers on a Ferris wheel are in translational motion along a circular path in a vertical plane; they do not rotate (unless the bearings lock!).

The fact that acceleration is relative, requires us to modify Newton's second law of motion when it is used in accelerating or rotating coordinate systems. Newton's second law $\mathbf{F} = m\mathbf{a}$ relates the resultant applied force $\mathbf{F}$ acting on a particle of mass m to the acceleration $\mathbf{a}$ of the particle. This law is valid only when the acceleration $\mathbf{a}$ of the particle is measured in a rigid coordinate system that is not itself accelerating or rotating. Any coordinate system in which Newton's second law applies is called an **inertial system** or an inertial frame of reference. A coordinate system that is accelerating or rotating is a **non-inertial system**. If we measure the acceleration of a particle using a non-inertial coordinate system (without correcting for the motion of the system) we will obtain a value $\mathbf{a}'$ and find that $\mathbf{F} \neq m\mathbf{a}'$, i.e. Newton's second law is not obeyed in a non-inertial coordinate system. We can however continue to use Newton's second law equation in the non-inertial system by writing it in the form $\mathbf{F}' = m\mathbf{a}'$ where $\mathbf{F}' = \mathbf{F} + \mathbf{f}$. The additional force $\mathbf{f}$ that we have to introduce as a result of the acceleration or rotation of our coordinate system is called an **inertial force**. Inertial forces are experienced as real forces. The force tending to fling you off a rotating roundabout is an example of an inertial force. Another example is the inertial force throwing you backwards, forwards or sideways in an accelerating, decelerating or cornering motor car.

Experience shows that inertial systems are those that do not rotate or accelerate relative to the stars.

We first consider the case of two observers in relative translational motion and derive the relative velocity equation quoted in Section 1.8.1. We then derive the inertial force on a particle when it is observed from an accelerating coordinate system. We then go on to the more interesting case of a rotating coordinate system. In this case there are several contributions to the inertial force. The vector algebra here is well worth studying carefully, not only for the intrinsic interest of the problem and the fascinating applications, but also because it provides an instructive illustration of the use of vector products, vector triple products and vector differentiation.

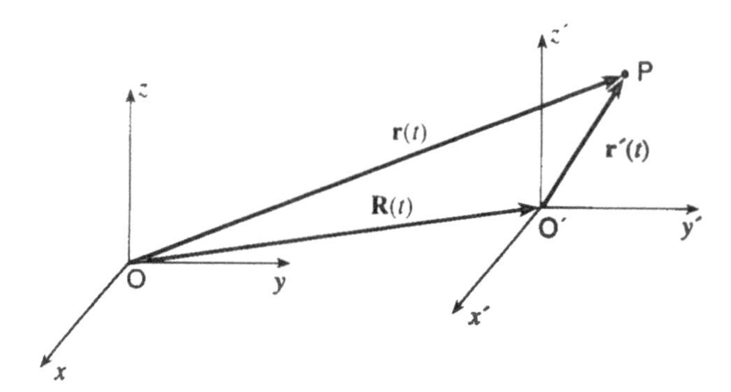

Fig 3.25
Two cartesian systems in relative translational motion.

3.6.1 Relative translational motion and inertial forces

Consider a coordinate system with origin O and refer to it as the system O. We introduce the term **observer** O to mean any person (or instrument such as a radar set) who uses the coordinate system O for position measurements. The observer O specifies the position of a moving point P by the position vector $\mathbf{r}(t) = \mathbf{OP}$ measured from the origin O of his coordinate system.

Now consider another observer using a coordinate system O′ in translational motion (i.e. moving but not rotating) relative to O. The position vector of the origin O′ as seen by observer O is $\mathbf{OO'} = \mathbf{R}(t)$ (Fig 3.25). The observer O′ specifies the position of the moving point P by the position vector $\mathbf{r}'(t) = \mathbf{O'P}$ measured from his origin O′. The vector addition rule tells us that these two measurements of the position of P are related at any instant by

$$\mathbf{r} = \mathbf{r}' + \mathbf{R} \tag{3.39}$$

where for convenience we have dropped the explicit reference to the time variable. Differentiating Eq (3.39) gives

$$\mathbf{v} = \mathbf{v}' + \mathbf{u} \tag{3.40}$$

where $\mathbf{v}$ and $\mathbf{v}'$ are the velocities of P measured by observers O and O′ respectively and $\mathbf{u} = \mathrm{d}\mathbf{R}/\mathrm{d}t$ is the velocity of O′ relative to O. Eq (3.40) is the relative velocity equation (see Eq (1.22)). When the velocities vary in time, Eq (3.40) is valid at each instant. Differentiating Eq (3.40) yields the relative acceleration equation

$$\mathbf{a} = \mathbf{a}' + \boldsymbol{\alpha} \tag{3.41}$$

where $\mathbf{a}$ and $\mathbf{a}'$ are the acceleration of P measured by observers O and O′ respectively and $\alpha = d\mathbf{u}/dt = d^2\mathbf{R}/dt^2$, the acceleration of O′.

Now let the system O be an inertial system – you can think of it as being at rest. Then an observer O using this system will find that Newton's second law $\mathbf{F} = m\mathbf{a}$ is obeyed, where $\mathbf{F}$ is the resultant force applied to a particle of mass m at P. Using Eq (3.41) in Newton's second law gives $\mathbf{F} = m\mathbf{a}' + m\alpha$ or $\mathbf{F} + \mathbf{f} = m\mathbf{a}'$ where

$$\mathbf{f} = -m\alpha \tag{3.42}$$

This shows that when the system O′ moves with an acceleration α relative to the inertial system, the observer O′ finds that an inertial force $\mathbf{f} = -m\alpha$ acts on the particle in addition to the applied force $\mathbf{F}$. Eq (3.42) describes the inertial force you feel in an accelerating vehicle.

Note that if the velocity $\mathbf{u}$ of O′ is constant then $\alpha = 0$ and $\mathbf{f} = 0$. Thus any system moving with constant velocity relative to an inertial system is also an intertial system. This explains why passengers in a high speed vehicle experience no inertial forces when the vehicle moves with constant velocity.

3.6.2 Relative rotational motion and inertial forces

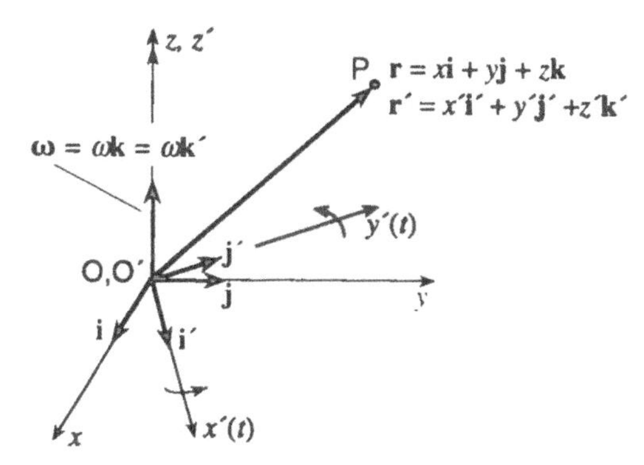

Fig 3.26
The stationary coordinate system Oxyz and the rotating coordinate system O′x′y′z′ with coincident origins and z-axes. The unit vectors i′ and j′ rotate with angular velocity $\omega = \omega\mathbf{k}' = \omega\mathbf{k}$

We now consider two coordinate systems designated O and O′ with the system O′ rotating relative to the stationary system O with angular velocity ω. In Fig 3.26 we choose the two systems to have a common z-axis lying on the axis of rotation of system O′ so that $\omega = \omega\mathbf{k} = \omega\mathbf{k}'$. We also let the two systems have a common origin point O = O′. The position vector $\mathbf{r}$ of a moving point P will then be the same in each system ($\mathbf{r} = \mathbf{OP} = \mathbf{O'P} = \mathbf{r}'$) but the coordinates and the cartesian unit vectors will be different. The cartesian form of $\mathbf{r}$ in system O is

$$\mathbf{r} = x\mathbf{i} + y\mathbf{j} + z\mathbf{k} \tag{3.43}$$

while in O′ it is

$$\mathbf{r} = \mathbf{r}' = x'\mathbf{i}' + y'\mathbf{j}' + z'\mathbf{k}' \tag{3.44}$$

with $z' = z$ and $\mathbf{k}' = \mathbf{k}$. The unit vectors $\mathbf{i}$, $\mathbf{j}$ and $\mathbf{k} = \mathbf{k}'$ are constant in time, but the unit vectors $\mathbf{i}'$ and $\mathbf{j}'$ rotate with angular velocity ω.

The velocity of the moving point P is found by differentiating its position vector. Differentiating Eq (3.43) gives

$$\mathbf{v} = \frac{d\mathbf{r}}{dt} = \frac{dx}{dt}\mathbf{i} + \frac{dy}{dt}\mathbf{j} + \frac{dz}{dt}\mathbf{k} \tag{3.45}$$

This same velocity vector, expressed in the rotating system, can be found by differentiating Eq (3.44), but here we must treat $\mathbf{i}'$ and $\mathbf{j}'$ as functions of time. Each of the first two terms on the right-hand side of Eq (3.44) is a product of a scalar component function (such as x') and a vector function (such as $\mathbf{i}'$). We therefore use the product rule 3.22 to differentiate these terms. The result of differentiating Eq (3.44) is

$$\mathbf{v} = \frac{dx'}{dt}\mathbf{i}' + x'\frac{d\mathbf{i}'}{dt} + \frac{dy'}{dt}\mathbf{j} + y'\frac{d\mathbf{j}'}{dt} + \frac{dz'}{dt}\mathbf{k}' \tag{3.46}$$

where we have used the fact that $\mathbf{k}' = \mathbf{k}$ is independent of time. We can find the derivatives of the rotating unit vectors in Eq (3.46) by using Eq (3.36),

$$\frac{d\mathbf{i}'}{dt} = \omega \times \mathbf{i}' = \omega\mathbf{k}' \times \mathbf{i}' = \omega\mathbf{j}', \qquad \frac{d\mathbf{j}'}{dt} = \omega \times \mathbf{j}' = \omega\mathbf{k}' \times \mathbf{j}' = -\omega\mathbf{i}'$$

Substituting into Eq (3.46) and rearranging terms gives

$$\mathbf{v} = \left\{\frac{dx'}{dt}\mathbf{i}' + \frac{dy'}{dt}\mathbf{j}' + \frac{dz'}{dt}\mathbf{k}'\right\} + (\omega x'\mathbf{j}' - \omega y'\mathbf{i}') \tag{3.47}$$

Notice that the three terms grouped together in the curly brackets are the three resolutes of a velocity vector measured in O′. They represent the velocity $\mathbf{v}'$ of P as seen by the observer O′. The last two terms in Eq (3.47) come from the time dependence of the rotating unit vectors, and may be recognised as a vector product. To see this consider

$$\omega \times \mathbf{r}' = \omega\mathbf{k} \times \mathbf{r}' = \omega\mathbf{k} \times (x'\mathbf{i}' + y'\mathbf{j}' + z'\mathbf{k}') = (\omega x'\mathbf{j}' - \omega y'\mathbf{i}')$$

Thus we can express Eq (3.47) as

$$\mathbf{v} = \mathbf{v}' + \omega \times \mathbf{r}' \tag{3.48}$$

where

$$\mathbf{v}'=\frac{dx'}{dt}\mathbf{i}'+\frac{dy'}{dt}\mathbf{j}'+\frac{dz'}{dt}\mathbf{k}' \tag{3.49}$$

Eq (3.48) is the relative velocity equation for two observers in relative rotational motion. Note that if the point P is at rest in the rotating system then $\mathbf{v}' = \mathbf{0}$ and Eq (3.48) yields $\mathbf{v} = \boldsymbol{\omega} \times \mathbf{r}' = \boldsymbol{\omega} \times \mathbf{r}$, which is Eq (3.34) for the velocity of a point in a rotating rigid body.

If the two systems are also in relative translational motion then the translational velocity $\mathbf{u}$ of O′ must be added to the right-hand side of Eq (3.48).

We now proceed to find the relative acceleration equation by differentiating Eq (3.48),

$$\mathbf{a}=\frac{d\mathbf{v}}{dt}=\frac{d\mathbf{v}'}{dt}+\frac{d(\boldsymbol{\omega}\times\mathbf{r}')}{dt} \tag{3.50}$$

Consider the first term on the right-hand side of Eq (3.50). Recall that $\mathbf{v}'$ (Eq (3.49)), like the position vector $\mathbf{r}'$ of Eq (3.44), is a vector measured in the rotating system. The differentiation of $\mathbf{v}'$ involves the differentiation of products of scalar component functions (such as dx'/dt) and rotating unit vectors (such as $\mathbf{i}'$). This is essentially the same mathematical problem as the differentiation of Eq (3.44) which led to Eq (3.48). We can therefore write directly

$$\frac{d\mathbf{v}'}{dt}=\mathbf{a}'+\boldsymbol{\omega}\times\mathbf{v}' \tag{3.51}$$

where

$$\mathbf{a}'=\frac{d^2x'}{dt^2}\mathbf{i}'+\frac{d^2y'}{dt^2}\mathbf{j}'+\frac{d^2z'}{dt^2}\mathbf{k}'$$

The vector $\mathbf{a}'$ is the acceleration of P as measured in the rotating system O′. We now consider the second term on the right-hand side of Eq (3.50). This derivative is found by using the vector product rule, Eq (3.21):

$$\frac{d(\boldsymbol{\omega}\times\mathbf{r}')}{dt}=\frac{d\boldsymbol{\omega}}{dt}\times\mathbf{r}'+\boldsymbol{\omega}\times\frac{d\mathbf{r}'}{dt}$$

But $\dfrac{d\mathbf{r}'}{dt}=\dfrac{d\mathbf{r}}{dt}=\mathbf{v}$, since $\mathbf{r}' = \mathbf{r}$ (Eq 3.44), and $\mathbf{v} = \mathbf{v}' + \boldsymbol{\omega} \times \mathbf{r}'$, from Eq (3.48). Hence

$$\frac{d(\boldsymbol{\omega}\times\mathbf{r}')}{dt}=\frac{d\boldsymbol{\omega}}{dt}\times\mathbf{r}'+\boldsymbol{\omega}\times\mathbf{v}'+\boldsymbol{\omega}\times(\boldsymbol{\omega}\times\mathbf{r}') \tag{3.52}$$

Using Eqs (3.51) and (3.52) in Eq (3.50), we obtain the relative acceleration equation for rotational motion

$$\mathbf{a} = \mathbf{a}' + \frac{d\boldsymbol{\omega}}{dt} \times \mathbf{r}' + 2\boldsymbol{\omega} \times \mathbf{v}' + \boldsymbol{\omega} \times (\boldsymbol{\omega} \times \mathbf{r}') \tag{3.53}$$

We see that the acceleration $\mathbf{a}'$ of P measured in the rotating system O′ differs from the acceleration $\mathbf{a}$ measured in the system O by the three $\boldsymbol{\omega}$ dependent terms on the right-hand side of Eq (3.53), the last two of which are in general non-zero even when the angular velocity $\boldsymbol{\omega}$ is constant.

Finally, to obtain the inertial forces, we assume that O is an inertial system and use Eq (3.53) in Newton's second law $\mathbf{F} = m\mathbf{a}$ to obtain

$$\mathbf{F} - m\frac{d\boldsymbol{\omega}}{dt} \times \mathbf{r}' - 2m\boldsymbol{\omega} \times \mathbf{v}' - m\boldsymbol{\omega} \times (\boldsymbol{\omega} \times \mathbf{r}') = m\mathbf{a}' \tag{3.54}$$

which is of the form $\mathbf{F} + \mathbf{f} = m\mathbf{a}'$. This shows that there are three inertial force terms observed in the rotating system,

$$\mathbf{f} = -m\frac{d\boldsymbol{\omega}}{dt} \times \mathbf{r}' - 2m\boldsymbol{\omega} \times \mathbf{v}' - m\boldsymbol{\omega} \times (\boldsymbol{\omega} \times \mathbf{r}')$$

in addition to the applied force $\mathbf{F}$. (Note that $\mathbf{r}'$ can be replaced by $\mathbf{r}$ in Eq (3.54) since $\mathbf{r}' = \mathbf{r}$ when the two coordinate systems have the same origin.)

The velocity-dependent inertial force term $-2m\boldsymbol{\omega} \times \mathbf{v}'$ is called the **Coriolis force**. The position-dependent term $-m\boldsymbol{\omega} \times (\boldsymbol{\omega} \times \mathbf{r}')$ is called the **centrifugal force**. You can feel both these forces on a rotating roundabout, the Coriolis force pushing you sideways when you move, and the centrifugal force tending to fling you off. If the roundabout is speeding up or slowing down then $d\boldsymbol{\omega}/dt$ is non-zero and you also feel the inertial force $-m(d\boldsymbol{\omega}/dt) \times \mathbf{r}'$ acting in a tangential direction.

Summary of section 3.6

- Newton's second law $\mathbf{F} = m\mathbf{a}$ applies in any **inertial system**. Observers using **non-inertial systems** (coordinate systems that are accelerating or rotating) find that **inertial forces f** act on bodies in addition to the applied force $\mathbf{F}$.

- For a coordinate system in translational motion (not rotating) the inertial force is $\mathbf{f} = -m\boldsymbol{\alpha}$ where m is the mass of the particle and $\boldsymbol{\alpha}$ is the acceleration of the system.

- The inertial force observed in a coordinate system rotating with angular velocity $\boldsymbol{\omega}$ is

$$\mathbf{f} = -m(d\boldsymbol{\omega}/dt) \times \mathbf{r}' - 2m\boldsymbol{\omega} \times \mathbf{v}' - m\boldsymbol{\omega} \times (\boldsymbol{\omega} \times \mathbf{r}')$$

The term $-2m\boldsymbol{\omega} \times \mathbf{v}'$ is the **Coriolis force** and the term $-m\boldsymbol{\omega} \times (\boldsymbol{\omega} \times \mathbf{r}')$ is the **centrifugal force**.

Example 6.1 (*Objective 8*) A playground roundabout rotates at 0.3 Hz anticlockwise. A child of mass 25 kg walks from the centre of the roundabout along a radial line at a speed of $0.5\ \mathrm{m\,s^{-1}}$ relative to the roundabout). Determine the Coriolis force and the centrifugal force acting on the child when he is 2 m from the centre. Give your answers in newtons and give the ratio of the magnitude of each inertial force to the magnitude of the child's weight.

Solution 6.1 We model the child as a particle in the x'–y' plane of a rotating coordinate system fixed on the roundabout with the z-axis lying on the axis of rotation and directed vertically upwards. The angular velocity of the roundabout is then $\boldsymbol{\omega} = 2\pi \times 0.3\mathbf{k}\ \mathrm{rad\ s^{-1}}$.

Let the radial line along which the child walks be the x'-axis. Then the Coriolis force is

$$\mathbf{f}_{cor} = -2m\boldsymbol{\omega} \times \mathbf{v}' = -2(25)(2\pi)(0.3\mathbf{k}) \times (0.5\mathbf{i}')\ \mathrm{N} = -47\mathbf{j}'\ \mathrm{N}$$

Thus the Coriolis force is of magnitude 47 N and is directed parallel to $-\mathbf{j}'$, i.e. it pushes the child to his right.

When the child is on the x'-axis 2 m from the centre we have $\mathbf{r}' = 2\mathbf{i}'$ m and the centrifugal force is

$$\mathbf{f}_{cen} = -m\boldsymbol{\omega} \times (\boldsymbol{\omega} \times \mathbf{r}') = -(25)(2\pi)(0.3\mathbf{k}) \times (2\pi)(0.3\mathbf{k} \times 2\mathbf{i}')\ \mathrm{N}$$

$$= -178\mathbf{k} \times (\mathbf{k} \times \mathbf{i}')\ \mathrm{N} = -178\mathbf{k} \times \mathbf{j}'\ \mathrm{N} = 178\mathbf{i}'\ \mathrm{N}$$

Thus the centrifugal force has magnitude 178 N and acts radially outwards.

The magnitude of the child's weight is $mg = 250$ N (using $g = 10\ \mathrm{ms^{-2}}$), and so

$$\frac{|\mathbf{f}_{cor}|}{mg} = 47/250 = 0.19, \qquad \frac{|\mathbf{f}_{cen}|}{mg} = 178/250 = 0.71$$

Example 6.2 (*Objective 8*) Show that when $\boldsymbol{\omega} = \omega\mathbf{k}$, the centrifugal force on a particle with position vector $\mathbf{r}' = \mathbf{r}$ always acts in a direction perpendicularly outwards from the z-axis and is of magnitude $|\mathbf{f}_{cen}| = m\omega^2\rho$ where ρ is the perpendicular distance of the particle from the z-axis.

Solution 6.2 Using the coordinate systems in the text, the centrifugal force is $\mathbf{f}_{cen} = -m\boldsymbol{\omega} \times (\boldsymbol{\omega} \times \mathbf{r}')$ or $-m\boldsymbol{\omega} \times (\boldsymbol{\omega} \times \mathbf{r})$ since $\mathbf{r}' = \mathbf{r}$ when the origins of the two coordinate systems coincide. We expand the vector triple product (Eq 2.37):

$$\mathbf{f}_{cen} = -m\boldsymbol{\omega} \times (\boldsymbol{\omega} \times \mathbf{r}) = -m[\boldsymbol{\omega}(\boldsymbol{\omega} \cdot \mathbf{r}) - \mathbf{r}(\boldsymbol{\omega} \cdot \boldsymbol{\omega})]$$

Now $\boldsymbol{\omega} \cdot \mathbf{r} = \omega r \cos\alpha = \omega z$ (Fig 3.27) and $\boldsymbol{\omega} \cdot \boldsymbol{\omega} = \omega^2$. Hence

$$\mathbf{f}_{cen} = -m[\omega\mathbf{k}(\omega z) - \omega^2\mathbf{r}] = m\omega^2(\mathbf{r} - \mathbf{k}z) = m\omega^2(x\mathbf{i} + y\mathbf{j})$$

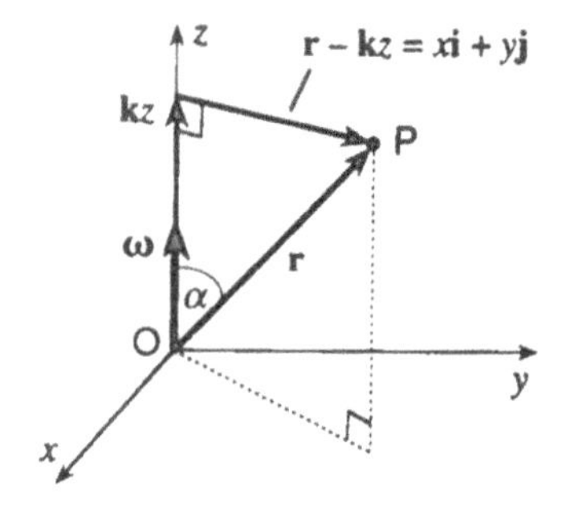

Fig 3.27
$\mathbf{f}_{cen} = m\omega(\mathbf{r} - \mathbf{k}z)$ where $\mathbf{r} = x\mathbf{i} + y\mathbf{j} + z\mathbf{k}$. (Note that $\mathbf{r} = \mathbf{r}'$, $z = z'$ and $\mathbf{k} = \mathbf{k}'$.)

Hence $\mathbf{f}_{cen}$ is in the direction of the vector $x\mathbf{i} + y\mathbf{j}$ which is perpendicularly outwards from the z-axis (Fig 3.27). The magnitude is $|\mathbf{f}_{cen}| = m\omega^2\rho$ where $\rho = |\mathbf{r} - \mathbf{k}z| = |x\mathbf{i} + y\mathbf{j}| = (x^2 + y^2)^{1/2}$, the perpendicular distance from the z-axis.

Example 6.3 (*Objective 8*) A proposed space station in gravity-free space consists of two modules each of "height" 10 m separated by a rigid communication tube 380 m long (Fig 3.28). In order to simulate Earth's gravity ($g = 10\text{ m s}^{-2}$) on a stationary astronaut at rest on the end wall of one of the modules, the station is made to spin at a constant angular speed ω about an axis through the centre of the tube perpendicular to its length.

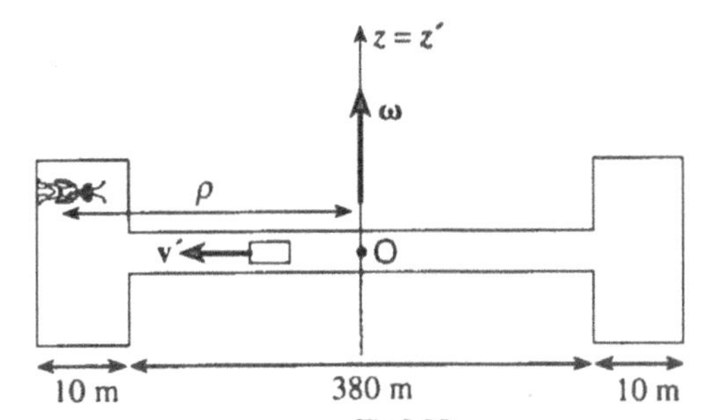

Fig 3.28
The rotating space station for Example 6.3. The shuttle moves with velocity $\mathbf{v}'$ along the communication tube.

(a) Determine the required value of ω.

(b) Transport between the modules is by shuttle of mass 1200 kg running along rails inside the communication tube. On which side or sides of the tube should the rails be mounted? What is the magnitude of the normal force between the shuttle and the rails when the shuttle is moving at a speed of 5 m s^{-1} along the tube?

Solution 6.3 We use a rotating coordinate system fixed in the space station with the z'-axis lying on the axis of rotation. Then the angular velocity is $\boldsymbol{\omega} = \omega\mathbf{k}'$.

(a) Gravity is simulated by the centrifugal force $\mathbf{f}_{cen} = -m\boldsymbol{\omega} \times (\boldsymbol{\omega} \times \mathbf{r}')$ which is directed perpendicularly outwards from the z-axis and has magnitude $|\mathbf{f}_{cen}| = m\omega^2\rho$, where ρ is the perpendicular distance from the axis (see Example 6.2). If this force is to simulate Earth's gravity, ω must satisfy $m\omega^2\rho = mg$ where $\rho = (380\text{ m})/2 + 10\text{ m}$ (Fig 3.28). Thus $\omega = (10\text{ ms}^{-2}/200\text{ m})^{1/2} = 0.224\text{ rad s}^{-1}$.

(b) The moving shuttle will experience a Coriolis force $\mathbf{f}_{cor} = -2m\boldsymbol{\omega} \times \mathbf{v}'$ as well as a centrifugal force. The latter is directed along the length of the tube and acts to accelerate or decelerate the shuttle. The Coriolis force is of magnitude $2m\omega v'\sin 90° = 2m\omega v'$ and is directed transversely. Referring to Fig 3.28, you can see that the direction of the Coriolis force on the shuttle ($-2m\boldsymbol{\omega} \times \mathbf{v}'$) is into the plane of the figure, i.e. towards the far wall of the tube, and so it is on this wall that the rails should be mounted. Rails on the opposite side are needed for the return journey when $\mathbf{v}'$ and the vector product $\boldsymbol{\omega} \times \mathbf{v}'$ change sign. The magnitude of this sideways force is $2m\omega v' = 2 \times 1200 \times 0.224 \times 5\text{ N} = 2690\text{ N}$.

Problem 6.1 (*Objective 8*) Determine the apparent weight of a 20 kg suitcase in a lift that is accelerating upwards at 2 ms^{-2}.

Problem 6.2 (*Objective 8*) Consider a stone falling vertically downwards with a speed of 40 m s^{-1} at a latitude of 50° north.

(a) Determine the magnitudes of the Coriolis force and the centrifugal force on the stone, due to the Earth's rotation. Give your answers as fractions of the magnitude of the stone's weight. Does a body released from rest fall vertically downwards? (Take the radius of the Earth as $R = 6.37 \times 10^6\text{ m}$.)

(b) Determine the directions of the two forces in part (a).

Problem 6.3 (*Objective 8*) Specify the directions of the Coriolis and centrifugal forces (if any) due to the Earth's rotation about its north-south axis,

(a) on a ship crossing the equator from north to south
(b) on a ship steaming along the equator from west to east
(c) on a snowcat as it crosses the south pole
(d) on St Paul's cathedral.

Problem 6.4 (*A problem for discussion*) Are inertial forces real forces?

4

Scalar and vector fields

Objectives

After you have studied this chapter you should be able to

- Describe the main characteristics of a scalar field from a given sketch of the contours (*Objective 1*).
- Describe the main characteristics of a vector field from a given sketch of the field lines (*Objective 2*).
- Determine the cartesian form of a scalar or vector field function given the coordinate-free form (*Objective 3*).
- Determine scalar field values at specified points in a given cartesian scalar field and obtain equations of the contours (*Objective 4*).
- Determine the magnitudes, directions and components of field vectors at specified points in a given cartesian vector field (*Objective 5*).
- Determine the equation of a vector field line (*Objective 6*).
- Convert coordinates between cylindrical polar, spherical polar and cartesian forms (*Objective 7*).
- Express the base vectors of a cylindrical or spherical polar coordinate system in terms of the cartesian base vectors (*Objective 8*).
- Convert between coordinate-free forms and coordinate forms of field functions, and evaluate field values (*Objective 9*).
- Determine the flux of a vector field across a surface when the normal component of the vector is constant (*Objective 10*).
- Determine the circulation of a vector field around a loop when the tangential component of the vector is constant (*Objective 11*).

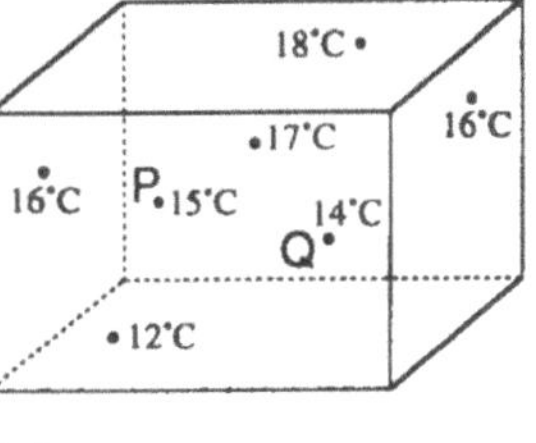

(a)

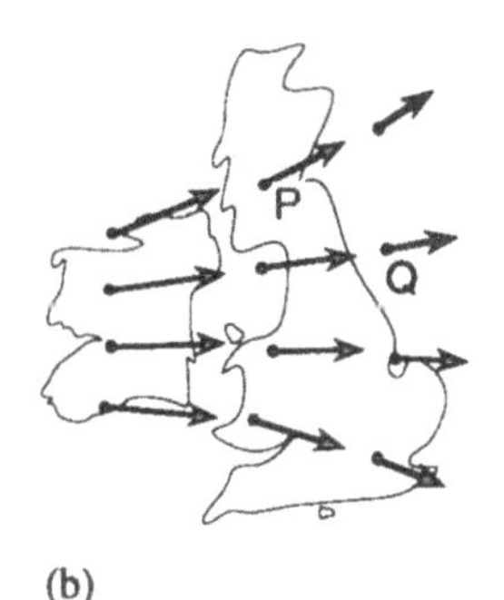

(b)

Fig 4.1
(a) A scalar field. The temperature inside the room is shown at a selection of points P, Q etc. (b) A vector field. The wind velocity is shown at a selection of points by drawing arrows.

The distribution of a physical quantity throughout a region of space is called a **field**. An example is the temperature in a room; we can associate a unique temperature with each point in the room at any instant (Fig 4.1a). This spatial distribution of temperatures is an example of a **scalar field** since temperature is a scalar quantity. When the physical quantity is a vector we have a vector field. An example of a **vector field** is the wind velocity in the atmosphere. We can associate a unique velocity vector with each point in the atmosphere at any instant (Fig 4.1b). These fields are obviously time-dependent when observed over say a 24 hour period but are approximately independent of time when observed over a short enough period, say an hour or so in the case of room temperature. We shall consider only fields that are independent of time.

Other examples of scalar fields are the barometric pressure in the atmosphere, the distribution of mass in a material body, and the electrostatic potential in the region near an electrically charged object. Other examples of vector fields are the gravitational field of a planet and the magnetic field produced by a magnet.

Most fields of interest are three-dimensional, i.e. the scalar or vector quantity is distributed throughout a region of three-dimensional space. Some fields however are restricted to a surface. An example is the static electric charge produced on the surface of a dry glass plate when it is rubbed with a silk cloth; this is a two-dimensional scalar field since the scalar (electric charge) is distributed over the surface of the glass. The water velocity on the surface of a river is an example of a two-dimensional vector field.

The concept of a field involves two sets of objects: the set of points in the region of space where the field exists; and the set of field values of the physical quantity, scalar or vector, that exists in the region. The field consists of an association of field values to points in a region of space. This leads to a mathematical description of fields by scalar or vector functions of position. Scalar and vector field functions are introduced and described in Sections 4.2 to 4.4 of this chapter. Section 4.5 introduces two quantities, flux and circulation, that encompass the physically significant properties of a vector field and play important roles in the description of many physical processes.

We begin in Section 4.1 by describing pictorial representations of fields which help us visualise their overall characteristics.

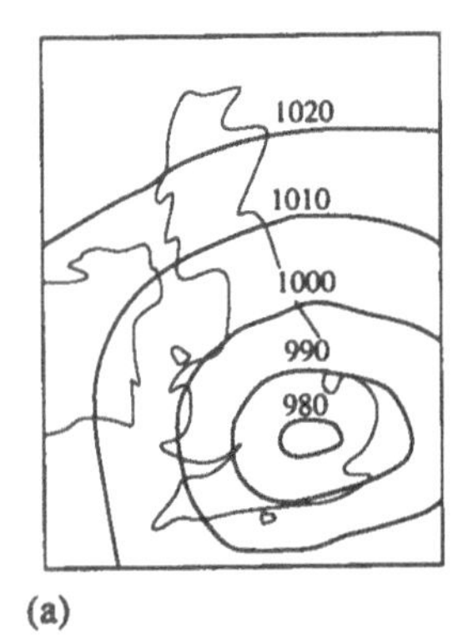

(a)

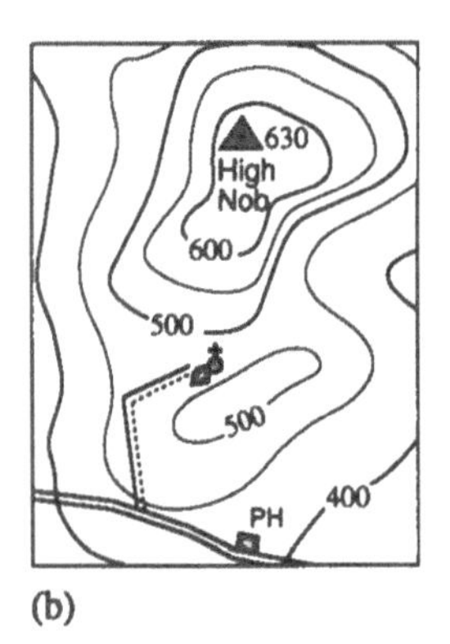

(b)

Fig 4.2
Contours. (a) Isobars drawn 10 millibars apart. (b) Land elevation contour lines at intervals of 50 m.

4.1 PICTORIAL REPRESENTATIONS OF FIELDS

The most direct way of showing scalar and vector fields pictorially is the rather obvious one of showing field values, numbers for scalar fields and arrows for vector fields, at a selection of points, as in Figs 4.1a and 4.1b. This section describes two other useful pictorial devices: *scalar field contours* and *vector field lines*.

4.1.1 Scalar field contours

The barometric pressure in the atmosphere is a scalar field commonly shown on weather maps by **isobars**. An isobar is a line joining points that have the same atmospheric pressure (Fig 4.2a) and is labelled by that pressure. Isobars are usually drawn at fixed pressure intervals, ten millibars in Fig 4.2a. The isobars show regions of high and low pressure indicating dry and wet conditions respectively. Regions where the pressure gradient is large and the winds strong are shown by the isobars being close together. Isobars are examples of **contour lines**, or contour curves. Another familiar example is land elevation above sea level shown on land maps by means of elevation contour lines. A contour line joins points of the same land elevation and is labelled by that value; adjacent lines are usually separated by some fixed elevation interval, 50 m in Fig 4.2b.

The contours enable you to identify summits, valleys, ridges, saddle points, steepness, etc.

The electrostatic potential (volts) in the vicinity of a positively charged conducting sphere is a scalar field shown in Fig 4.3 by circular contour lines, each line being labelled by the value of the potential. This scalar field is three-dimensional and the contours are in fact concentric spherical **contour surfaces** (called equipotentials) surrounding the charged sphere, the circular lines shown in Fig 4.3 being the intersections of these surfaces with the plane of the figure. Most of the fields we have to deal with are three-dimensional and the contours are surfaces often shown as lines in plane figures. Contour surfaces are sometimes called **level surfaces**.

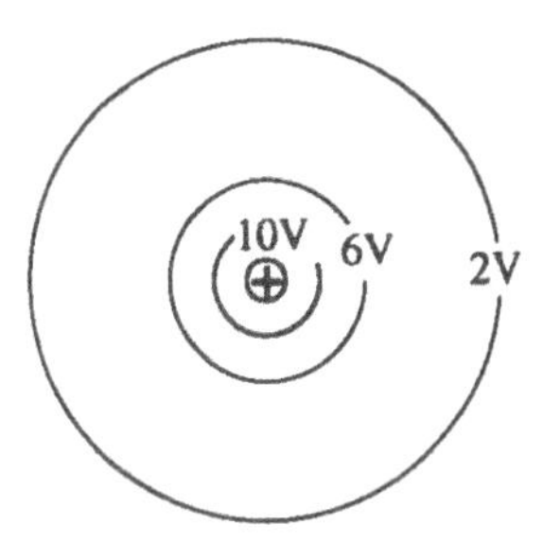

Fig 4.3
Eqipotentials around a positively charged sphere. The circles are the intersections of spherical contour surfaces with the plane of the figure.

4.1.2 Vector field lines

The electrostatic field near an electrically charged object is an example of a vector field, the field vector at any point being the electrostatic force that would act on a unit positive charge placed at that point. Fig 4.4a depicts the electrostatic field in the vicinity of a positively charged conducting sphere. The field vector at a point such as P is shown as an arrow with its beginning point (tail end) at P. In this example the magnitude of the field vector obeys an inverse square law and the direction is radially outward, as indicated by the lengths and directions of the arrows. Fig 4.4b shows the same vector field by means of electrostatic field lines.

Field lines are directed lines drawn such that the direction of the field line passing through any point gives the direction of the field vector at that point. There is a field line passing through each point where the field vector is non-zero, but we can of course show only a limited number of field lines in any diagram. No two field lines can cross since there is always a unique field vector at any point. If the vector at a point is the zero vector then no field line can pass through that point since the direction of the zero vector is undefined.

We can sometimes indicate the magnitudes of the field vectors by the density of field lines, as in Fig 4.4b where the field lines have the largest density, i.e. are closest together, in the region near the charge where the field is strongest. Let N be the number of field lines passing through a small imaginary surface of area $A_\perp$ oriented at right-angles to the direction of the field (Fig 4.4c). Then the **density of field lines** is $N/A_\perp$ and we can take this as an indication of the relative magnitude of the field in that region. Thus we can see that the magnitude of the field depicted in Fig 4.4c is greater at the point P than at Q since more field lines pass through the small normally-oriented surface at P than through the same small surface at Q. It is often quite difficult to draw field lines to show the relative magnitude of a field at all accurately. Field line diagrams are normally used to show the direction of the field and to indicate roughly where the field magnitude is large or small.

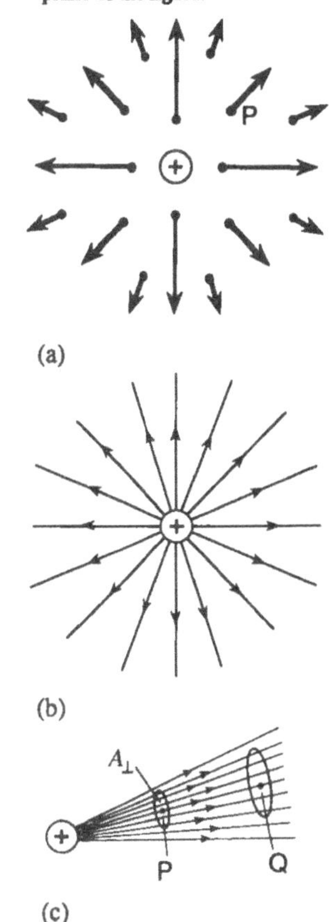

Fig 4.4
The electrostatic field in the region outside a uniformly charged sphere shown (a) by arrows at a selection of points and (b) by field lines. (c) The magnitude of the vector field is proportional to the density of field lines.

Summary of section 4.1

- A **contour line** or a **contour surface (level surface)** connects points in a scalar field where the scalar has the same value. Contours are labelled by the scalar values and are usually shown at fixed intervals of the scalar; regions where the scalar value varies strongly with distance are then shown by the contours being close together.

- Vector **field lines** are drawn to show the direction of a vector field. The relative magnitudes of the field vectors can sometimes be indicated by the **density of field lines**. The total number of field lines drawn in any diagram is arbitrary.

Example 1.1 (*Objectives 1,2*) Describe, in a sentence or two, the main characteristics of the two-dimensional scalar fields depicted in Figs 4.5(a) and (b), and the three-dimensional vector fields depicted in Fig 4.6(a) and in Fig 4.4(b).

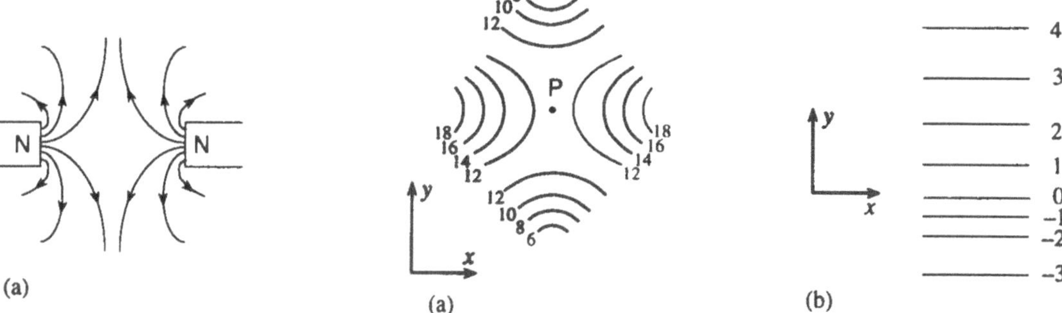

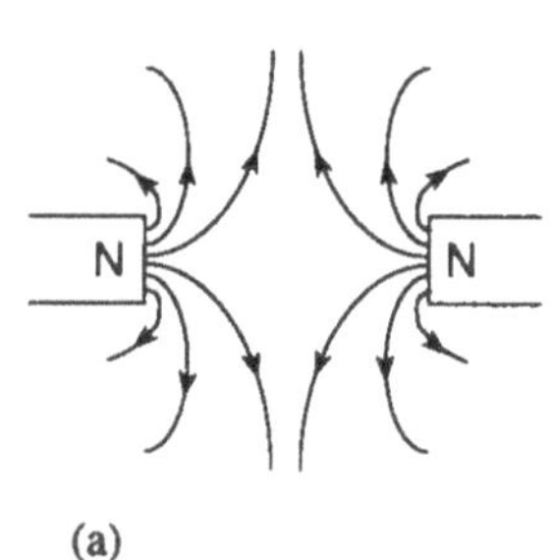

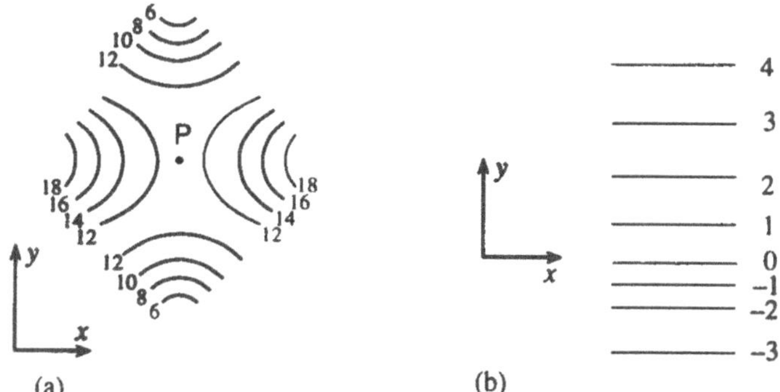

Fig 4.5
Contour lines for two scalar fields.

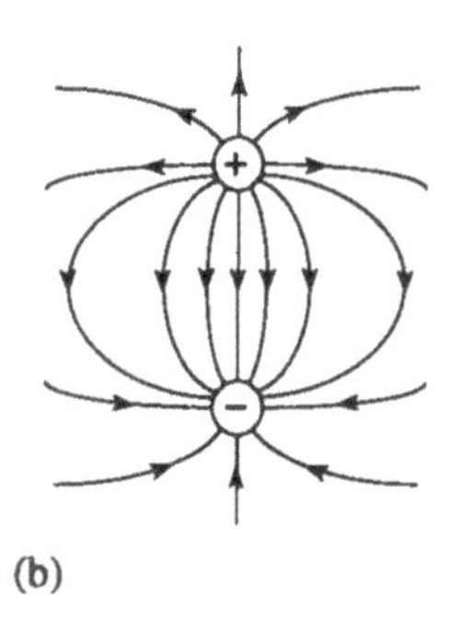

Fig 4.6
Examples of vector field lines.
(a) Magnetic field lines near the north poles of two similar magnets. (Magnetic field lines are always closed loops; only segments of the lines are shown in the figure.)
(b) Electrostatic field lines near an electric dipole.

Solution 1.1 The contours in Fig 4.5(a) have the form of a saddle (or col) with a scalar value of about 12 at the saddle point P from which the scalar values decrease in the positive and negative y directions and increase in the positive and negative x directions.

In Fig 4.5(b) the scalar values are independent of x but increase in the y direction, the rate of increase with y being greatest near the -1 contour where the contour spacing is least.

The directions of the magnetic field lines in Fig 4.6(a) show that the field vectors near a north pole are directed away from the pole. The field magnitude is largest near the poles where the field lines are closest together, i.e. of greatest density. The symmetry of the figure indicates that the field vector at the midpoint between the two poles is a zero vector (a **null point**).

Fig 4.4(b) shows the electric field lines outside a positively charged sphere. The lines show the direction of the field to be radially outwards from the centre. The increasing separation of the lines with radial distance shows that the field magnitude is largest near the surface of the sphere where the lines are closest together and becomes weaker with radial distance. In fact we can quantify the field line density by noting that all N field lines pass at right angles through any imaginary spherical surface of radius r surrounding the charged sphere. The area of the spherical surface is $4\pi r^2$ and so the field line density is $N/4\pi r^2$. This shows that the magnitude of the electrostatic field depicted in Fig 4.4(b) falls off with the inverse of the square of the distance from the centre.

Example 1.2 (*Objectives 1,2*) State where the magnitudes of each of the two-dimensional fields shown in Figs 4.7(a) and 4.7(b) are greatest and least.

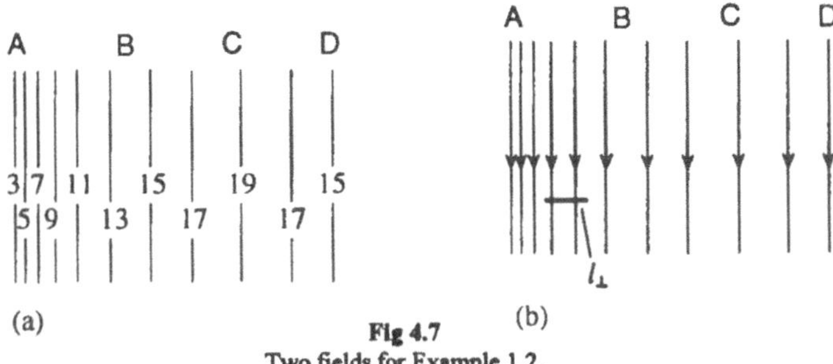

Fig 4.7
Two fields for Example 1.2.

Solution 1.2 The field depicted in Fig 4.7(a) is a scalar field, the vertical lines being contour lines labelled by the scalar values. The greatest scalar value is 19, on the contour near C; the least is 3, on the contour near A.

The field depicted in Fig 4.7(b) is a vector field directed downwards everywhere. The magnitude of the field vector is greatest near A where the field lines are closest together (greatest density) and least near C where the separation of field lines is largest.

The density of field lines in a two-dimensional vector field is $n/l_\perp$ where n is the number of field lines crossing a short line of length $l_\perp$ drawn at right angles to the direction of the field (Fig 4.7b).

Example 1.3 (*Objective 1*) The light intensity (as measured by a photographer's light meter) in the region near a small source of light (a light bulb say) is an example of a three-dimensional scalar field. What shape would you expect the contour surfaces to have and how would they be shown in a plane sketch? Assume that the light source is isotropic, i.e. it appears equally bright from all directions.

Solution 1.3 The light intensity is the same at all points at the same distance from the source and so the contours are concentric spherical surfaces centred on the source. A sketch in a plane containing the source would show the circles of intersection of the contour surfaces with the plane of the sketch.

Problem 1.1 (*Objectives 1,2*) Describe, in a sentence each, the main features of the fields shown in Fig 4.8.

Problem 1.2 (*Objective 2*) Fig 4.9 shows two vector fields by means of arrows drawn at a selection of points. Sketch roughly some vector field lines for each of these fields.

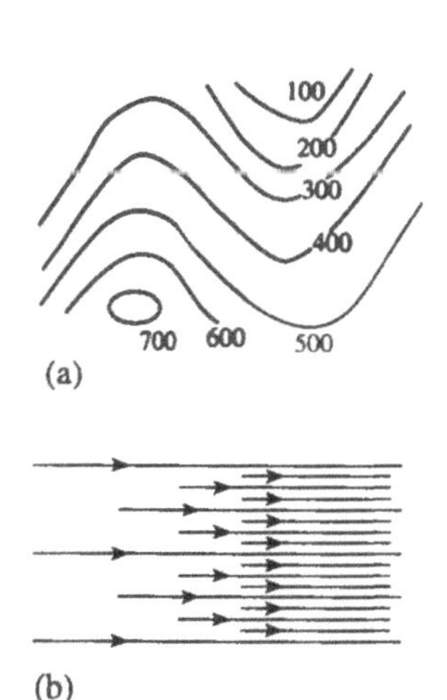

Fig 4.8
Two fields for Problem 1.1.

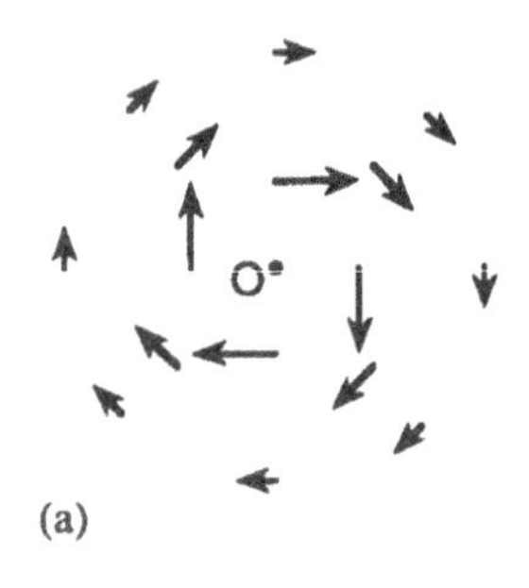

(a)

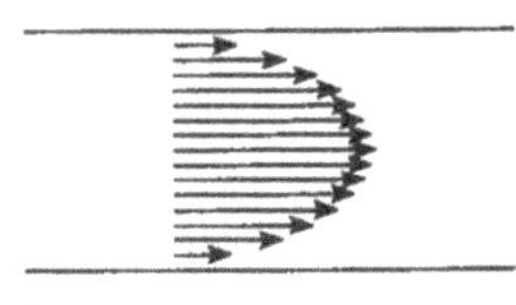

(b)

Fig 4.9
Two vector fields shown by arrows. Each arrow is drawn with its beginning point at the corresponding field point.
(a) Some magnetic field vectors outside a wire carrying an electric current into the plane of the page at O. (b) Some surface velocity vectors on a line across a river.

Problem 1.3 (*Objective 1*) Make rough sketches to indicate the land elevation contours that would describe

(a) a conical volcano,
(b) a hemispherical depression in the ground,
(c) a hillside of uniform slope rising towards the north.

Problem 1.4 (*Objective 2*) Describe in a sentence the main features of the electrostatic field produced by the electric dipole shown in Fig 4.6b.

Problem 1.5 (*Objective 2*) The magnitude of the radially-directed electrostatic field outside an elecrically charged conducting sphere is inversely proportional to the square of the distance from the centre. Comment on the difficulty of indicating the magnitude of this field by field lines in a plane section as in Fig 4.4b.

4.2 SCALAR FIELD FUNCTIONS

Scalar fields are represented mathematically by scalar functions of position. We show how these functions are defined, and describe how they are used to specify scalar values at given points and to derive equations for the contours.

4.2.1 Specifying scalar field functions

Let the scalar quantity be denoted by the symbol U and suppose points in the region of the field are specified by their position vectors $\mathbf{r}$ relative to some fixed reference point. Then the value of the scalar at a point with position vector $\mathbf{r}$ is denoted by the symbol $U(\mathbf{r})$. The point $\mathbf{r}$ is called the **field point** and the scalar $U(\mathbf{r})$ is called the **field value** at $\mathbf{r}$. The set of all field points $\mathbf{r}$, together with the corresponding field values $U(\mathbf{r})$, constitute a **scalar field function** or simply a scalar field. The function is specified by a **rule** for determining the scalar field value $U(\mathbf{r})$ at each field point $\mathbf{r}$, and a statement of the **domain** of the function, i.e. the region of space in which the field exists and in which the rule is valid. The symbol $U(\mathbf{r})$ serves to denote either a field value at a field point $\mathbf{r}$ or the scalar field function itself, depending on the context. The scalar field function may also be denoted simply by the symbol U.

A simple example of a three-dimensional scalar field function is

$$U(\mathbf{r}) = \frac{1}{|\mathbf{r}|} \qquad \text{(}\mathbf{r}\text{ is any vector with the exclusion of the zero vector.)} \qquad (4.1\text{a})$$

The rule of this function is the equation $U(\mathbf{r}) = 1/|\mathbf{r}|$; it tells us that the scalar field value $U(\mathbf{r})$ at the field point $\mathbf{r}$ is equal to the reciprocal of the distance of the field point from the reference point. The statement in parentheses following the rule specifies the domain of the function as being the whole of space excluding the reference point from which the position vectors are measured.

This exclusion is obviously necessary here since the rule cannot be evaluated when $\mathbf{r} = 0$. In practice we would normally abbreviate this statement of the domain by writing simply ($\mathbf{r} \neq 0$).

4.2.2 Cartesian scalar fields

The scalar field function (4.1a) is expressed in terms of position vectors and makes no reference to a coordinate system; it is a **coordinate-free** description. Coordinate-free descriptions are concise and convenient for algebraic manipulations, but for numerical work it is often useful to introduce a coordinate system and to express the function in terms of the coordinates of points rather than position vectors.

In a cartesian coordinate system, a point in space with position vector $\mathbf{r}$ is specified by its cartesian coordinates (x,y,z), and we can write $\mathbf{r} = (x,y,z)$. A scalar field $U(\mathbf{r})$ can then be denoted by $U(x,y,z)$. Thus the symbol $U(x,y,z)$ denotes the scalar field value at the field point with coordinates (x,y,z). The cartesian form of the scalar field function of Eq (4.1a) is

$$U(x,y,z) = \frac{1}{(x^2+y^2+z^2)^{1/2}} \qquad ((x,y,z) \neq (0,0,0)) \qquad (4.1b)$$

Eqs (4.1a) and (4.1b) represent the same function. Eq (4.1b) is the cartesian form and is a function of the three cartesian variables. We can use the cartesian form of the rule to evaluate, for example, the field value at the field point with coordinates $(1,2,-3)$: thus $U(1,2,-3) = 1/(1^2+2^2+(-3)^2)^{1/2} = (1/14)^{1/2} = 0.267$.

As another example, consider the temperature distribution in a room. Let the room have dimensions 8 m by 6 m and height 5 m, and suppose the temperature is the same at all points on the same horizontal level and increases uniformly with height from 12°C at floor level to 18°C at ceiling level. We introduce a cartesian coordinate system (Fig 4.10) with the floor in the x-y plane and the z axis increasing upwards. Then the temperature field is described by the cartesian field function

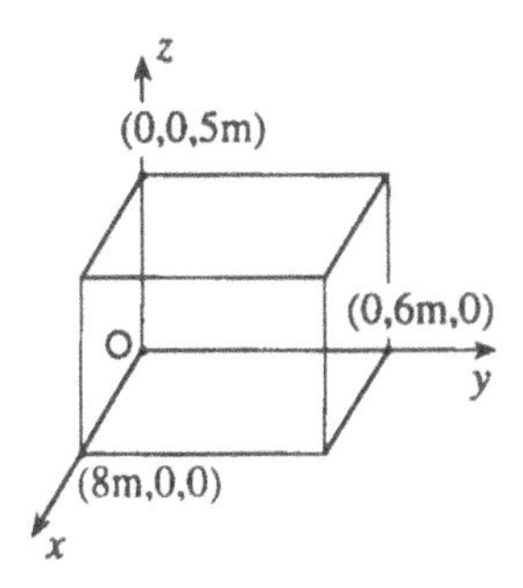

Fig 4.10
A cartesian system for the temperature field in the room.

$$T(x,y,z) = 12\left(1+\frac{z}{10\ \text{m}}\right)\ {}^\circ\text{C} \quad (0 \le x \le 8\ \text{m},\ 0 \le y \le 6\ \text{m},\ 0 \le z \le 5\ \text{m}) \quad (4.2)$$

The domain is specified by the three intervals which define the interior region of the room in which the rule applies. You can see that the rule generates the given temperatures at floor level ($z = 0$) and at ceiling level ($z = 5$ m) and gives, for example, a temperature of 15°C half way up (at $z = 5$ m/2). The absence of x and y from the rule is a result of the temperature being independent of horizontal position and the chosen horizontal orientation of the x-y plane.

We often loosely refer to a function such as Eq (4.2) as a function of one variable (z in this case) even though it is actually a function of all three cartesian variables as indicated by the domain statement.

The domain of a field function is sometimes restricted by mathematical requirements, as in Eq (4.1a) (or Eq (4.1b)) where the rule cannot be evaluated at the point where the denominator is zero; while in other examples the domain is restricted to a region of space for physical reasons as in Eq (4.2) where the rule is to be applied only to points inside the room. In practice a function is often specified simply by giving its algebraic rule with the domain understood from the context; physical units are also often omitted when it is convenient to do so.

A two-dimensional scalar field on a plane surface can be represented in cartesian form by a function of two cartesian variables. Suppose a small drop of coloured liquid falls on a large sheet of absorbing paper lying in the x-y plane. The intensity distribution of the stain produced on the paper can often be described by a function of the form

The domain of the function (4.3) may alternatively be stated as $(x,y) \in \Re^2$ where the symbol $\Re^2$ denotes the set of points in the plane.

$$f(x,y) = A\exp\left[-\frac{x^2+y^2}{a^2}\right] \qquad (x \in \Re, y \in \Re) \qquad (4.3)$$

where the constant A is the intensity of the stain at the centre (0,0) and a is the radial distance at which the intensity falls to $A/e = A/2.718$. The domain statement specifying the whole of the x-y plane expresses the fact that the dimensions of the sheet are very large compared to the extent of the stain.

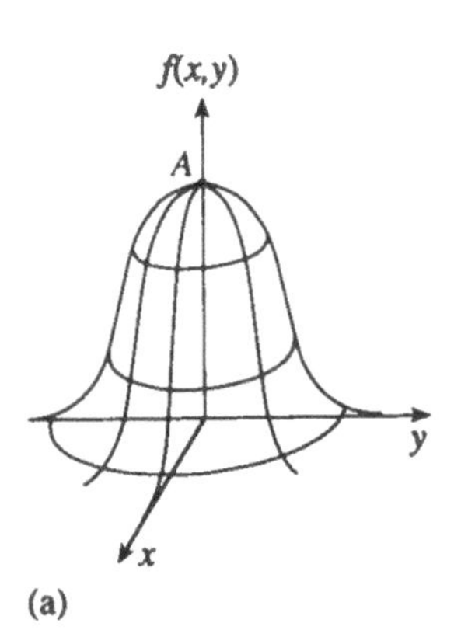

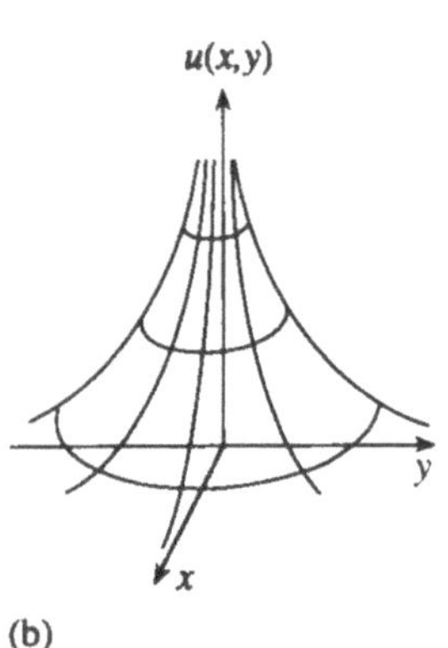

Fig 4.11
(a) The bell-shaped surface is the graph of the two-dimensional scalar field function $f(x,y)$. (b) The graph of the section function $u(x,y) = 1/(x^2+y^2)^{1/2}$ $(x,y) \neq (0,0)$.

4.2.3 Graphs and contours

We can sketch the graph of a two-dimensional scalar field by using the redundant z-axis to represent the field values $f(x,y)$. Thus we put $z = f(x,y)$ and plot the z values above or below the corresponding field points (x,y); then the graph is the surface connecting all points (x,y,z). We shall usually assume that the function is "well-behaved", meaning that the graph is a continuous smooth surface. The graph of the two-dimensional function of Eq (4.3) is the bell-shaped surface shown in Fig 4.11a.

We can't draw the graph of a three-dimensional scalar field, such as Eq (4.1b), since all three spatial dimensions are needed to show the field points (x,y,z); there is no fourth dimension of space for showing the field values! We can however show the graphs of two-dimensional sections. Suppose we fix the z-coordinate in Eq (4.1b) at the value $z = 0$. Then we have a function of two variables, $u(x,y) = U(x,y,0) = 1/(x^2+y^2)^{1/2}$ $(x,y) \neq (0,0)$. This section function gives the scalar field values on the plane $z = 0$ (the x-y plane). The surface shown in Fig 4.11b is the graph of this section function. By taking other fixed values of z we can generate a sequence of section functions and their representative surfaces, and so build up a picture of the three-dimensional field.

The general difficulty of sketching curved surfaces is one reason why the contour representation of scalar fields is so useful. We now show how to obtain equations for contour lines and contour surfaces.

In Section 4.1.1 we defined the contour surface or level surface of a three-dimensional scalar field as a surface on which the scalar has a constant value. For a scalar field $U(x,y,z)$, the equation of the contour surface on which the scalar has some constant value U_1 is obtained simply by putting $U(x,y,z)$ equal to U_1. Thus

$$U(x,y,z) = U_1 \qquad \text{(equation of the } U_1 \text{ contour)} \qquad (4.4)$$

As an example consider the 15°C contour of the temperature field described by Eq (4.2). The equation of this contour is, according to Eq (4.4),

$$12\left(1+\frac{z}{10}\right) = 15 \qquad \text{(equation of the 15 contour)}$$

where the units have been dropped for convenience. This equation is easily rearranged to give the $z = 2.5$. Thus the 15° contour is the plane surface 2.5 m above the floor of the room (Fig 4.12).

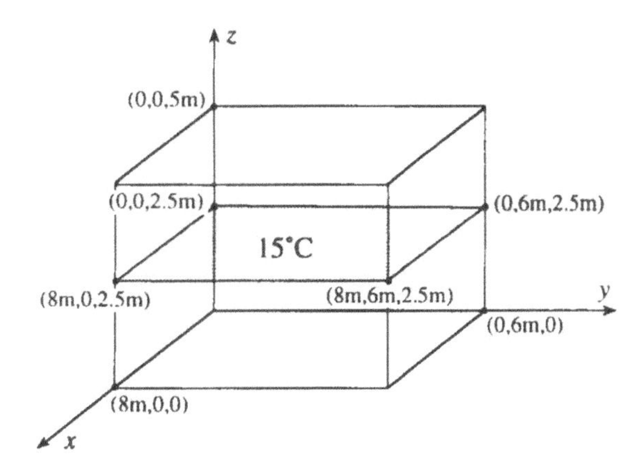

Fig 4.12
The 15°C contour surface for the room temperature field.

In the case of a two-dimensional scalar field $f(x,y)$ the contours are lines given by equations $f(x,y) = f_1$ where f_1 is the constant scalar value on the contour. For example, the two-dimensional function of Eq (4.3) (with $A = a = 1$ for convenience) has the constant value 0.5 on the contour line defined by the equation $\exp[-(x^2 + y^2)] = 0.5$. By taking logs to the base e of both sides of this equation we obtain $x^2 + y^2 = \log_e 2$ which is the equation of a circle of radius $(\log_e 2)^{1/2} = 0.833$ centred on the origin; this circle is the 0.5 contour of f. This and other contours of the function are shown in Fig 4.11c.

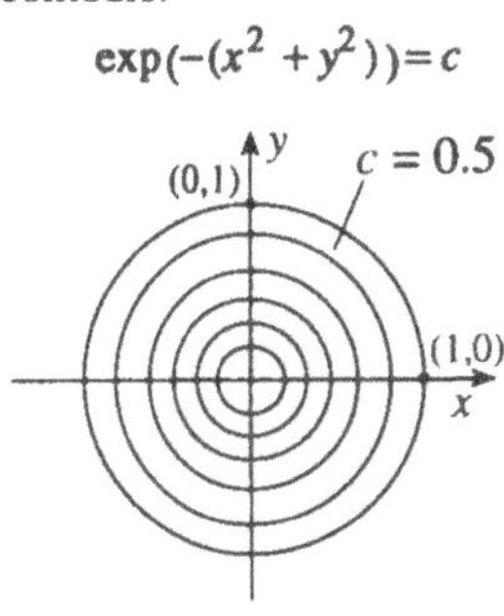

Fig 4.11c
(c) Contours of the function $f(x,y)$ (with $A = a = 1$).

Summary of section 4.2

- A **scalar field function** U is specified by a **rule** and a **domain**. The rule specifies the scalar **field value** $U(\mathbf{r})$ at each **field point** $\mathbf{r}$, and the domain gives the region of space in which the field exists and in which the rule is valid. A three-dimensional field function U is given in cartesian form by a function of the three cartesian variables $U(x,y,z)$.

- The graph of a two-dimensional scalar field f (or a two-dimensional **section function** of a three-dimensional scalar field) is the surface containing all field values $f(x,y)$ when these are plotted as z values above or below the x-y plane, i.e. it is the surface of points (x,y,z) where $z = f(x,y)$.

- The contour surface on which a three-dimensional scalar field $U(x,y,z)$ has the constant value U_1 is given by the equation

  $$U(x,y,z) = U_1 \qquad \text{(equation of the } U_1 \text{ contour)}$$

  A contour line of a two-dimensional field $f(x,y)$ is given by $f(x,y) = f_1$ where f_1 is the constant value of the scalar on the contour line.

Example 2.1 (*Objective 3*) A scalar field g is given in coordinate-free form by

$$g(\mathbf{r}) = \frac{1}{|\mathbf{r} - \mathbf{a}|} \qquad (\mathbf{r} \neq \mathbf{a})$$

where $\mathbf{a}$ is a constant vector. Write down the cartesian form of g in a system in which the origin is at $\mathbf{r} = \mathbf{0}$ and $\mathbf{a} = (\alpha,\beta,\gamma)$.

Solution 2.1 We write $\mathbf{r} = (x,y,z)$ and $\mathbf{r} - \mathbf{a} = (x-\alpha,\ y-\beta,\ z-\gamma)$. The cartesian form of g is therefore

$$g(x,y,z) = \frac{1}{[(x-\alpha)^2 + (y-\beta)^2 + (z-\gamma)^2]^{1/2}} \qquad (x,y,z) \neq (\alpha,\beta,\gamma)$$

Example 2.2 (*Objective 4*) A scalar field function is defined by

$$f(x,y,z) = 2x - \frac{x^2}{y} \qquad (y \neq 0)$$

Find the field value at the point $(1,-2,5)$ and the field values at points $(0,1,z)$. State the location and shape of the region of space excluded from the domain of the function.

Solution 2.2 $f(1,-2,5) = 2\times 1 - 1^2/(-2) = 5/2$.
$f(0,1,z) = 2\times 0 - 0^2/1 = 0$.

The domain of the function is the whole of space with the exclusion of the x-z plane $(y = 0)$.

Example 2.3 (*Objective 4*) Consider the two-dimensional scalar field specified by

$$s(x,y) = \cos(5x)\sin(5y) \qquad (x \in \Re, y \in \Re)$$

This domain statement can alternatively be written as $(x,y) \in \Re^2$.

(a) In what region of space does the field exist?

(b) Determine the value of s at

(i) the origin,
(ii) the point $(0,\pi/10)$ and
(iii) the point $(0.2,0.2)$.
(iv) What is the value of s at points on the y-axis at unit distance from the origin?

Solution 2.3

(a) The domain statement $(x \in \Re, y \in \Re)$ is a formal way of saying that both x and y can have any values. Thus the field exists everywhere in the x-y plane.

(b) (i) The field value at the origin is $s(0,0) = \cos(5\times0)\sin(5\times0) = 0$.

(ii) $s(0,\pi/10) = \cos(5\times0)\sin(\pi\times5/10) = 1$.

(iii) $s(0.2,0.2) = \cos(5\times0.2)\sin(5\times0.2) = \cos(1)\sin(1) = 0.455$.

(iv) There are two points on the y-axis at unit distance from the origin; they have coordinates $(0,1)$ and $(0,-1)$. The values of s are $s(0,1) = \cos(5\times0)\sin(5\times1) = -0.959$ and $s(0,-1) = 0.959$.

Example 2.4 (*Objective 4*) Fig 4.13 shows a section of a long hot straight wire of radius a enclosed by insulation of outer radius b. In the steady state, the temperature field in the insulation is described by the scalar field function

$$T_{ins}(x,y,z) = T_1 - (T_1 - T_2)\frac{\log_e\left[\dfrac{(x^2+y^2)^{1/2}}{a}\right]}{\log_e\left(\dfrac{b}{a}\right)}$$

$$(z \in \Re;\ a \le (x^2+y^2)^{1/2} \le b) \qquad (4.5)$$

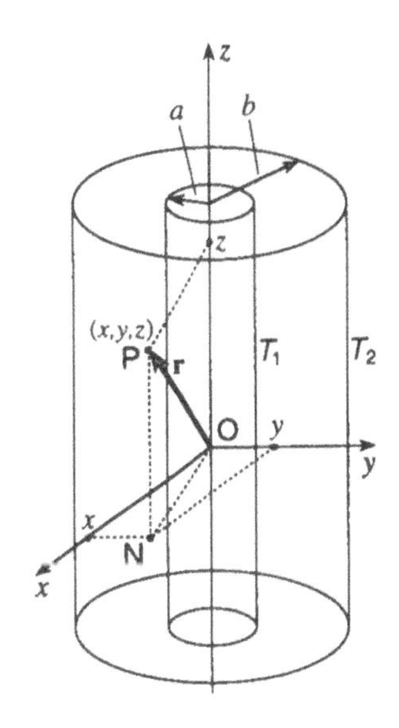

Fig 4.13
Segment of a very long straight wire of radius a surrounded by insulation of external radius b. The z-axis of the cartesian coordinate system is along the axis of the wire. A point P in the insulation has position vector **r** and cartesian coordinates (x,y,z).

where T_1 is the temperature of the inner surface of the insulation which is in contact with the wire and T_2 is the temperature of the outer surface. The axis of the wire lies on the z-axis and the wire is assumed to be infinitely long

($z \in \Re$); this assumption ensures that there are no so-called "end effects" and the temperature is independent of z.

(a) Confirm that the function T_{ins} gives the temperatures at the inner and outer surfaces to be T_1 and T_2 respectively.

(b) Given that $a = 0.5$ mm, $b = 1.5$ mm, $T_1 = 80°C$ and $T_2 = 20°C$, use the function to determine, where possible, the temperatures at the following points: (0.7 mm, 0, 12 mm), (0.5 mm, 0.5 mm, 3 mm), (0, 1 mm, z) and the origin (0,0,0).

Solution 2.4

(a) The quantity $(x^2 + y^2)^{1/2}$ is the perpendicular distance out from the axis. Thus for a point on the inner surface of the insulation we put $(x^2 + y^2)^{1/2} = a$, and the rule gives a temperature (in °C) of $T = T_1 - (T_1 - T_2)[\log_e(a/a)]/\log_e(b/a)$. But $\log_e(a/a) = \log_e 1 = 0$ and so we have $T = T_1$ as required. Points on the outer surface have $(x^2 + y^2)^{1/2} = b$ and so the temperature there is $T = T_1 - (T_1 - T_2)[\log_e(b/a)]/\log_e(b/a) = T_2$.

(b) Note that all distances are given in millimetres (mm) and appear in the rule of Eq (4.5) only as ratios. Thus, omitting units, we have

$$T(0.7,\ 0,\ 12) = 80 - 60\frac{\log_e\left[\dfrac{(0.7^2+0^2)^{1/2}}{0.5}\right]}{\log_e\left(\dfrac{1.5}{0.5}\right)}$$

$$= 80 - 18.4 = 61.6$$

i.e. the temperature is 61.6°C. Similarly we find $T(0.5, 0.5, 3) = 61.1$ and $T(0,1,z) = 42.1$. $T(0,0,0)$ cannot be evaluated using the given function since the origin point is not in the domain.

Problem 2.1 (*Objective 4*) f is a scalar field function defined by

$$f(x,y,z) = z + x^2 - y^2 \qquad (z \geq 0,\ (x^2 + y^2) \leq 9)$$

State the shape and location of the region of space in which the field is defined and evaluate where possible the scalar values at the following points: (1,2,3), (–1,–2,0), (–1,–2,–3), the origin point, the point three units up on the z-axis.

Problem 2.2 (*Objective 4*) The pressure P (including atmospheric pressure) below the surface of a lake is described by the function

$$P(x,y,z) = 10^4(10 - z)\ \mathrm{Nm}^{-2} \qquad (z \le 0)$$

where the z-axis points vertically upwards and $z = 0$ on the surface. Determine the pressure at the surface of the lake and at a depth of 12 m, and specify the location and shape of the contour surface on which $P = 22 \times 10^4\ \mathrm{Nm}^{-2}$.

Problem 2.3 (*Objectives 1,3,4*) A scalar field is defined by

$$\phi(\mathbf{r}) = \boldsymbol{\eta} \cdot \mathbf{r} + \delta \qquad (\text{any } \mathbf{r})$$

where $\mathbf{r}$ is a position vector measured from a fixed reference point, and the vector $\boldsymbol{\eta}$ and the scalar δ are constant.

(a) Determine the value of the scalar ϕ at

(i) the reference point,
(ii) $\mathbf{r} = (3\pi/\eta)\hat{\boldsymbol{\eta}}$ where $\eta = |\boldsymbol{\eta}|$ and $\hat{\boldsymbol{\eta}} = \boldsymbol{\eta}/\eta$,
(iii) points for which $\mathbf{r}$ is normal to $\boldsymbol{\eta}$.

(b) Now suppose the constant vector $\boldsymbol{\eta}$ is directed along the z-axis of a cartesian coordinate system so that $\boldsymbol{\eta} = \eta\mathbf{k}$. Write down the scalar field function in cartesian form and specify the shapes of the contour surfaces. If contours are drawn at scalar intervals of $\pi/2$, what is the distance between adjacent contours?

You may recognise that the scalar $\phi(\mathbf{r})$ describes the spatial variation of the phase (at a fixed instant of time) of a plane wave with propagation vector $\boldsymbol{\eta}$ and wavelength $\lambda = 2\pi/\eta$.

Problem 2.4 (*Objectives 1,4*) Each of the six options in Fig 4.14 shows some contour lines of a two-dimensional scalar field. Each contour line is labelled by the corresponding scalar value and some field points are also shown. Match the options to the following scalar field functions:

$$f(x,y) = 2 + x, \quad g(x,y) = y - 1, \quad h(x,y) = x + y, \quad p(x,y) = x^2 + y^2,$$

$$q(x,y) = x^2 - y^2, \quad s(x,y) = x^2$$

In each case assume that the domain is the whole of the x-y plane.

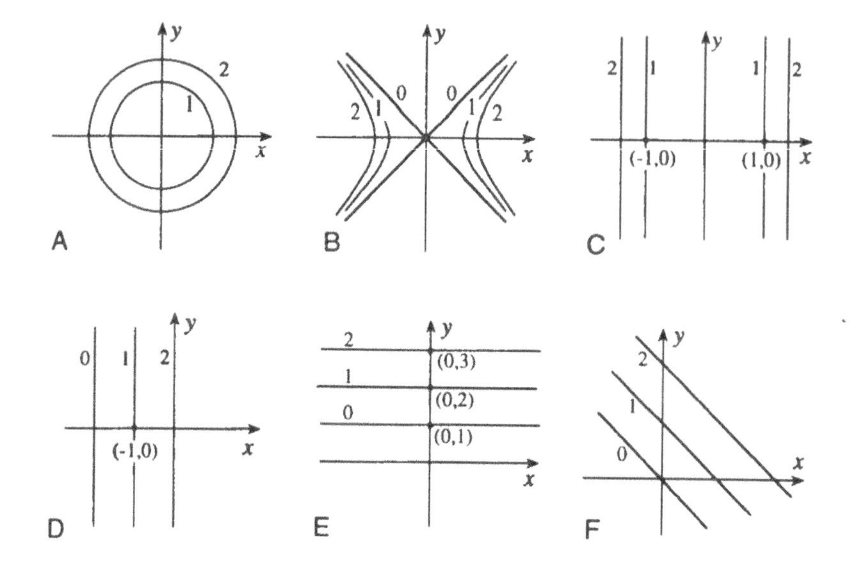

Fig 4.14
Contour lines for Problem 2.4.

Problem 2.5 (*Objective 4*) The light intensity (Wm^{-2}) outside a spherical isotropic lamp of radius 0.02 m is described by the inverse-square-law function

$$I(\mathbf{r}) = \frac{0.01\,\mathrm{W}}{|\mathbf{r}|^2} \qquad (|\mathbf{r}| \geq 0.02)$$

where the reference point for position vectors is at the centre of the source.

(a) What is the light intensity

(i) at a point 5 m from the centre of the source, and
(ii) at a point on the surface of the lamp?

(b) Determine the intensity contour surfaces for intensities (Wm^{-2}) of 11, 9, 7, 5, 3 and 1. Show these contours as curves on a rough sketch of a section through the origin.

4.3 VECTOR FIELD FUNCTIONS

Vector fields are represented mathematically by vector functions of position. We show how these functions are specified, and describe how they are used to calculate the magnitudes, directions and components of the field vectors at given field points. We show also how the equations of the field lines can be derived from the field function.

4.3.1 Specifying vector field functions

Let the vector quantity be denoted by the symbol **F**. The value of the vector at a field point **r** is denoted by **F**(**r**). The set of field points **r** together with the corresponding **field vectors F**(**r**) constitutes a **vector field function** or simply a vector field. The function is specified by giving a rule from which the field vector can be found at each field point, and a domain which gives the region of space in which the rule applies. An example of a three-dimensional vector field function is

$$\mathbf{F}(\mathbf{r}) = \frac{C\mathbf{r}}{|\mathbf{r}|^3} \qquad (\mathbf{r} \neq \mathbf{0};\ C \text{ is a constant}) \qquad (4.6a)$$

The rule associates a field vector $C\mathbf{r}/|\mathbf{r}|^3$ with each field point **r**. Thus the field vector has a magnitude proportional to the inverse of the square of the distance from the reference point ($\mathbf{r} = \mathbf{0}$) and a direction given by the direction of $C\mathbf{r}$, i.e. a radial direction, inward or outward depending on the sign of the constant C. The domain, specified by the statement $\mathbf{r} \neq \mathbf{0}$, is the whole of space with the exclusion of the reference point. An example of such a field is the electrostatic field produced by a point charge Q at $\mathbf{r} = \mathbf{0}$, which is given by Eq (4.6a) with $C = Q/4\pi\epsilon_0$.

4.3.2 Cartesian vector fields

When the position vector of the field point is expressed in cartesian form, $\mathbf{r} = (x,y,z)$, we have a **cartesian vector field** denoted by $\mathbf{F}(x,y,z)$. The field vector at any point is a linear combination of the cartesian unit vectors. Thus we can express the vector field in terms of three cartesian component fields

$$\mathbf{F}(x,y,z) = \mathbf{i}F_x(x,y,z) + \mathbf{j}F_y(x,y,z) + \mathbf{k}F_z(x,y,z) \qquad (4.7a)$$

which can also be written as an ordered triple of scalar fields

$$\mathbf{F}(x,y,z) = (F_x(x,y,z),\ F_y(x,y,z),\ F_z(x,y,z)) \qquad (4.7b)$$

The component fields are found by projecting **F** onto the cartesian unit vectors, thus $F_x(x,y,z) = \mathbf{i} \cdot \mathbf{F}(x,y,z)$, etc.

The cartesian form of the vector field (4.6a), with the origin of coordinates chosen to coincide with $\mathbf{r} = \mathbf{0}$, is

$$\mathbf{F}(x,y,z) = \frac{C(x\mathbf{i}+y\mathbf{j}+z\mathbf{k})}{(x^2+y^2+z^2)^{3/2}} \qquad ((x,y,z) \neq (0,0,0)) \qquad (4.6b)$$

where the position vector **r** is expressed in its cartesian form, $\mathbf{r} = x\mathbf{i} + y\mathbf{j} + z\mathbf{k}$. The y-component of the field, for example, is the scalar field $F_y(x,y,z) = \mathbf{j} \cdot \mathbf{F}(x,y,z)$, i.e.

$$F_y(x,y,z) = \frac{Cy}{(x^2+y^2+z^2)^{3/2}} \qquad ((x,y,z) \neq (0,0,0))$$

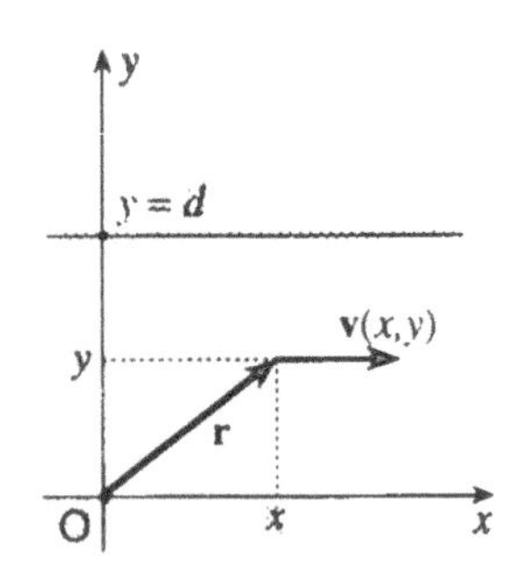

Fig 4.15
The surface velocity of a river. The field point (x,y) is fixed relative to the river bank and does not move with the river. The surface velocity $\mathbf{v}(x,y)$ is the velocity with which a small floating object would pass through the point (x,y).

An example of a two-dimensional vector field is the surface velocity on a river. Let the water surface define the x-y plane with the x-axis lying along one river bank and the origin point fixed on this bank (Fig 4.15), the other bank being at $y = d$. Let $\mathbf{v}(x,y)$ denote the velocity of the water surface at the point (x,y). This velocity field can often be represented by a vector field function of the form

$$\mathbf{v}(x,y) = ky(d-y)\mathbf{i} \qquad (0 \le y \le d) \qquad (4.8)$$

where k is a constant determined by the flow rate. The surface velocity is everywhere parallel to the x-axis except at the river banks where the velocity is zero. The maximum surface speed is $kd^2/4$ in mid-stream ($y = d/2$). Once the parameters k and d are specified, the function gives the surface velocity at any point on the river. For example, with $k = 0.05$ and $d = 10$ (SI units) the surface velocity at points with coordinates $(x,7)$ is $\mathbf{v}(x,7) = (0.05)7(10-7)\mathbf{i} = 1.05\mathbf{i}$. That is, the velocity at any point 7 m out from the bank is $1.05\,\mathrm{m\,s^{-1}}$ in the x-direction.

4.3.3 Equation of a field line

Vector field lines were introduced in Section 4.1.2. We now show how the equation of the field line passing through a given point can be calculated.

Consider a point $P(x,y)$ in the domain of a two-dimensional vector field **F**. Fig 4.16 shows the field line of **F** passing through P and the field vector $\mathbf{F}(x,y)$ at P. The direction of $\mathbf{F}(x,y)$ is along the tangent to the field line. This direction is specified by the angle θ_x where $\tan\theta_x = F_y(x,y)/F_x(x,y)$. Now the field line passing through P is the graph of some equation $y = y(x)$ and so the slope or gradient of the field line at P is the derivative $\mathrm{d}y/\mathrm{d}x$ at P. Thus we can equate $\mathrm{d}y/\mathrm{d}x$ with $\tan\theta_x$ and obtain the differential equation

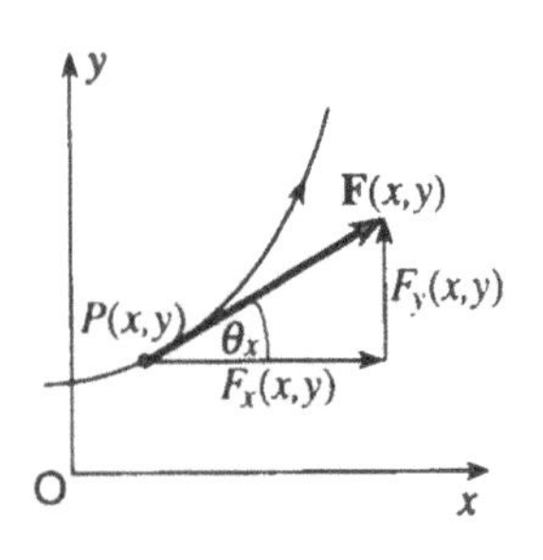

Fig 4.16
The direction of the field vector $\mathbf{F}(x,y)$ at the point $P(x,y)$ is along a tangent to the field line at P and is specified by the angle θ_x where $\tan\theta_x = F_y(x,y)/F_x(x,y)$.

$$\frac{\mathrm{d}y}{\mathrm{d}x} = \frac{F_y(x,y)}{F_x(x,y)} \qquad \text{(gradient of field line)} \qquad (4.9)$$

which is sometimes written as $\mathrm{d}x/F_x = \mathrm{d}y/F_y$. The equation of the field line through a given point is found by integrating the differential equation (4.9) using the coordinates of the given point to determine the constant of integration. For example, the velocity field $\mathbf{v}(x,y)$ of Eq (4.8) has cartesian components $v_y(x,y) = 0$ and $v_x(x,y) = ky(d-y)$, and so Eq (4.9) gives $\mathrm{d}y/\mathrm{d}x = 0$. Integration yields the solution $y = C$ where C is a constant. To find the velocity field line passing through, say, the mid-stream point $M(0,d/2)$, we determine the constant C by putting $y = d/2$ in the solution $y = C$. This gives $C = d/2$ and so the field line passing through M is the straight line $y = d/2$ (Fig 4.17).

Field lines in a three-dimensional vector field obey a set of three differential equations commonly written as

$$dx/F_x = dy/F_y = dz/F_z \qquad (4.10)$$

These equations have two independent solutions each of which represents a surface. The curve of intersection of the two surfaces is a field line of **F**. We shall not consider the details.

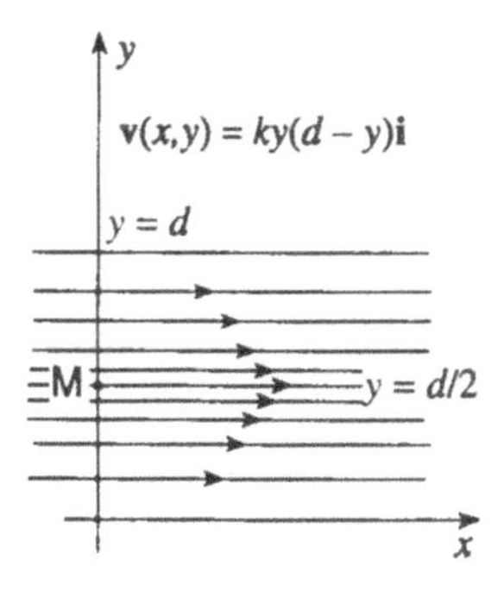

Fig 4.17
Field lines of the surface velocity $\mathbf{v}(x,y)$ on a river.

Summary of section 4.3

- A **vector field function F** is specified by a rule which gives the field vector $\mathbf{F}(\mathbf{r})$ at each field point $\mathbf{r}$, and a domain statement which specifies the region of space in which the rule applies. A **cartesian vector field** $\mathbf{F}(x,y,z)$ can be expressed in terms of three cartesian component fields

$$\mathbf{F}(x,y,z) = \mathbf{i}F_x(x,y,z) + \mathbf{j}F_y(x,y,z) + \mathbf{k}F_z(x,y,z) \qquad (4.6a)$$

- The equation of the field line passing through a given point in a two-dimensional vector field $\mathbf{F}(x,y)$ is found by integrating the differential equation $dy/dx = F_y(x,y)/F_x(x,y)$ using the coordinates of the given field point to fix the constant of integration.

Example 3.1 (*Objective 5*) A vector field **A** is specified by

$$\mathbf{A}(x,y,z) = (xy^2 + 2)\mathbf{i} + xy(\mathbf{j} + \mathbf{k}) \qquad (\text{all } (x,y,z))$$

(a) Write down the three cartesian component fields.

(b) Specify the vector $\mathbf{A}(1,2,-3)$.

(c) Determine the cartesian components and the magnitude of the field vector at the field point (5,1,3).

(d) Specify the field vectors at points on the z-axis.

Solution 3.1

(a) The x component of the field **A** is the scalar field

$$A_x(x,y,z) = \mathbf{i} \cdot \mathbf{A}(x,y,z) = xy^2 + 2$$

Similarly the y and z component fields are $A_y(x,y,z) = A_z(x,y,z) = xy$.

(b) $\mathbf{A}(1,2,-3) = (1\times2^2 + 2)\mathbf{i} + 1\times2(\mathbf{j} + \mathbf{k}) = 6\mathbf{i} + 2\mathbf{j} + 2\mathbf{k}$.

(c) $\mathbf{A}(5,1,3) = (5\times1^2 + 2)\mathbf{i} + 5\times1(\mathbf{j} + \mathbf{k}) = 7\mathbf{i} + 5\mathbf{j} + 5\mathbf{k}$. Thus the components of $\mathbf{A}(5,1,3)$ are $A_x(5,1,3) = 7$ and $A_y(5,1,3) = A_z(5,1,3) = 5$. The magnitude is $|\mathbf{A}(5,1,3)| = (7^2 + 2\times5^2)^{1/2} = 9.95$.

(d) At points on the z-axis, $x = y = 0$ and so the field vectors on the z-axis are $\mathbf{A}(0,0,z) = 2\mathbf{i}$.

Example 3.2 (*Objectives 3,5*) We can define a velocity field $\mathbf{v}(\mathbf{r})$ for a rigid body rotating at constant angular velocity ω about a fixed axis of rotational symmetry, as follows: Let $\mathbf{r}$ be the position vector of a point fixed in a stationary frame of reference with the positive z-axis in the direction of ω and the origin on the axis of rotation. Then $\mathbf{v}(\mathbf{r})$ is the velocity with which elements of the rotating body pass through the fixed point $\mathbf{r}$.

The velocity field of a solid sphere of radius a rotating uniformly about a fixed axis through its centre with angular velocity ω is $\mathbf{v}(\mathbf{r}) = \boldsymbol{\omega} \times \mathbf{r}$ ($|\mathbf{r}| \leq a$). Express this field in cartesian form and identify the x, y and z component fields. Determine the x and y components of the velocity at points $(a/2,a/3,z)$ in the sphere. Find the speed at these points if $\omega = 3$ rad s^{-1} and $a = 0.2$ m.

Solution 3.2 The cartesian form of $\mathbf{v}(\mathbf{r}) = \boldsymbol{\omega} \times \mathbf{r}$ is

$$\mathbf{v}(x,y,z) = \omega\mathbf{k} \times (x\mathbf{i} + y\mathbf{j} + z\mathbf{k}) = \omega(-y\mathbf{i} + x\mathbf{j}) \qquad (4.11)$$

The component fields are: $v_x = -\omega y$, $v_y = \omega x$ and $v_z = 0$. Hence the velocity components at points $(a/2,a/3,z)$ are

$$v_x(a/2,a/3,z) = -\omega a/3, \quad v_y(a/2,a/3,z) = \omega a/2, \quad v_z = 0$$

The required speed is

$$v(a/2,a/3,z) = (v_x^2 + v_y^2 + v_z^2)^{1/2}$$

$$= [(-3\times0.2/3)^2 + (3\times0.2/2)^2 + 0^2]^{1/2} \text{ ms}^{-1} = 0.36 \text{ ms}^{-1}.$$

Example 3.3 (*Objective 5*) The magnetic field in the space outside a very long thin straight wire carrying a current I in the z-direction is described by the function

$$\mathbf{B}(x,y,z) = \frac{\mu_0 I(x\mathbf{j} - y\mathbf{i})}{2\pi(x^2 + y^2)} \qquad (x,y,z) \neq (0,0,z) \qquad (4.12)$$

where μ_0 is a constant (permeability of free space), and the line of the wire (the z-axis) is outside the domain.

(a) Express the magnitude of the magnetic field in terms of the perpendicular distance ρ from the wire?

(b) What is the direction of the magnetic field at points on the x-axis?

(c) Does the magnetic field have a z-component anywhere?

Solution 3.3

(a) $$|\mathbf{B}(x,y,z)| = \frac{\mu_0 I|x\mathbf{j} - y\mathbf{i}|}{2\pi(x^2 + y^2)}$$

$$= \frac{\mu_0 I(x^2 + y^2)^{1/2}}{2\pi(x^2 + y^2)} = \frac{\mu_0 I}{2\pi(x^2 + y^2)^{1/2}}$$

But $(x^2 + y^2)^{1/2} = \rho$, the perpendicular distance from the wire (see Fig 4.18a). Hence

$$|\mathbf{B}(x,y,z)| = \frac{\mu_0 I}{2\pi\rho} \qquad (\rho \neq 0)$$

(b) For points on the x-axis we put $y = z = 0$ in Eq (4.12) to obtain

$$\mathbf{B}(x,0,0) = \frac{\mu_0 I\mathbf{j}}{2\pi x} \qquad (x \neq 0)$$

which is in a direction parallel or antiparallel to $\mathbf{j}$ when x is positive or negative respectively.

(c) No. $B_z = \mathbf{k} \cdot \mathbf{B} = 0$.

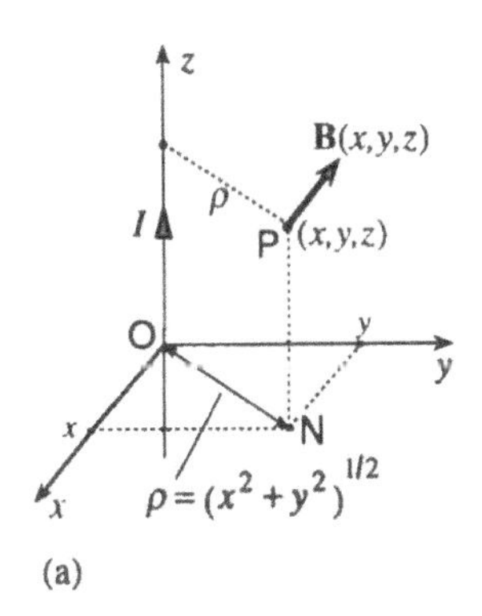

Fig 4.18a
The magnetic field $\mathbf{B}(x,y,z)$ produced by an electric current I flowing along the z-axis. The perpendicular distance ρ of a point $P(x,y,z)$ from the z-axis is $\rho = (x^2 + y^2)^{1/2}$.

Example 3.4 (*Objective 6*) Consider the magnetic field **B** given by Eq (4.12).

(a) Write down an expression for the gradient (dy/dx) of any field line of **B**.

(b) Derive the equation of the field line passing through the point (3,4,0).

(c) Sketch roughly a few field lines in the x-y plane to indicate the magnitude and direction of the field.

Solution 3.4

(a) The field **B** has no z-component and so the field vectors and the field lines all lie in planes parallel to the x-y plane. (The field is effectively two-dimensional.) The gradient of any field line is given by Eq (4.9) (or the first of Eqs (4.10)):

$$\frac{dy}{dx} = \frac{B_y}{B_x} = \frac{\mu_0 Ix}{(-\mu_0 Iy)} = -\frac{x}{y}$$

(b) The differential equation dy/dx = $-x/y$ can be integrated by the method of separation of variables to give the equation of any field line. Thus we have $\int y\mathrm{d}y = -\int x\mathrm{d}x$, which gives $x^2 + y^2 = C$, where C is a constant of integration. This is the equation of a circle of radius $\sqrt{C}$ centred on the z-axis. For a particular field line passing through a point (3,4,z) we have $C = 3^2 + 4^2 = 25$, and so the equation of the field line is $x^2 + y^2 = 25$. Thus the field line passing through a point (3,4,z) is a circle of radius 5.

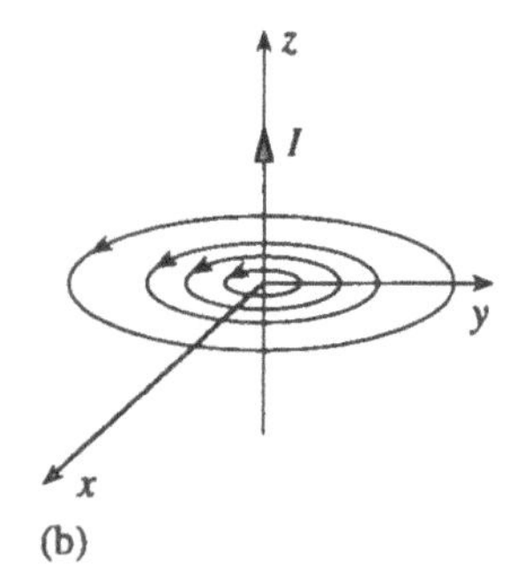

Fig 4.18b
Magnetic field lines in the x-y plane.

(c) Fig 4.18b shows circular field lines in the x-y plane. The separation of adjacent lines increases in proportion to distance ρ from the origin, i.e. the line density is proportional to $1/\rho$ in accordance with the magnitude of the field vectors found in part (a) of Example 3.3. The direction of any field line is that of a positive rotation about the direction of the current (z-axis); this can be deduced from the directions of the field vectors on the x-axis found in part (b) of Example 3.3.

Problem 3.1 (*Objective 5*) A vector field is specified by

$$\mathbf{A}(x,y,z) = xy\mathbf{i} + (y^2 - x^2)\mathbf{j} + z\mathbf{k}$$

for all space. Determine the vector **A**(−1,2,3), the scalar **i** . **A**(3,1,0) and the scalar field $A_y(x,y,z)$. Find also the magnitude of **A** at the point (9,9,81).

Problem 3.2 (*Objective 5*) Given the vector field $\mathbf{R}(\mathbf{r}) = 3\mathbf{r}$, determine the magnitudes of the field vectors at distances 1, 2, 3, 4 and 5 from the reference point $\mathbf{r} = \mathbf{0}$. What is the direction of the field vector at a point on the z–axis of a coordinate system with origin at $\mathbf{r} = \mathbf{0}$?

Problem 3.3 (*Objectives 5,6*) Each of the six options in Fig 4.19 shows some field lines corresponding to one of the two-dimensional vector fields specified below. Match the options to the fields below. (The field line diagrams in Fig 4.19 show the directions but not necessarily the magnitudes of the fields. Note that field lines cannot pass through points where the field vector is the zero vector.)

$$\mathbf{P}(x,y) = x\mathbf{i} + y\mathbf{j}, \quad \mathbf{Q}(x,y) = x\mathbf{i} - y\mathbf{j}, \quad \mathbf{S}(x,y) = -y\mathbf{i} + x\mathbf{j},$$

$$\mathbf{T}(x,y) = (x\mathbf{j} + y\mathbf{i}), \quad \mathbf{U}(x,y) = (\mathbf{i} + \mathbf{j})/\sqrt{2}, \quad \mathbf{V}(x,y) = x\mathbf{i} + x\mathbf{j}$$

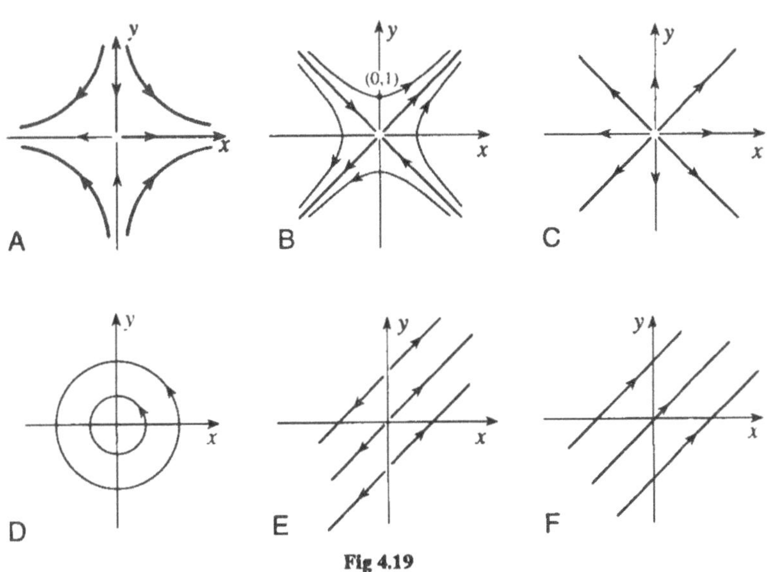

Fig 4.19
Vector field lines for Problem 3.3.

Problem 3.4 (*Objective 5*) A vector field **H** is defined by

$$\mathbf{H}(x,y,z) = (y, (x + 2), 7) \qquad \text{(all } (x,y,z))$$

Determine the x-component of the field vector at the origin and the magnitude of the field vector at the point on the positive x-axis three units from the origin.

Problem 3.5 (*Objective 6*) Determine the equation of the field line of $\mathbf{u}(x,y) = x\mathbf{i} - y\mathbf{j}$ passing through the point (1,1).

4.4 POLAR COORDINATE SYSTEMS

Cartesian coordinate systems are not always convenient. Two other types of coordinate system are in common use and are described in this section. The most suitable coordinate system for any particular problem often depends on the spatial symmetry of the field.

4.4.1 Symmetries and coordinate systems

The subject of symmetry is of fundamental importance in science and mathematics and the classification of symmetries is a complex mathematical subject which we shall not develop in this book. However, some basic considerations of the spatial symmetry of a field can help us to choose a convenient coordinate system. This may involve choosing the most appropriate origin and orientation of a cartesian system or it may suggest using non-cartesian systems.

We have seen examples of three-dimensional field functions where the rule for calculating the field values involves only one or two of the three cartesian variables, as in Eq (4.2) where the room temperature depends only on the z-coordinate. A reduction in the number of variables appearing in the rule of a function has obvious advantages and can happen whenever the field exhibits spatial symmetry of some kind and we choose a coordinate system that exploits the symmetry. The temperature field in the room exhibits a horizontal **translational symmetry** since the temperature is the same everywhere on any horizontal plane, i.e. the contours are plane surfaces (Fig 4.12). To exploit this symmetry we choose a cartesian coordinate system with the x-y plane horizontal so that the temperature is independent of the x- and y- coordinates and varies with z only (Eq (4.2)). If instead we choose a coordinate system with the x-y plane tilted with respect to the horizontal then the temperature would vary across the x-y plane and the rule of the field function in this system would depend on x and y as well as on z.

An example of a field exhibiting a different kind of symmetry is the temperature field in the insulation of the heated wire (Eq (4.5) and Fig 4.13). Here the temperature is the same everywhere on any cylindrical surface coaxial with the wire, i.e. the temperature contours are concentric cylinders. This is a special case of rotational symmetry about an axis called **cylindrical symmetry**. Another example of cylindrical symmetry is the magnetic field around a very long straight current-carrying wire (Eq (4.12)). This vector field has the same magnitude everywhere on any cylindrical surface which has its axis on the wire, and the direction of the vector is everywhere tangential to the surface and normal to the axis.

Another kind of symmetry that occurs commonly is **spherical symmetry**. An example of a spherically symmetric scalar field is illustrated by the scalar inverse distance function of Eq (4.1a) and the inverse square law function in Problem 2.5. Here the scalar value is the same everywhere on a spherical surface centred on the origin point, i.e. the contours are concentric spheres. An example of a vector field exhibiting spherical symmetry is the field **F** specified in

Eq (4.6) where the field vector is directed radially outwards everywhere and has the same magnitude at all points on the surface of a sphere.

We have seen that for a field with horizontal translational symmetry an appropriate orientation of the cartesian axes can give a field function that varies with one spatial coordinate only (Eq (4.2)). For fields exhibiting cylindrical symmetry it is possible to reduce the number of cartesian variables in the rule of the function from three to two by taking the z-axis along the axis of symmetry, as in Eqs (4.5) and (4.11), but there is no cartesian system that would reduce the number of variables further. In the case of spherical symmetry, exemplified by Eqs (4.1b) and (4.6), the rule of the function depends on all three cartesian variables and there is no cartesian system that would reduce this number. However, in these cases the symmetries themselves suggest new non-cartesian variables in which to express the fields. Consider for example the spherically symmetric field described by Eq (4.1b). The rule for calculating the field values depends on each of the three cartesian variables in exactly the same way. You can see this by making any permutation of the three cartesian variables (interchange x and y for example); you end up with the same rule. In fact the rule depends on the cartesian variables only in the combination $(x^2 + y^2 + z^2) = |\mathbf{r}|^2$, as shown explicitly in Eq (4.1a). This suggests that spherically symmetric fields are most simply described in a coordinate system in which the distance from the origin, $r = |\mathbf{r}|$, plays the role of a coordinate. Such a system is the *spherical polar coordinate system* described in Section 4.4.3.

Similar considerations apply to fields with cylindrical symmetry such as Eq (4.5) where the rule depends on each of the two cartesian variables x and y in exactly the same way; in fact you can see that it depends only on the combination $(x^2 + y^2) = \rho^2$ where the non-cartesian variable ρ is the perpendicular distance out from the wire. In Section 4.4.2 we describe the *cylindrical polar coordinate* system in which ρ is one of three coordinates.

4.4.2 Cylindrical polar coordinate systems

The **cylindrical polar coordinates** of a point P are denoted by (ρ,ϕ,z) and shown in Fig 4.20. The coordinate ρ is the perpendicular distance of P from the z-axis (or **polar axis**); ρ is also equal to the length ON where N is the foot of the perpendicular from P onto the x-y plane. The angle ϕ is the angular displacement of the line ON from the positive x-axis measured in the anticlockwise sense, and z is the z-coordinate. The angle ϕ is commonly called the **azimuthal angle** or **longitude** and is restricted to values within one complete cycle $(0 \le \phi < 2\pi)$; the distance ρ is restricted to non-negative values $(\rho \ge 0)$ and z can be any number.

The three coordinates (ρ,ϕ,z) are sufficient to specify the position of any point in space, and conversely any point in space has a unique set of cylindrical polar coordinates with the exception of points on the z-axis for which the angle ϕ is undetermined.

Surfaces called **coordinate surfaces** are generated by fixing one coordinate while allowing the others to vary. For example, suppose we fix $\rho = 2$ and allow ϕ and z to vary. The set of points $(2,\phi,z)$ lies on a cylindrical surface of radius 2

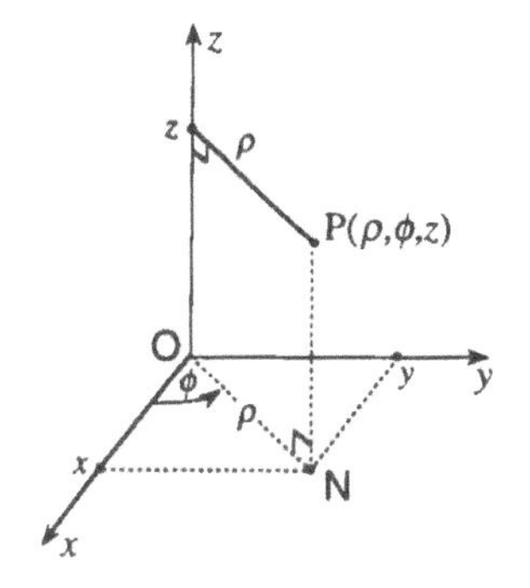

Fig 4.20
Cylindrical polar coordinates of a point P.

In some books the cylindrical polar coordinate $\rho = (x^2 + y^2)^{1/2}$ is denoted by r which should not then be confused with the magnitude of the position vector $|\mathbf{r}| = r = (x^2 + y^2 + z^2)^{1/2}$. Also, the azimuthal angle in cylindrical polar coordinates is often denoted by θ rather than ϕ. Thus you may see the cylindrical polar coordinates of a point written as (r, θ, z) rather than (ρ,ϕ,z).

(Fig 4.21a). We refer to this coordinate surface as the surface $\rho = 2$ or the surface $(2,\phi,z)$. The coordinate surfaces generated by fixing ϕ are half-planes normal to the x-y plane (Fig 4.21b) while those generated by fixing z are planes parallel to the x-y plane (Fig 4.21c).

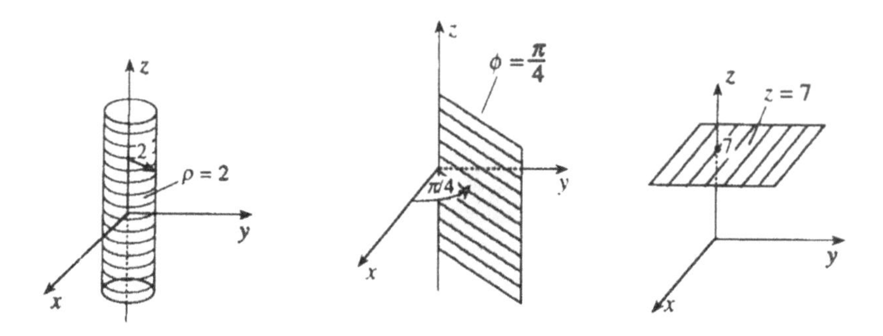

Fig 4.21
Coordinate surfaces: (a) $\rho = 2$, (b) $\phi = \pi/4$, (c) $z = 7$.

You can see from the geometry of Fig 4.20 that the cylindrical polar coordinates (ρ,ϕ,z) of a point P are related to the cartesian coordinates of the same point by

$$x = \rho\cos\phi, \quad y = \rho\sin\phi, \quad z = z, \tag{4.13}$$

To find ϕ uniquely we use $\sin\phi = y/\rho$ and $\cos\phi = x/\rho$.

$$\rho = (x^2 + y^2)^{1/2}, \quad \tan\phi = \frac{y}{x} \tag{4.14}$$

Cylindrical polar coordinates can be adapted to specify points on a plane surface simply by suppressing the redundant z-coordinate. This is illustrated in Fig 4.20 where the **plane polar coordinates** of the point N in the x-y plane are (ρ,ϕ).

Cylindrical polar systems are used not only for describing the coordinates of points but also for specifying field vectors. For this purpose we introduce a set of three mutually orthogonal unit vectors at each point called the **cylindrical polar base vectors** (Fig 4.22a):

$$\mathbf{e}_\rho, \quad \mathbf{e}_\phi, \quad \mathbf{e}_z, \quad (\mathbf{e}_\rho = \mathbf{e}_\phi \times \mathbf{e}_z) \tag{4.15}$$

The unit vector $\mathbf{e}_\rho$ at P is directed perpendicularly away from the z-axis and therefore lies in a plane parallel to the x-y plane. $\mathbf{e}_\phi$ lies in the same plane but points in the direction of increasing ϕ, i.e. the direction in which a small positive (anticlockwise) rotation about the z-axis would take the point P. $\mathbf{e}_z = \mathbf{k}$, the unit vector in the z-direction. Note that the unit vectors are mutually orthogonal at any point but the directions of $\mathbf{e}_\rho$ and $\mathbf{e}_\phi$ depend on the value of the azimuthal coordinate ϕ. You can see from Fig 4.22b that the cartesian components of these unit vectors are $\mathbf{i}\cdot\mathbf{e}_\rho = \cos\phi$, $\mathbf{j}\cdot\mathbf{e}_\rho = \sin\phi$, $\mathbf{i}\cdot\mathbf{e}_\phi = -\sin\phi$ and $\mathbf{j}\cdot\mathbf{e}_\phi = \cos\phi$. Hence

$$\left.\begin{aligned}\mathbf{e}_\rho &= \mathbf{i}\cos\phi + \mathbf{j}\sin\phi = \frac{(x\mathbf{i}+y\mathbf{j})}{(x^2+y^2)^{1/2}}\\ \mathbf{e}_\phi &= -\mathbf{i}\sin\phi + \mathbf{j}\cos\phi = \frac{(x\mathbf{j}-y\mathbf{i})}{(x^2+y^2)^{1/2}}\\ \mathbf{e}_z &= \mathbf{k}\end{aligned}\right\} \qquad (4.16)$$

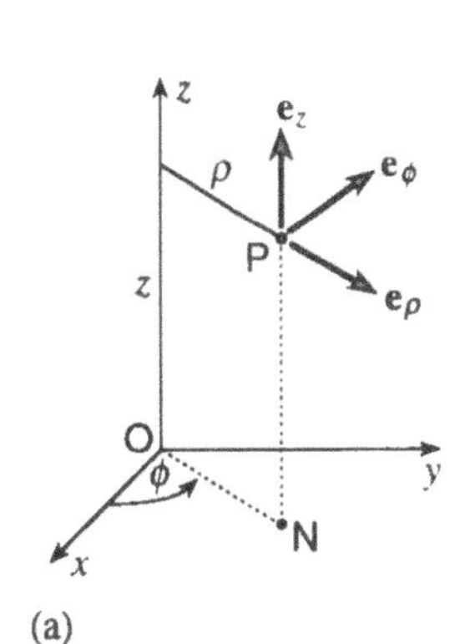

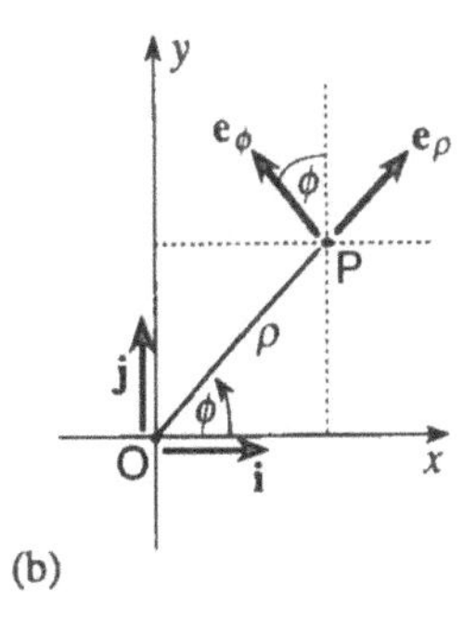

Fig 4.22
(a) The cylindrical polar base vectors at P. (b) The unit vectors $\mathbf{e}_\rho$ and $\mathbf{e}_\phi$ lie in a plane parallel to the x-y plane.

We can represent any vector field **A** as a linear combination of the cylindrical polar base vectors

$$\mathbf{A}(\rho,\phi,z) = A_\rho(\rho,\phi,z)\,\mathbf{e}_\rho + A_\phi(\rho,\phi,z)\,\mathbf{e}_\phi + A_z(\rho,\phi,z)\,\mathbf{e}_z \qquad (4.17a)$$

where the scalar functions A_ρ, A_ϕ and A_z are the **cylindrical polar components** of the field. They are found by projecting the field **A** onto the base vectors. Thus

$$A_\rho = \mathbf{e}_\rho\cdot\mathbf{A}, \quad A_\phi = \mathbf{e}_\phi\cdot\mathbf{A}, \quad A_z = \mathbf{e}_z\cdot\mathbf{A} \qquad (4.17b)$$

As an example of how cylindrical polar coordinates can be used to advantage, consider the magnetic field **B** of Eq (4.12). This field has cylindrical symmetry; its magnitude depends only on $x^2 + y^2 = \rho^2$ and it points in the direction of $\mathbf{e}_\phi$. You can see from inspection of Eq (4.12), or formally by projecting both sides of Eq (4.12) onto the cylindrical polar unit vectors, that $B_\rho = B_z = 0$ and $B_\phi = \mu_0 I/2\pi\rho$, i.e. we can express the field in the cylindrical polar form

$$\mathbf{B}(\rho,\phi,z) = \frac{\mu_0 I\mathbf{e}_\phi}{2\pi\rho} \qquad (\rho \neq 0)$$

You may wish at this point to study Examples 4.1a, 4.2a, 4.3a and 4.4 before going on to the next section.

4.4.3 Spherical polar coordinate systems

The **spherical polar coordinates** of a point P (Fig 4.23) are denoted by (r,θ,ϕ). Here $r = |\mathbf{r}| = (x^2 + y^2 + z^2)^{1/2}$, the magnitude of the position vector from the origin to the point P. θ is the angle between the positive z-axis (polar axis) and the position vector and is called the **polar angle** or **colatitude**, and ϕ is the azimuthal angle. The coordinate r is of course non-negative, θ is restricted to the range $(0 \le \theta \le \pi)$ and ϕ to $(0 \le \phi < 2\pi)$. The three coordinates (r,θ,ϕ) are sufficient to specify any point in space, and conversely each point in space has a unique set of spherical polar coordinates with the exception of points on the z-axis on which ϕ is undetermined; θ is also undetermined at the origin.

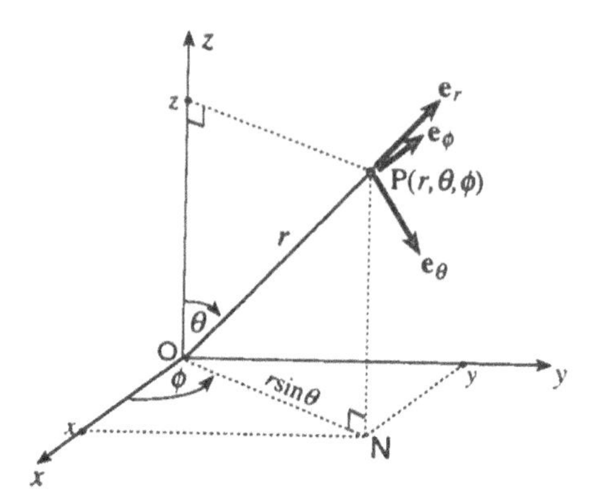

Fig 4.23
The spherical polar coordinates of a point P and the spherical polar base vectors $\mathbf{e}_r$, $\mathbf{e}_\theta$ and $\mathbf{e}_\phi$.

Coordinate surfaces are generated by fixing one coordinate and allowing the other two to vary. For example, the coordinate surface specified by $\theta = \pi/2$ is the x-y plane.

The spherical polar coordinates of a point are seen from Fig 4.23 to be related to the cartesian coordinates by

$$x = r\sin\theta\cos\phi, \quad y = r\sin\theta\sin\phi, \quad z = r\cos\theta,$$

$$r = (x^2 + y^2 + z^2)^{1/2}, \quad \cos\theta = \frac{z}{r}, \quad \tan\phi = \frac{y}{x} \tag{4.18}$$

A simple example of a scalar field expressed in spherical polar coordinates is the spherically symmetric inverse distance field

$$U(r,\theta,\phi) = \frac{1}{r} \qquad (r \neq 0)$$

Here only the r coordinate appears in the rule of the function while all three cartesian coordinates appear in the cartesian representation (Eq (4.1b)).

The **spherical polar base vectors** are the mutually orthogonal unit vectors (Fig 4.23)

$$\mathbf{e}_r, \quad \mathbf{e}_\theta, \quad \mathbf{e}_\phi \qquad (\mathbf{e}_r = \mathbf{e}_\theta \times \mathbf{e}_\phi) \tag{4.19}$$

where $\mathbf{e}_r = \mathbf{r}/|\mathbf{r}|$ is directed radially outwards from the origin, $\mathbf{e}_\theta$ is in the plane of $\mathbf{r}$ and the z-axis and points normal to $\mathbf{r}$ in the direction of increasing θ, and $\mathbf{e}_\phi$ lies in a plane parallel to the x-y plane and points normal to $\mathbf{r}$ in the direction of increasing ϕ. Note that the directions of all three unit vectors depend on the position of the point P. It can be deduced from the geometry of Fig 4.23 that

$$\left.\begin{aligned} \mathbf{e}_r &= \mathbf{i}\sin\theta\cos\phi + \mathbf{j}\sin\theta\sin\phi + \mathbf{k}\cos\theta \\ \mathbf{e}_\theta &= \mathbf{i}\cos\theta\cos\phi + \mathbf{j}\cos\theta\sin\phi - \mathbf{k}\sin\theta \\ \mathbf{e}_\phi &= -\mathbf{i}\sin\phi + \mathbf{j}\cos\phi \end{aligned}\right\} \quad (4.20)$$

An arbitrary vector field can be expressed as a linear combination of the spherical polar base vectors

$$\mathbf{A}(r,\theta,\phi) = A_r(r,\theta,\phi)\mathbf{e}_r + A_\theta(r,\theta,\phi)\mathbf{e}_\theta + A_\phi(r,\theta,\phi)\mathbf{e}_\phi \quad (4.21)$$

The scalar functions A_r, A_θ and A_ϕ are the **spherical polar components** of the field obtained by projecting **A** onto the base vectors

$$A_r = \mathbf{e}_r \cdot \mathbf{A}, \quad A_\theta = \mathbf{e}_\theta \cdot \mathbf{A}, \quad A_\phi = \mathbf{e}_\phi \cdot \mathbf{A} \quad (4.22)$$

As an example of a vector field expressed in a spherical polar system we consider the radially-directed *isotropic* inverse square law field given in coordinate-free form by Eq (4.6a). This field is spherically symmetric since it depends on $r = (x^2 + y^2 + z^2)^{1/2}$ and $\mathbf{e}_r$ only, and has the spherical polar form

The term **isotropic** means having the same properties in all directions. Thus an isotropic field varies with r only and is independent of θ and ϕ.

$$\mathbf{F}(r,\theta,\phi) = \frac{C\mathbf{e}_r}{r^2} \qquad (r \neq 0) \quad (4.23)$$

Summary of section 4.4

- The **cylindrical polar coordinates** of a point are (ρ,ϕ,z) and are shown in Fig 4.20; the coordinate ϕ is called the **azimuthal angle**. They are related to the cartesian coordinates by

$$x = \rho\cos\phi, \quad y = \rho\sin\phi, \quad z = z; \quad \rho = (x^2 + y^2)^{1/2}, \quad \tan\phi = \frac{y}{x}$$

  The **cylindrical polar base vectors** are the mutually orthogonal unit vectors $\mathbf{e}_\rho$, $\mathbf{e}_\phi$, $\mathbf{e}_z$, ($\mathbf{e}_\rho = \mathbf{e}_\phi \times \mathbf{e}_z$) shown in Fig 4.22a and specified by Eq (4.16). **Plane polar coordinates** (ρ,ϕ) are used to specify points in the x-y plane.

- The **spherical polar coordinates** of a point are (r,θ,ϕ), as shown in Fig 4.23; ϕ is the azimuthal angle and the coordinate θ is called the **polar angle**. They are related to the cartesian coordinates by

$$x = r\sin\theta\cos\phi,\quad y = r\sin\theta\sin\phi,\quad z = r\cos\theta;\quad r = (x^2 + y^2 + z^2)^{1/2}$$

- The **spherical polar base vectors** are the mutually orthogonal unit vectors $\mathbf{e}_r$, $\mathbf{e}_\theta$, $\mathbf{e}_\phi$, ($\mathbf{e}_r = \mathbf{e}_\theta \times \mathbf{e}_\phi$) shown in Fig 4.23 and specified by Eqs (4.20).

Example 4.1 (*Objective 7*)

(a) Specify the azimuthal angle ϕ for

(i) any point on the positive x-axis,
(ii) any point on the negative x-axis,
(iii) any point on the positive y-axis,
(iv) the point with cartesian coordinates (–1,–1,0).

(b) What is the polar angle θ of the points referred to in part (a)? What is the polar angle for any point on the positive z-axis and for any point on the negative z-axis?

Solution 4.1

(a) In cylindrical and spherical polar coordinate systems ϕ is the angular displacement about the z-axis measured anticlockwise from the positive x-axis. Thus (i) points on the positive x-axis have $\phi = 0$, (ii) points on the negative x-axis have $\phi = \pi$ (= 180°), (iii) any point on the positive y-axis has $\phi = \pi/2$, and (iv) the cartesian point (–1,–1,0) has $\phi = \pi + \pi/4 = 5\pi/4$.

(b) All the points referred to in part (a) are in the x-y plane and so the polar angle is $\theta = \pi/2$ for all of them. The polar angle θ is 0 and π respectively for points on the positive and negative z-axes.

Example 4.2 (*Objective 7*)

(a) Make sketches to show the positions of the two points P and Q whose cylindrical polar coordinates are, respectively, $(2, 3\pi/4, 3)$ and $(3,\pi,-1)$, and give the cartesian coordinates of these points.

(b) Repeat part (a) for two points R and S whose spherical polar coordinates are, respectively, $(\sqrt{3},\pi/4, \pi/2)$ and $(1,\pi,0)$.

Solution 4.2

(a) Cylindrical polar coordinates (ρ,ϕ,z) are illustrated in Fig 4.20. The coordinates $(2, 3\pi/4, 3)$ identify the point P as being 2 units perpendicularly out from the z-axis ($\rho = 2$), 3 units up from the x-y plane ($z = 3$), and rotated from the positive x-axis by $3\pi/4$ ($\phi = 3\pi/4$), as shown in Fig 4.24a. The cartesian coordinates of P are (Eqs (4.13)):

$$x = 2\cos(3\pi/4) = -\sqrt{2} = -1.41,$$

$$y = 2\sin(3\pi/4) = \sqrt{2} = 1.41, \quad z = 3.$$

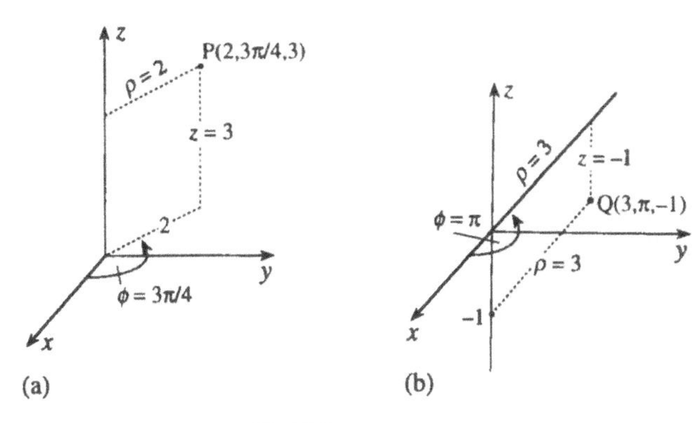

Fig 4.24
Cylindrical polar coordinates of the points P and Q of Example 4.2a.

The point Q is 3 units out from the point $z = -1$ on the negative z-axis, with the azimuthal angle $\phi = \pi$ placing it directly below the negative x-axis, as shown in Fig 4.24b. The cartesian coordinates of Q are:

$$x = 3\cos\pi = -3, \quad y = 3\sin\pi = 0, \quad z = -1$$

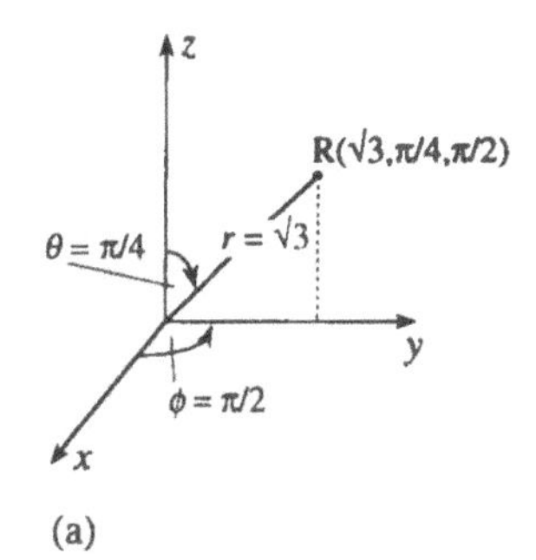

(b) Spherical polar coordinates (r,θ,ϕ) are illustrated in Fig 4.23. The point R is at a distance $r = \sqrt{3}$ from the origin, has a polar angle of $\pi/4$ and azimuthal angle $\pi/2$, as shown in Fig 4.25a. The cartesian coordinates of R (see Eq (4.18)) are:

$$x = \sqrt{3}\sin(\pi/4)\cos(\pi/2) = 0,$$

$$y = \sqrt{3}\sin(\pi/4)\sin(\pi/2) = \sqrt{3} \times \frac{1}{\sqrt{2}} = 1.22,$$

$$z = \sqrt{3}\cos(\pi/4) = 1.22$$

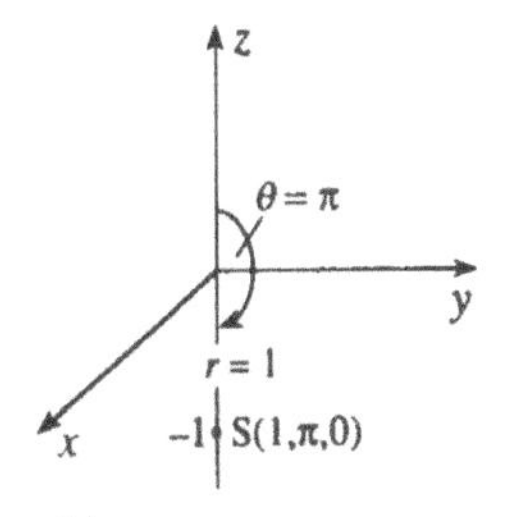

Fig 4.25
Spherical polar coordinates of the points R and S of Example 4.2b.

The point S is at a distance $r = 1$ from the origin and has polar angle π, which places it on the negative z-axis at $z = -1$, as shown in Fig 4.25b. Note that the azimuthal angle for points on the z-axis is undefined and is usually taken to be $\phi = 0$. The cartesian coordinates of S are $(x,y,z) = (0,0,-1)$.

Example 4.3 (Objective 7) Describe the location and shape of each of the following six coordinate surfaces:

(a) $\phi = \pi$; $\rho = 4$; $(\rho,\phi,7)$

(b) $r = 3$; $\theta = \pi/4$; $(r,\theta,0)$.

Solution 4.3

(a) $\phi = \pi$ specifies the x-z half plane with $x < 0$; $\rho = 4$ is the cylindrical surface of radius 4 with its axis on the z-axis; $(\rho,\phi,7)$ is the set of points with $z = 7$, i.e. the plane 7 units above the x-y plane.

(b) $r = 3$ describes the spherical surface of radius 3 centred on the origin; $\theta = \pi/4$ is a conical surface with apex on the origin and axis lying on the positive z axis; $(r,\theta,0)$ is the x-z half plane with $x \geq 0$.

Example 4.4 (*Objective 8*) Specify, where possible, the cylindrical polar base vectors in terms of the cartesian unit vectors at

(a) points on the positive x-axis,
(b) points on the positive y-axis,
(c) points on the negative x-axis,
(d) points on the negative y-axis,
(e) a point on the z-axis,
(f) the point with cylindrical polar coordinates $(1,\pi/2,0)$.

Solution 4.4 We have $\mathbf{e}_z = \mathbf{k}$ everywhere. To find $\mathbf{e}_\rho$ and $\mathbf{e}_\phi$ we use Eqs (4.16).

(a) For any point on the positive x-axis, $\phi = 0$ and so $\mathbf{e}_\rho = \mathbf{i}$ and $\mathbf{e}_\phi = \mathbf{j}$.
(b) On the positive y-axis we have $\phi = \pi/2$ and so $\mathbf{e}_\rho = \mathbf{j}$, $\mathbf{e}_\phi = -\mathbf{i}$.
(c) On the negative x-axis $\phi = \pi$, $\mathbf{e}_\rho = -\mathbf{i}$ and $\mathbf{e}_\phi = -\mathbf{j}$.
(d) On the negative y-axis $\phi = 3\pi/2$, $\mathbf{e}_\rho = -\mathbf{j}$, and $\mathbf{e}_\phi = \mathbf{i}$.
(e) ϕ is not defined on the z-axis and so $\mathbf{e}_\rho$ and $\mathbf{e}_\phi$ are also not defined there.
(f) The same as in part (b) since this point is on the positive y-axis.

Example 4.5 (*Objectives 7,8*)

(a) A point in the x-y plane has cartesian coordinates $(-3,4)$. Determine the plane polar coordinates (ρ,ϕ) of the point and express the unit vectors $\mathbf{e}_\rho$ and $\mathbf{e}_\phi$ at this point in terms of the cartesian unit vectors $\mathbf{i}$ and $\mathbf{j}$.

(b) Express the cartesian vector $\mathbf{a} = (1,2)$ in terms of vectors $\mathbf{e}_\rho$ and $\mathbf{e}_\phi$.

Solution 4.5

(a) Using Eqs (4.14) we find $\rho = ((-3)^2 + 4^2)^{1/2} = 5$, and ϕ is given by $\sin\phi = 4/5$ and $\cos\phi = -3/5$ from which $\phi = 2.214$ rad. Hence the plane polar coordinates of the point are (5, 2.214). Now use Eqs (4.16) for the unit vectors at this point and we find:

$$\mathbf{e}_\rho = \frac{(-3\mathbf{i}+4\mathbf{j})}{5}, \quad \mathbf{e}_\phi = \frac{-(4\mathbf{i}+3\mathbf{j})}{5}$$

(b) The given cartesian vector is

$$\mathbf{a} = (1,2) = \mathbf{i} + 2\mathbf{j}$$

and we are asked to express this as

$$\mathbf{a} = a_\rho \mathbf{e}_\rho + a_\phi \mathbf{e}_\phi$$

The plane polar component a_ρ is the projection of $\mathbf{a}$ onto $\mathbf{e}_\rho$, and similarly for a_ϕ (see Eq 4.17). Thus $a_\rho = \mathbf{e}_\rho \cdot \mathbf{a} = (-3\mathbf{i} + 4\mathbf{j}) \cdot (\mathbf{i} + 2\mathbf{j})/5 = 1$ and $a_\phi = \mathbf{e}_\phi \cdot \mathbf{a} = -(4\mathbf{i} + 3\mathbf{j}) \cdot (\mathbf{i} + 2\mathbf{j})/5 = -2$, and so we have $\mathbf{a} = \mathbf{e}_\rho - 2\mathbf{e}_\phi$.

Example 4.6 (*Objective 8*)

(a) Express the cartesian unit vector **i** as a linear combination of the spherical polar base vectors.

(b) Confirm that $\mathbf{e}_\theta$ is a unit vector and show that it is orthogonal to $\mathbf{e}_\phi$, where $\mathbf{e}_\theta$ and $\mathbf{e}_\phi$ are defined by Eqs (4.20).

Solution 4.6

(a) We put $\mathbf{i} = \mathbf{A}$ in Eqs (4.21) and (4.22). Then

$$\mathbf{i} = (\mathbf{e}_r \cdot \mathbf{i})\mathbf{e}_r + (\mathbf{e}_\theta \cdot \mathbf{i})\mathbf{e}_\theta + (\mathbf{e}_\phi \cdot \mathbf{i})\mathbf{e}_\phi$$

where the components $\mathbf{e}_r \cdot \mathbf{i}$ etc. are found by using Eqs (4.20). Thus

$$\mathbf{i} = (\sin\theta \cos\phi)\mathbf{e}_r + (\cos\theta \cos\phi)\mathbf{e}_\theta - (\sin\phi)\mathbf{e}_\phi$$

(b)
$$|\mathbf{e}_\theta|^2 = (\mathbf{e}_\theta \cdot \mathbf{e}_\theta)$$
$$= \cos^2\theta \cos^2\phi + \cos^2\theta \sin^2\phi + (-\sin\theta)^2$$
$$= (\cos^2\phi + \sin^2\phi)\cos^2\theta + \sin^2\theta$$

$$= \cos^2\theta + \sin^2\theta = 1$$

Hence $\mathbf{e}_\theta$ is a unit vector. To show that it is orthogonal to $\mathbf{e}_\phi$ we calculate $\mathbf{e}_\phi \cdot \mathbf{e}_\theta = -\sin\phi\cos\theta\cos\phi + \cos\phi\cos\theta\sin\phi = 0$. Therefore $\mathbf{e}_\theta$ and $\mathbf{e}_\phi$ are orthogonal.

Example 4.7 (*Objective 9*) A three-dimensional scalar field is defined by

$$f(\mathbf{r}) = \frac{C\mathbf{p}\cdot\mathbf{r}}{|\mathbf{r}|^3} \qquad (|\mathbf{r}| \neq 0) \tag{4.24}$$

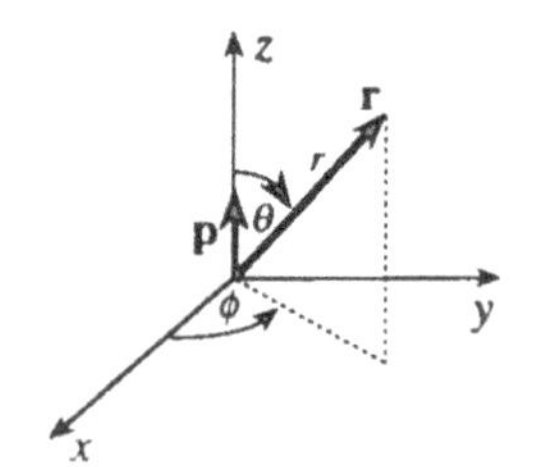

Fig 4.26
The spherical polar coordinate system for Example 4.7.

where C is a constant scalar, $\mathbf{p}$ is a constant vector and $\mathbf{r}$ is the position vector. Choose a spherical polar coordinate system with origin at $\mathbf{r} = \mathbf{0}$ and the z-axis aligned along the direction of the vector $\mathbf{p}$. See Fig 4.26. Give the spherical polar representation of the field and find the field value at the point with spherical polar coordinates $(6,\pi,0)$. (With $C = 1/4\pi\epsilon_0$, the field f gives the electrostatic potential field at large distances from a small electric dipole of dipole moment $\mathbf{p}$ at $\mathbf{r} = \mathbf{0}$.)

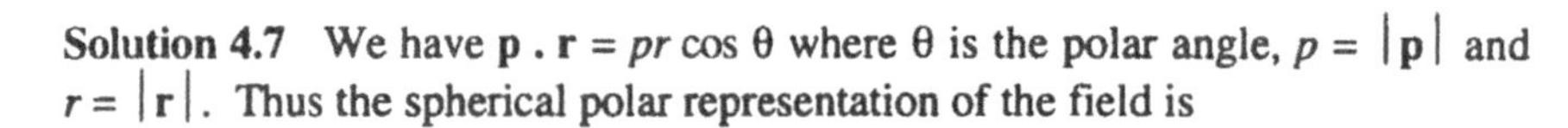

Solution 4.7 We have $\mathbf{p}\cdot\mathbf{r} = pr\cos\theta$ where θ is the polar angle, $p = |\mathbf{p}|$ and $r = |\mathbf{r}|$. Thus the spherical polar representation of the field is

$$f(r,\theta,\phi) = \frac{C(pr\cos\theta)}{r^3} = \frac{Cp\cos\theta}{r^2} \qquad (r \neq 0)$$

Hence $f(6,\pi,0) = \dfrac{Cp\cos\pi}{6^2} = -\dfrac{Cp}{36}$.

Problem 4.1 (*Objective 7*) The three entries of each row of Table 4.1 should specify the coordinates of the same point in the coordinate system indicated by the column headings. Fill in the blank entries.

Table 4.1

cartesian	cylindrical polar	spherical polar
(0,1,0)	1,π/2,0)	(1,π/2,π/2)
(1,0,0)		
(0,0,1)		
		$(\sqrt{2},\pi/4,\pi/2)$
	$(\sqrt{2},0,-1)$	
		(1,π,0)

Problem 4.2 (*Objective 8*)

(a) Express the cartesian unit vector **j** as a linear combination of the cylindrical polar base vectors at an arbitrary point.

(b) Refer to Eqs (4.20). Show that the spherical polar unit vector $\mathbf{e}_r$ has unit modulus and show that $\mathbf{e}_r$ and $\mathbf{e}_\theta$ are orthogonal.

Problem 4.3 (*Objective 9*) The density of a spherically symmetric planet of radius a is modelled in spherical polar coordinates by the function

$$\rho(r,\theta,\phi) = \left[5 + \frac{2(a-r)}{a}\right] \times 10^3 \text{ kg m}^{-3} \qquad (r \leq a)$$

What is the density of the planet at

(i) its centre,
(ii) its surface, and
(iii) the point with spherical polar coordinates $(a/3, \pi/6, \pi)$?

Problem 4.4 (*Objective 9*) Consider the vector field

$$\mathbf{F} = \mathbf{a} \times \frac{\mathbf{r}}{|\mathbf{r}|^3} \qquad (|\mathbf{r}| \neq 0) \tag{4.25}$$

where **a** is a constant vector and **r** is the position vector. Express this field in cartesian form and in cylindrical polar form. (Choose a direction of the z axis that will lead to a simple expression of the function.)

Problem 4.5 (*Objective 9*) A linear **electric quadrupole** consists of two equal electric dipoles of the same magnitude but opposite directions placed end to end on a line. When the quadrupole is placed at the origin and aligned along the z-axis (Fig 4.27) the spherical polar form of the electrostatic potential due to the quadrupole is given (for very large distances compared with the size of the quadrupole) by the function

$$V(r,\theta,\phi) = \frac{M}{4\pi\epsilon_0 r^3}(3\cos^2\theta - 1) \qquad (r \neq 0) \tag{4.26}$$

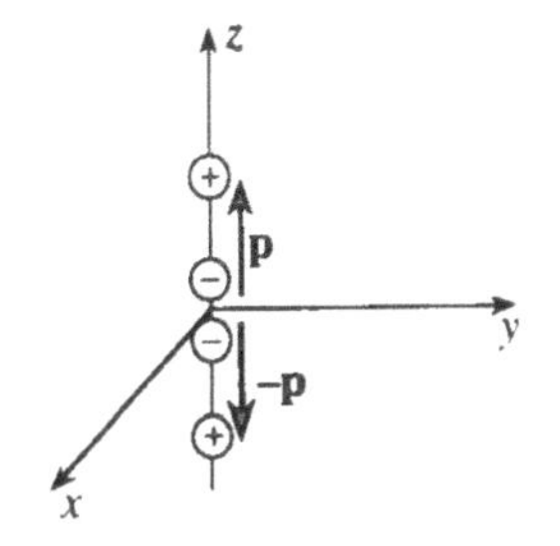

Fig 4.27
A linear electric quadrupole aligned along the z-axis.

where M and ϵ_0 are constant scalars (M is called the quadrupole moment). Determine the value of V at the following points:

(a) a point on the z-axis at a distance d from the origin,
(b) a point in the x-y plane at a distance d from the origin,
(c) the point with spherical polar coordinates $(d, \pi/4, 0.1\pi)$,
(d) the point with spherical polar coordinates (d,α,β) where $\cos\alpha = 1/\sqrt{3}$,

(e) the point with cartesian coordinates $(s,s,0)$,
(f) a point "at infinity".

Problem 4.6 (*Objective 9*) The electrostatic field produced by an electric dipole of dipole moment **p** is given (for distances that are large compared to the size of the dipole) by

$$\mathbf{E}(\mathbf{r}) = C\left[\frac{3(\mathbf{p}\cdot\mathbf{r})\mathbf{r}}{|\mathbf{r}|^5} - \frac{\mathbf{p}}{|\mathbf{r}|^3}\right] \tag{4.27}$$

where **r** is the position vector from the dipole and $C = 1/4\pi\epsilon_0$.

(a) Show that the spherical polar form of this field is

$$\mathbf{E}(r,\theta,\phi) = \frac{Cp}{r^3}(2\mathbf{e}_r\cos\theta + \mathbf{e}_\theta\sin\theta) \tag{4.28}$$

where the dipole is at the origin with **p** pointing in the z-direction (Fig 4.26).

(b) Find the ratio of the magnitudes of the field vectors at two points equidistant from the dipole, one on the z-axis and one in the x-y plane.

(c) Show that the angle α between the field vector and the radial direction $\mathbf{e}_r$ at any point is given by $\tan\alpha = (1/2)\tan\theta$.

(d) Make a rough sketch of the field lines in any plane containing the dipole.

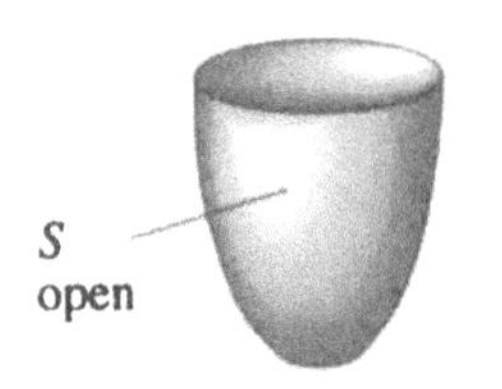

(a)

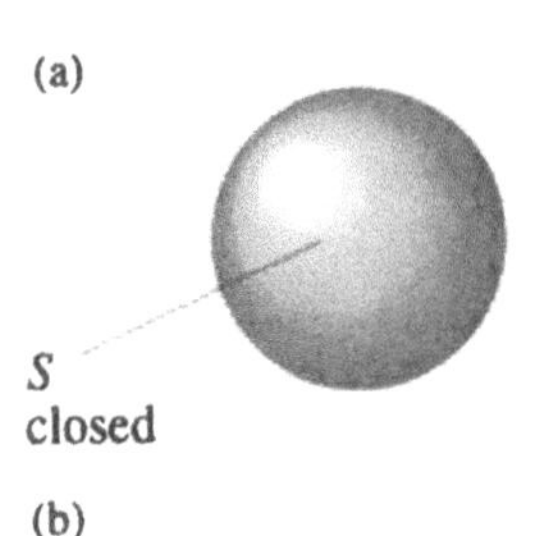

(b)

Fig 4.28
(a) An open surface (such as an egg cup) and (b) a closed surface (such as a soap bubble).

4.5 INTRODUCING FLUX AND CIRCULATION

Sections 4.1 to 4.4 of this chapter have been concerned with pictorial and mathematical descriptions of fields. We now turn our attention to some of the properties of vector fields and introduce two quantities that lie at the heart of vector calculus: the *flux* of a vector field and the *circulation* of a vector field. These quantities embody physically important features of a vector field and play an important role in the expression of many physical laws. Flux and circulation are in fact certain integrals of a vector field and will be introduced as such in Chapter 6. We introduce them here in simple contexts where it is not necessary to consider integrals.

4.5.1 Flux of a vector field

Consider a surface labelled S of area A in the domain of a three-dimensional vector field **F**. The surface may be an **open surface** or a **closed surface** (Fig 4.28). We arbitrarily choose one side of the surface to be the "positive" side, the other side being the "negative" side. In the case of a closed surface we usually choose the outside to be the positive side. Let **n** be a unit vector perpendicular to the surface S at a point P on the surface, and let **n** point from the negative side to the positive side (Fig 4.29). **n** is called the **unit surface normal** at P. Now let **F** be the field vector at P, and θ the angle between **n** and **F**. Then the scalar $\mathbf{n} \cdot \mathbf{F} = F\cos\theta = F_n$ is the component of **F** normal to the surface at P. We now restrict our attention to cases where **n . F** is the same everywhere on the surface, i.e. **n . F** is constant on S. Then the **flux** of **F** across the surface S of area A is defined to be

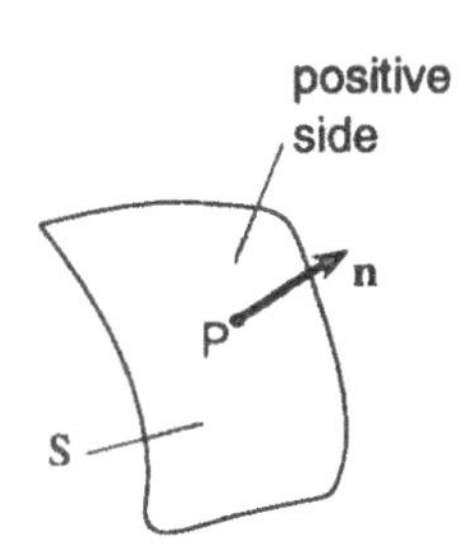

Fig 4.29
The unit surface normal n at a point P on a surface.

$$\Phi = \mathbf{n} \cdot \mathbf{F}A = F_n A = FA\cos\theta \qquad (\mathbf{n} \cdot \mathbf{F} \text{ is constant on S}) \qquad (4.29)$$

Note that flux is a scalar quantity. A more general definition of flux applicable to cases where **n . F** varies across the surface will be given in Chapter 6.

The condition (**n . F** is constant on S) is satisfied in a number of important cases, for example: when **F** is a constant vector field and S is a plane surface (Fig 4.30a); when **F** is an isotropic radial field (i.e. a spherically symmetric vector field) and S is the surface of a sphere (Fig 4.30b), and when **F** is cylindrically symmetric and S is the curved surface of a cylinder (Fig 4.30c). The condition is also approximately satisfied whenever the dimensions of S are very small compared with the distances over which **n . F** varies significantly.

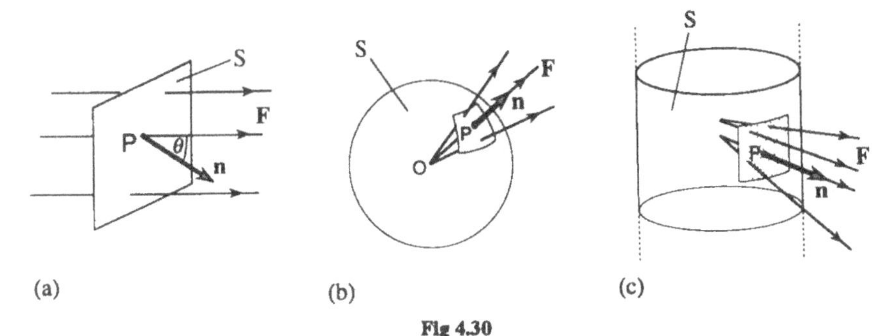

Fig 4.30
Cases where **n . F** is constant over a surface. (a) S is a plane surface and **F** is a constant vector field; $\mathbf{n}.\mathbf{F} = |\mathbf{F}|\cos\theta$. (b) S is the surface of a sphere and **F** is an isotropic radial field; $\mathbf{n} \cdot \mathbf{F} = |\mathbf{F}|$, a constant on S. (c) S is the curved surface of a cylinder and the cylindrically symmetric field **F** is normal to it, $\mathbf{n} \cdot \mathbf{F} = |\mathbf{F}|$, a constant on S.

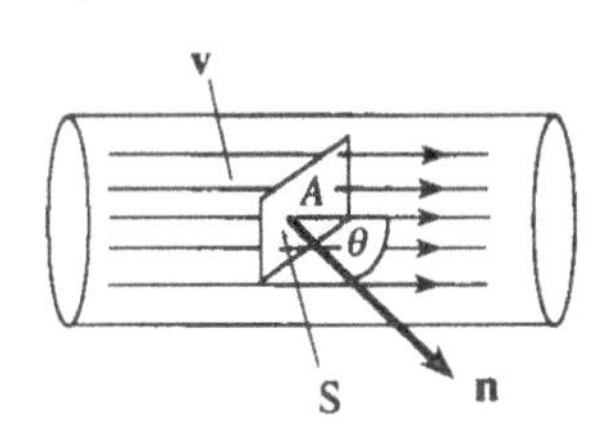

Fig 4.31
The surface normal on the plane surface of area A is inclined at an angle θ to the velocity flow lines; flux $\Phi = \mathbf{n} \cdot \mathbf{v}A = |\mathbf{v}|A\cos\theta$.

In physical terms, flux can often be thought of as a "flow" across a surface. In the case of a three-dimensional velocity field **v** describing the flow of water through a pipe (Fig 4.31) the flux of **v** across a surface S is the volume of water flowing across S per second. This flow rate obviously depends on the orientation of the surface relative to the direction of **v**. When a plane surface is oriented at right-angles to the direction of flow, with **n** and **v** parallel, the flux has its maximum value of $\mathbf{n} \cdot \mathbf{v}A = vA$; when the surface lies along the direction of flow, **n** and **v** are perpendicular and the flux is zero.

The flux of a velocity field, $\Phi = \mathbf{n} \cdot \mathbf{v}A$, has units of velocity times area, i.e. $(\mathrm{ms}^{-1}) \times \mathrm{m}^2 = \mathrm{m}^3\mathrm{s}^{-1}$, which is a volume flow rate. The unit vector **n** does not carry physical units.

The flux of a vector field across a surface cannot always be interpreted literally as a flow rate. The flux of a magnetic field across a surface does not describe a physical flow of any kind, although we may speak figuratively of the flow of the field across the surface.

By the total area of a closed surface we mean the total area of one side. Both sides, the inside and the outside, have the same total area.

Of special interest is the net **outward flux** of a vector field across a closed surface such as the surface of a sphere or the closed surface formed by the six faces of a box. The net outward flux is $\Phi_o = \mathbf{n} \cdot \mathbf{F}A$ where A is the total area of the closed surface and $\mathbf{n}$ is the outward directed normal. Net outward flux plays an important role in many physical laws, such as Gauss's law of electrostatics (Example 5.3).

It is sometimes useful to think of flux in terms of field lines. Suppose we have a field line diagram in which the magnitude $|\mathbf{F}|$ of the field is proportional to the density of field lines, as described in Section 4.1.2. Then $|\mathbf{F}| \propto N/A_\perp = N/(A\cos\theta)$ where N is the number of field lines crossing a plane surface S of area A, and $A_\perp = A\cos\theta$ is the area of the projection of S onto a plane at right angles to the field (Fig 4.32). Thus we have $N \propto |\mathbf{F}|A\cos\theta = \Phi$, i.e. the flux Φ across any surface S is simply proportional to the number of field lines crossing it, and is positive when the lines cross from the negative side to the positive side. The net outward flux Φ_o across a closed surface is therefore proportional to the number of field lines crossing the surface from the inside to the outside minus the number crossing from the outside to the inside. If the field lines are continuous in the enclosed region, i.e. have no beginning points or terminating points, then any line that enters through one point P on the surface must leave through another point Q (Fig 4.33) and so the net outward flux across the surface is zero. Thus the continuity of field lines is the pictorial expression of the net outward flux across a closed surface being zero. Any vector field $\mathbf{F}$ in which $\Phi_o = 0$ across all possible closed surfaces is said to be **solenoidal**. All magnetic fields, for example, are solenoidal.

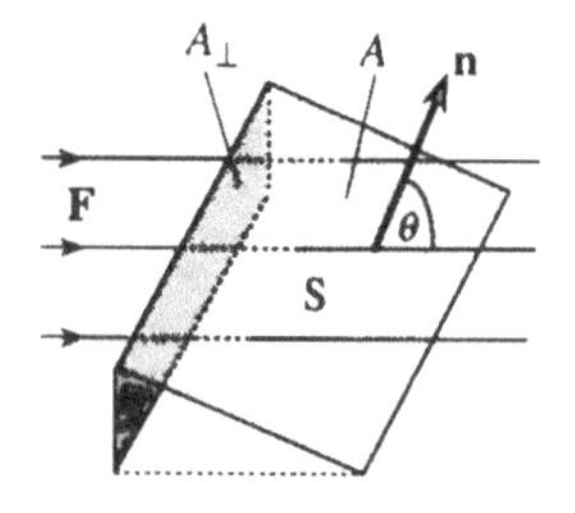

Fig 4.32
The surface S has unit surface normal n and area A. The projection of S onto a plane normal to F has area $A_\perp = A\cos\theta$ where θ is the angle between n and F.

An example of a non-solenoidal vector field is $\mathbf{G}(x,y,z) = x\mathbf{i}$ $(x > 0)$. We show in Example 5.4 that the net outward flux of this field across a closed cube of side length a is $\Phi_o = a^3$, which is positive. Thus more field lines of $\mathbf{G}$ leave the cube than enter it. Field lines of $\mathbf{G}$ are shown in Fig 4.34a. As the magnitude x of the field $\mathbf{G}$ increases from left to right we introduce more field lines to show a closer packing (larger density). The field lines are not continuous; there are beginning points, or **sources**, from which new field lines emerge.

If the net outward flux of a vector field across a closed surface is negative then more field lines enter the surface than leave it. This is illustrated in Fig 4.34b for the vector field $\mathbf{F} = (c - x)\mathbf{i}$ $(x < c)$. This field has **sinks** where field lines terminate.

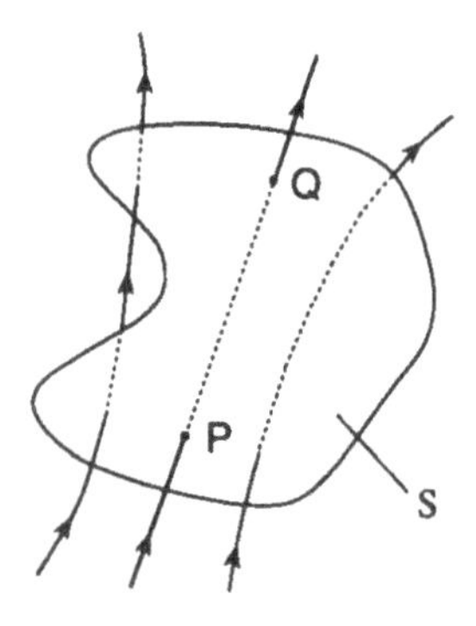

Fig 4.33
Continuous field lines across a closed surface. A field line enters the enclosed volume at P and leaves at Q.

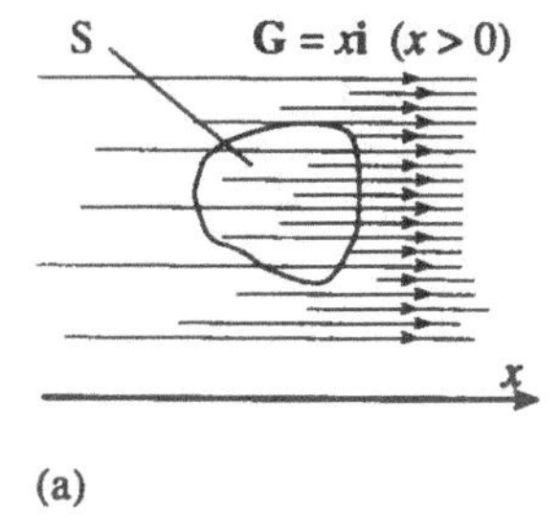

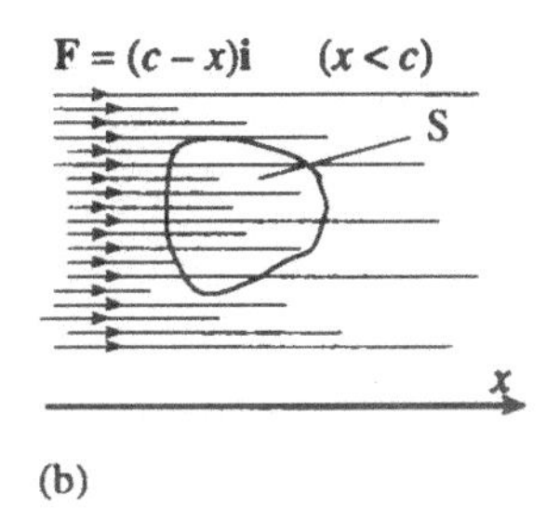

Fig 4.34

(a) Field lines representing the magnitude and direction of $\mathbf{G} = x\mathbf{i}$ $(x > 0)$ are not continuous. There are beginning points (sources) inside any closed surface S, and the net outwards flux is positive.

(b) Field lines representing the magnitude and direction of $\mathbf{F} = (c - x)\mathbf{i}$ $(x < c)$. The magnitude of $\mathbf{F}$ decreases progressively with x and so field lines terminate inside any closed surface S and the net outward flux is negative.

4.5.2 Circulation of a vector field

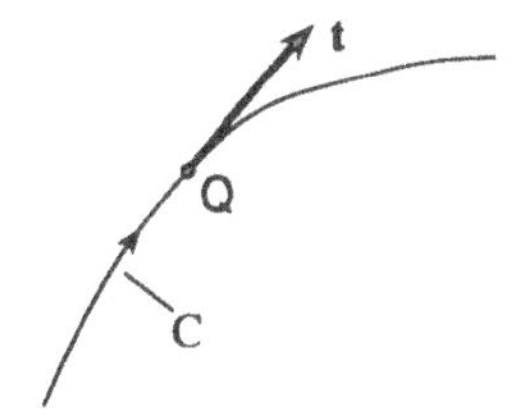

Fig 4.35
A unit tangent vector **t** at a point Q on a curve C.

Consider a curve C of length L in the domain of a vector field $\mathbf{F}(\mathbf{r})$. Choose a positive direction along the curve, as indicated by the arrow on curve C in Fig 4.35. Draw a tangent to the curve at Q and let **t** be the unit vector at Q lying on the tangent line and pointing in the positive direction. **t** is called the **unit tangent vector** at Q and $\mathbf{t} \cdot \mathbf{F} = F_t$ is the tangential component of **F** at Q. We now restrict our attention to cases where $\mathbf{t} \cdot \mathbf{F}$ is the same everywhere on the curve C, i.e. $\mathbf{t} \cdot \mathbf{F}$ is constant on C. Consider the scalar quantity

$$W = \mathbf{t} \cdot \mathbf{F}L = F_t L = FL\cos\theta \qquad (\mathbf{t} \cdot \mathbf{F} \text{ is constant on C}) \qquad (4.30)$$

where θ is the angle between **F** and **t** at Q. If $\mathbf{F}(\mathbf{r})$ represents the force on a particle at **r**, then W is the work done by the field when the particle moves in the positive direction along the length L of the curve C. The condition ($\mathbf{t} \cdot \mathbf{F}$ is constant on C) is satisfied in many important cases the simplest of which is when **F** is a constant vector field and C is a straight line. Other cases are considered in the Examples and Problems.

When the curve is a **closed curve** or **loop**, the value of W given by Eq (4.30) is called the **circulation** of **F** around the loop, which we shall denote by W_o. The positive direction around a loop is usually taken by convention to be the anticlockwise direction, i.e. the enclosed area is on the left as the loop is traversed. If the circulation of **F** around every possible loop in the domain of the field **F** is zero, the field is said to be a **conservative** field. Conservative fields are discussed further in Chapters 5 and 6.

As an example of the use of Eq (4.30) we calculate the anticlockwise circulation of the two-dimensional vector field

$$\mathbf{u}(x,y) = y\mathbf{i} \qquad (y > 0)$$

around the square loop ABCD with sides of length 2 shown in Fig 4.36. The field **u** is everywhere directed parallel to the x-axis and is therefore normal to the sides BC and DA. The tangential component of **u** on these lines is therefore zero and the only contribution to the circulation comes from the sides AB and CD which are segments of the lines $y = 3$ and $y = 5$ respectively. The field vector at any point on the side AB is $\mathbf{u} = 3\mathbf{i}$ and the unit tangent vector there is $\mathbf{t} = \mathbf{i}$, while the field vector on CD is $\mathbf{u} = 5\mathbf{i}$ and the unit tangent vector is $\mathbf{t} = -\mathbf{i}$. The anticlockwise circulation, found by using Eq (4.30) on each of these two sides, is therefore

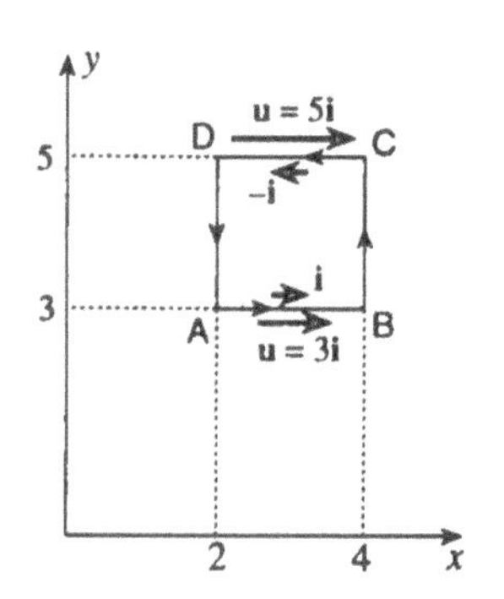

Fig 4.36
A square loop of side length 2 in the x-y plane.

$$W_0 = (\mathbf{i} \,.\, 3\mathbf{i}) \times 2 + (-\mathbf{i} \,.\, 5\mathbf{i}) \times 2 = -4$$

We can conclude from this that the vector field **u** is not a conservative field. Note however that if we find that the circulation of a vector field around a particular loop is zero we cannot conclude from this alone that the field is conservative since there may be other loops in the domain of the field on which the circulation is nonzero (Problem 5.5).

Flux and circulation are the key ideas behind two concepts "divergence" and "curl" which are introduced in Chapters 5 and 6 where they play a central role in the calculus of vector fields.

Summary of section 4.5

- The flux Φ of a vector field $\mathbf{F}(\mathbf{r})$ across a surface S of area A is

$$\Phi = \mathbf{n} \,.\, \mathbf{F}A = F_n A = FA\cos\theta \qquad \text{(for } \mathbf{n} \,.\, \mathbf{F} \text{ constant on S)} \qquad (4.29)$$

where **F** is the field vector at a point on the surface where the **unit surface normal** is **n**. For a **closed surface** of area A, **n** is the outward directed normal and $\Phi_0 = \mathbf{n} \,.\, \mathbf{F}A$ is the net outward flux.

- The **circulation** of a vector field around a **closed curve** or **loop** C of length L is

$$W_0 = \mathbf{t} \,.\, \mathbf{F}\,L = F_t L = FL\cos\theta \qquad \text{(for } \mathbf{t} \,.\, \mathbf{F} \text{ constant on C)}$$

where **F** is the field vector at a point on the curve where the **unit tangent vector** is **t**.

Example 5.1 (*Objective 10*) Find the flux of the vector field

$$\mathbf{h}(x,y,z) = xy(\mathbf{i} + \mathbf{j}) + (z + 5)\mathbf{k}$$

across a surface of unit area of the x-y plane.

Solution 5.1 The surface S is in the x-y plane ($z = 0$) and so the field vectors on S are

$$\mathbf{h}(x,y,0) = xy(\mathbf{i} + \mathbf{j}) + 5\mathbf{k}$$

A unit surface normal is $\mathbf{n} = \mathbf{k}$ or $\mathbf{n} = -\mathbf{k}$. Choosing the top of the x-y plane to be the positive side we have $\mathbf{n} = \mathbf{k}$. The normal component of $\mathbf{h}(x,y,0)$ is then $\mathbf{k} \cdot \mathbf{h}(x,y,0) = 5$. This is the same everywhere on the surface and so we can use Eq (4.29). Thus the flux Φ of $\mathbf{h}$ across a surface S of unit area ($A = 1$) is $\Phi = \mathbf{k} \cdot \mathbf{h} \times 1 = 5$.

Example 5.2 (*Objective 10*) A garden shed has a vee-shaped roof of total area $12\ \mathrm{m}^2$ inclined at 20° to the horizontal (Fig 4.37). Find the flux of the constant vector field $\mathbf{F} = \mathbf{g}$ across the roof, where $\mathbf{g}$ is directed vertically downwards and is of magnitude $10\ \mathrm{m\,s}^{-2}$. Take the top face of the roof to be the positive side.

Determine the flux of $\mathbf{F}$ across the floor of the shed, taking the positive surface normal to be pointing downwards.

What is the net outward flux of $\mathbf{F}$ across the shed?

Solution 5.2 The unit normals $\mathbf{n}_1$ and $\mathbf{n}_2$ on the two halves of the roof are shown in Fig 4.37. Then using Eq (4.29) we have

$$\Phi = (\mathbf{n}_1 \cdot \mathbf{g}) \times 6\ \mathrm{m}^2 + (\mathbf{n}_2 \cdot \mathbf{g}) \times 6\ \mathrm{m}^2 = 10 \times 12 \cos(160°)\ \mathrm{m}^3\mathrm{s}^{-2} = -113\ \mathrm{m}^3\mathrm{s}^{-2}.$$

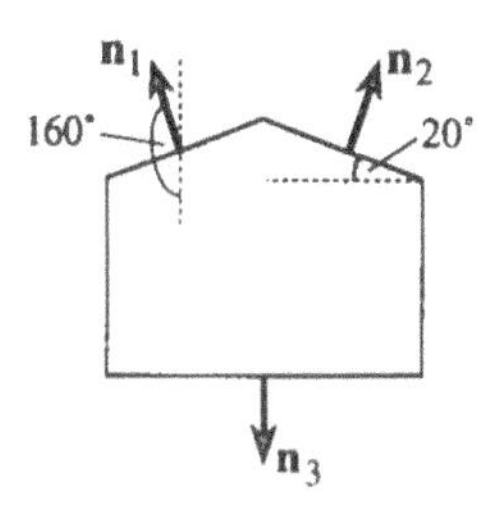

Fig 4.37
A garden shed and the outward unit surface normals.

Now consider the floor which has an area of $(12\ \mathrm{m}^2)\cos(20°) = 11.3\ \mathrm{m}^2$. With the unit surface normal $\mathbf{n}_3$ directed vertically downwards we find the flux across the floor to be $\Phi = \mathbf{n}_3 \cdot \mathbf{g} \times 11.3\ \mathrm{m}^2 = 10\ \mathrm{m\,s}^{-2} \times 11.3\ \mathrm{m}^2 = 113\ \mathrm{m}^3\mathrm{s}^{-2}$.

To find the net outward flux across the shed we take the unit surface normals to be the outward directed ones: $\mathbf{n}_1$ and $\mathbf{n}_2$ for the two halves of the roof and $\mathbf{n}_3$ for the floor (Fig 4.37). There is no flux of $\mathbf{g}$ across the vertical walls of the shed since the surface normals on the walls are perpendicular to $\mathbf{g}$. The net outward flux Φ_0 is therefore the sum of $-113\ \mathrm{m}^3\mathrm{s}^{-2}$ across the roof and $113\ \mathrm{m}^3\mathrm{s}^{-2}$ across the floor, which gives $\Phi_0 = 0$.

Example 5.3 (*Objective 10*) **Gauss's law of electrostatics** states that the net outward flux of an electrostatic field $\mathbf{E}$ across any closed surface is equal to Q/ϵ_0 where Q is the total electric charge in the region enclosed by the surface.

The stated form of Gauss's law assumes that there are no dielectric materials present.

Suppose an amount of electric charge Q is uniformly distributed in a spherical region of radius R. Use Gauss's law to determine the electrostatic field in the region (a) outside the sphere ($r > R$) and (b) inside the sphere ($r < R$).

Solution 5.3 Since the charge distribution is spherically symmetric we shall assume that the electrostatic field also has spherical symmetry, i.e. it is an isotropic radial field. Then we can write $\mathbf{E}(r,\theta,\phi) = E(r)\mathbf{e}_r$ where the scalar $E(r)$ is a function of r only. To find E we apply Gauss's law to an imaginary spherical surface of radius r_1 centred at the centre of symmetry. Then at any point on this

The subscript 1 on r_1 is just a reminder that r_1 is the radius of the imaginary spherical surface; it is optional and can be omitted.

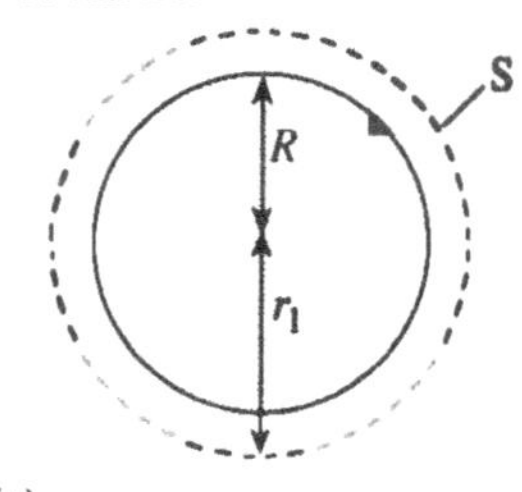

(a)

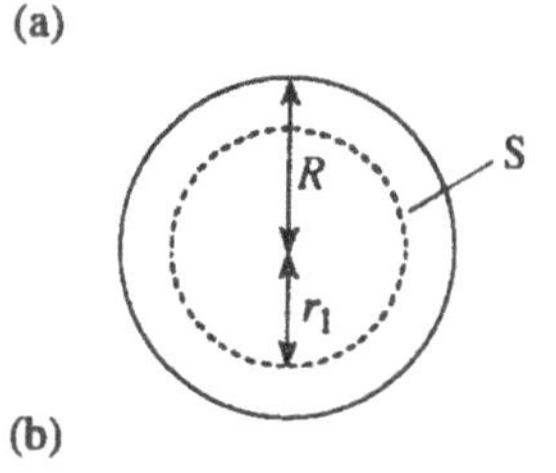

(b)

Fig 4.38
A spherical distribution of electric charge of radius R and imaginary "Gaussian" surfaces S of radius r_1. (a) $r_1 > R$ and (b) $r_1 < R$.

"Gaussian" surface, the field vector is $E(r_1)\mathbf{e}_r$ and the outward directed surface normal at the same point is $\mathbf{n} = \mathbf{e}_r$. The outward normal component of the field at this point is $\mathbf{e}_r \cdot E(r_1)\mathbf{e}_r = E(r_1)$. This is the same at all points on the "Gaussian" surface and so the net outward flux of the field across the surface is found from Eq (4.29) to be $\Phi_o = E(r_1)4\pi r_1^2$.

We now consider separately the two regions specified in the question:

(a) Outside the sphere ($r > R$).

Let the imaginary "Gaussian" surface have radius $r_1 > R$ (Fig 4.38a). The total charge enclosed by the surface is the entire charge Q and so Gauss's law yields $E(r_1)4\pi r_1^2 = Q/\epsilon_0$ which gives $E(r_1) = Q/4\pi\epsilon_0 r_1^2$. Hence we conclude that the electric field in the region outside the charge distribution is

$$\mathbf{E}(r,\theta,\phi) = E(r)\mathbf{e}_r = \frac{Q\,\mathbf{e}_r}{4\pi\epsilon_0 r^2} \qquad (r > R)$$

(b) Inside the sphere ($r < R$).

Let the imaginary "Gaussian" surface have radius $r_1 < R$ (Fig 4.38b). The total charge enclosed by this surface is $Q(4\pi r_1^3/3)/(4\pi R^3/3) = Q(r_1^3/R^3)$ since we are told that the charge is distributed uniformly. Hence Gauss's law yields $E(r_1)4\pi r_1^2 = Qr_1^3/\epsilon_0 R^3$ and we conclude that the electrostatic field inside the charge distribution is

$$\mathbf{E}(r,\theta,\phi) = E(r)\mathbf{e}_r = \frac{Q\,r\,\mathbf{e}_r}{4\pi\epsilon_0 R^3} \qquad (r < R)$$

Example 5.4 (*Objective 10*) Calculate the net outward flux of the vector field $\mathbf{G}(x,y,z) = x\mathbf{i}$ ($x > 0$) across the closed surface comprising the six faces of a cube of side length a (Fig 4.39). The cube is located with its centre in the plane $x = x_1$ ($x_1 > a/2$) and is aligned with two faces S_1 and S_2 perpendicular to the direction of the field and in the planes $x = x_1 + a/2$ and $x = x_1 - a/2$ respectively.

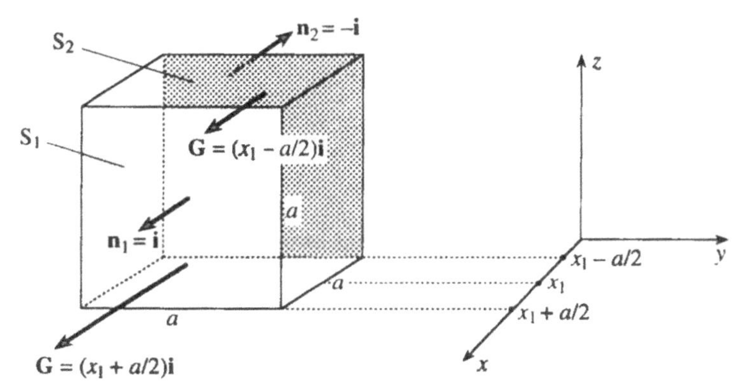

Fig 4.39
A cube of side a in the vector field $\mathbf{G} = x\mathbf{i}$.

Solution 5.4 The net outward flux is the sum of the outward fluxes across the six faces of the cube. The field vector **G** points in the x-direction everywhere and so there is no flux across any of the four faces that are parallel to the x-y plane or the x-z plane. The net outward flux is therefore the sum of the outward fluxes across the two faces labelled S_1 and S_2 in Fig 4.39. The field values are $\mathbf{G} = (x_1 + a/2)\mathbf{i}$ everywhere on S_1 and $\mathbf{G} = (x_1 - a/2)\mathbf{i}$ everywhere on S_2. The net outward flux is therefore

$$\Phi_o = \mathbf{n}_1 \cdot (x_1 + a/2)\mathbf{i}a^2 + \mathbf{n}_2 \cdot (x_1 - a/2)\mathbf{i}a^2$$

where $\mathbf{n}_1 = \mathbf{i}$ and $\mathbf{n}_2 = -\mathbf{i}$, and so

$$\Phi_o = (x_1 + a/2)a^2 - (x_1 - a/2)a^2 = a^3$$

Hence the net outward flux is a^3, the volume of the cube.

Example 5.5 (*Objective 11*) A constant two-dimensional force field acting on a particle is described by

$$\mathbf{F}(x,y) = C(2\mathbf{i} - \mathbf{j}) \qquad (C \text{ is a constant})$$

(a) Find the work done by the field when the particle moves along the x-axis from $x = 2$ to $x = 7$.

(b) Calculate the anticlockwise circulation of **F** around the rectangular loop (Fig 4.40) defined by the points A(2,0), B(7,0), C(7,3) and D(2,3).

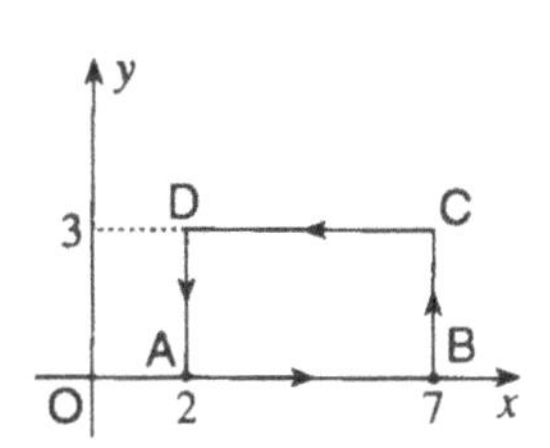

Fig 4.40
A rectangular loop ABCD.

Solution 5.5

(a) Use Eq (4.30) with $\mathbf{t} = \mathbf{i}$ and $L = 7 - 2 = 5$. Then

$$W = \mathbf{i} \cdot \mathbf{F} \times 5 = 2C \times 5 = 10C$$

(b) We use Eq (4.30) for each side of the rectangle and add the results:

$$W_{AB} = 10C, \text{ from part (a)}$$
$$W_{BC} = \mathbf{j} \cdot \mathbf{F}(3 - 0) = -3C, \text{ since } \mathbf{t} = \mathbf{j} \text{ along BC}$$
$$W_{CD} = -\mathbf{i} \cdot \mathbf{F} \times 5 = -10C$$
$$W_{DA} = -\mathbf{j} \cdot \mathbf{F} \times 3 = 3C$$

The circulation is therefore $W_o = 10C - 3C - 10C + 3C = 0$.

The stated form of Ampère's circuital law assumes that there are no magnetic materials present.

Example 5.6 (*Objective 11*) **Ampère's circuital law** for steady currents states that the circulation of the magnetic field vector **B** around any closed loop is equal to $\mu_0 I$ where I is the total electric current enclosed by the loop, the sense of the circulation being given by the right-hand rule with respect to the direction of the current.

A very long straight cylindrical wire of radius R with its axis lying on the z-axis carries a steady current I in the positive z direction. The current is distributed uniformly in the wire and produces a magnetic field of the form $\mathbf{B}(\rho,\phi,z) = B(\rho)\mathbf{e}_\phi$ where the scalar B is a function of ρ only. Use Ampère's law to find $B(\rho)$ in the region (a) outside the wire ($\rho > R$) and (b) inside the wire ($\rho < R$).

Solution 5.6 Choose a circular "Amperian" loop of radius ρ_1 in any plane parallel to the x-y plane and with its centre on the z-axis (Fig 4.41). To apply Ampere's law when the current is in the positive z direction we calculate the anticlockwise circulation of **B** around the loop as required by the right-hand rule. We have $\mathbf{t} = \mathbf{e}_\phi$ on the loop, and so $\mathbf{t} \cdot \mathbf{B} = B(\rho_1)$ which is the same everywhere on the loop. The circulation around the loop is then found from Eq (4.30) to be $W_o = B(\rho_1) \times 2\pi\rho_1$. Ampère's law requires us to equate this anticlockwise circulation of **B** to μ_0 times the total current enclosed by the loop. We now consider the two cases:

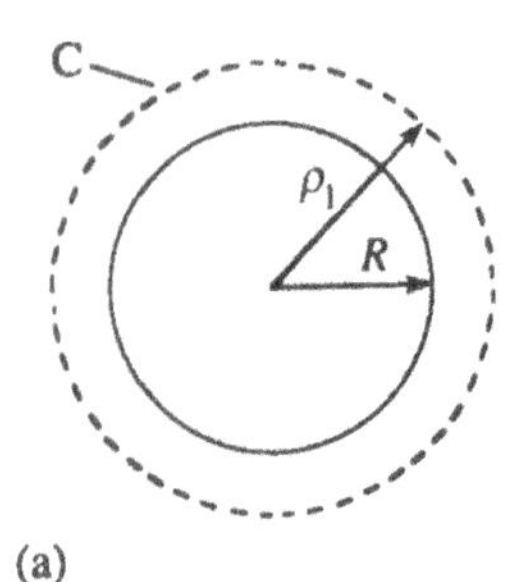

(a)

(a) Outside the wire ($\rho > R$).

The total current enclosed by an "Amperian" loop of radius $\rho_1 > R$ is the entire current I (Fig 4.41a) and so we have $B(\rho_1) \times 2\pi\rho_1 = \mu_0 I$ which gives $B(\rho_1) = \mu_0 I/2\pi\rho_1$. Thus the magnetic field outside the wire is given by

$$\mathbf{B} = \frac{\mu_0 I \mathbf{e}_\phi}{2\pi\rho} \qquad (\rho > R)$$

(b)

Fig 4.41
A section of the cylindrical current-carrying wire of radius R and an imaginary "Amperian" loop C of radius ρ_1. (a) $\rho_1 > R$ and (b) $\rho_1 < R$.

(b) Inside the wire ($\rho < R$).

The "Amperian" loop is shown in Fig 4.41b. The total enclosed current is $I(\pi\rho_1^2/\pi R^2)$ since we are told that the current is uniformly distributed in the wire. Thus Ampère's law gives $B(\rho_1) \times 2\pi\rho_1 = \mu_0 I(\rho_1^2/R^2)$ from which

$B(\rho_1) = \mu_0 I\rho_1/2\pi R^2$. We conclude that the magnetic field anywhere inside the wire is

$$\mathbf{B} = \frac{\mu_0 I\rho \mathbf{e}_\phi}{2\pi R^2} \qquad (\rho < R)$$

Problem 5.1 *(Objective 10)* Determine the flux of the vector field $\mathbf{V}(x,y,z) = 2y\mathbf{i} - x\mathbf{j} - 3\mathbf{k}$ across a surface of unit area of the x-y plane.

Problem 5.2 *(Objective 10)* Consider a vector field of the form $\mathbf{F}(r,\theta,\phi) = \lambda\mathbf{e}_r/r^n$ $(r \neq 0)$ where λ is a constant and n is an integer. Find the net outward flux of $\mathbf{F}$ across a spherical surface of radius r centred at the origin, and state the value of n for which the net outward flux is independent of r.

Problem 5.3 *(Objective 10)* A very long thin plastic fibre lying on the z-axis carries a uniform distribution of static electric charge. The density of charge (charge per unit length) is a constant λ. Consider a closed cylindrical "Gaussian" surface S of radius ρ greater than the radius of the fibre, and of height l, with its axis on the axis of the fibre (Fig 4.42). Use Gauss's law of electrostatics (see Example 5.3) to find the electrostatic field $\mathbf{E}$ at a point a distance ρ from the axis of the fibre. Assume that $\mathbf{E} = E(\rho)\mathbf{e}_\rho$ where E is a function of ρ only.

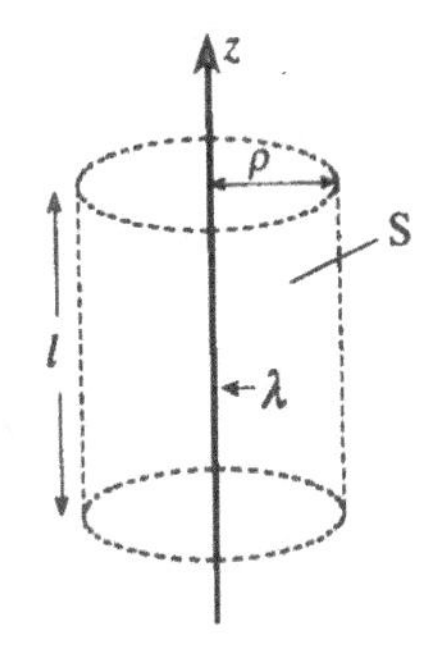

Fig 4.42
A cylindrical "Gaussian" surface S of length l coaxial with a long thin plastic fibre carrying a static electric charge per unit length of λ.

Problem 5.4 *(Objective 11)* Calculate the anticlockwise circulation of the constant vector field $\mathbf{g}(x,y) = \mu\mathbf{i} + \lambda\mathbf{j}$ (λ and μ are constants) around the square with corners at the points (0,0), (5,0), (5,5) and (0,5).

Problem 5.5 *(Objective 11)* The surface velocity field (in $\mathrm{m\,s^{-1}}$) on a river of width 20 m is $\mathbf{v}(x,y) = 0.1y(20 - y)\mathbf{i}$ $(0 \leq y \leq 20\ \mathrm{m})$.

(a) Calculate the circulation of $\mathbf{v}$ anticlockwise around the closed rectangular path with corners at A(3,3), B(4,3), C(4,5) and D(3,5). See Fig 4.43.

(b) Repeat part (a) for the rectangular path with corners at E(3,9), F(4,9), G(4,11) and H(3,11). Is the field conservative?

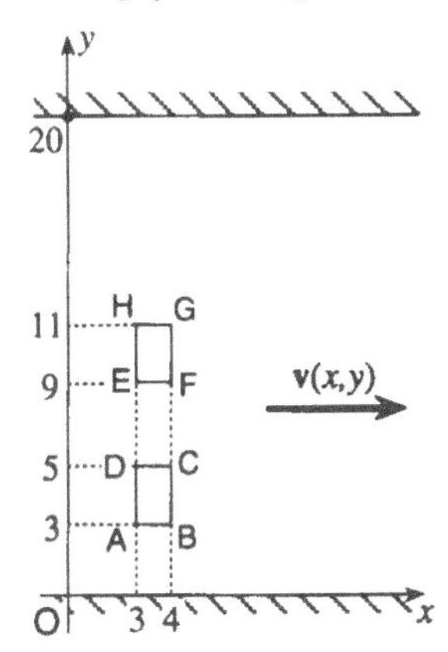

Fig 4.43
Two rectangular loops in the surface of a river.

Problem 5.6 *(Objectives 10,11)*

(a) In Problem 5.3 it was assumed that $\mathbf{E}$ has no azimuthal component $(E_\phi = 0)$. Use a symmetry argument and the fact that any electrostatic field is conservative to justify this assumption.

(b) In Example 5.6 it was assumed that the field $\mathbf{B}$ has no radial component. Use a symmetry argument and the fact that any magnetic field is solenoidal to justify this assumption.

Problem 5.7 (*Objective 11*) Consider the two-dimensional vector field defined by

$$\mathbf{u}(x,y) = g(y)\mathbf{i} + f(x)\mathbf{j}$$

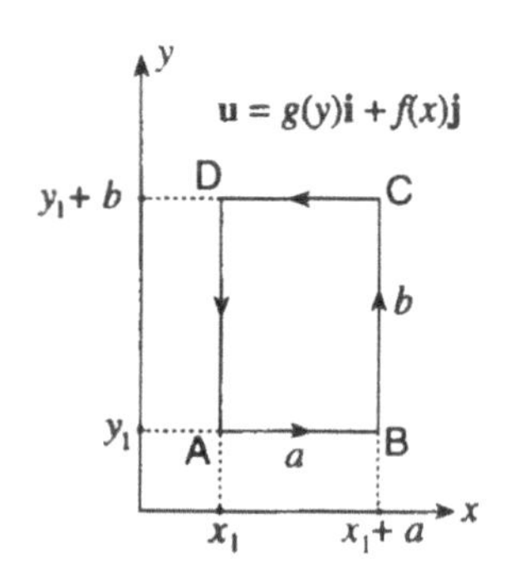

Fig 4.44
A rectangular loop in the x-y plane with sides of length a and b.

where g is a function of y only and f is a function of x only.

(a) Calculate the anticlockwise circulation of **u** around the rectangular loop ABCD in Fig 4.44.

(b) Determine the circulation per unit of area enclosed by the loop.

(c) What is the limit of the circulation per unit area as the dimensions of the loop go to zero? (Hint: see Eq (3.13) for the definition of the derivative of a scalar function.)

Problem 5.8 (*Objective 11*) Find the anticlockwise circulation of the constant vector field $\mathbf{f}(x,y) = \mathbf{i} - 2\mathbf{j} + 6\mathbf{k}$ around the unit square defined by the points O(0,0,0), A(1,0,0), B(1,1,0) and C(0,1,0).

5

Differentiating fields

After you have studied this chapter you should be able to

Objectives

- Calculate partial derivatives of functions of two and three variables (*Objective 1*).
- Calculate the gradient vector field from a given scalar field, and use the gradient to calculate directional derivatives (*Objective 2*).
- Calculate the divergence of a vector field (*Objective 3*).
- Calculate the curl of a vector field (*Objective 4*).
- Use the del (or nabla) notation ∇ for grad, div and curl (*Objective 5*).

An important property of any field is the way in which the field value changes from one field point to another. These spatial variations can be quite complex even for a scalar field since the change of field value from $\Phi(\mathbf{r})$ at a point P to $\Phi(\mathbf{r} + \Delta\mathbf{r})$ at a nearby point Q may depend on the direction as well as the magnitude of the displacement $\Delta\mathbf{r} = \mathbf{PQ}$. This is illustrated in Fig 5.1 where the change in Φ is $7 - 6 = 1$ when the displacement $\mathbf{PQ}_1$ made, and about $5.5 - 6 = -0.5$ for a displacement $\mathbf{PQ}_2$ of the same magnitude but different direction. We shall see that the way in which a scalar field varies in any direction from a point P can be found from a certain vector at P called the *gradient vector*. There is a gradient vector at each point in a scalar field and so the gradient vector is in fact a vector field. This gives the neat result that for any scalar field Φ there exists a gradient vector field from which all the spatial variations of Φ can be found.

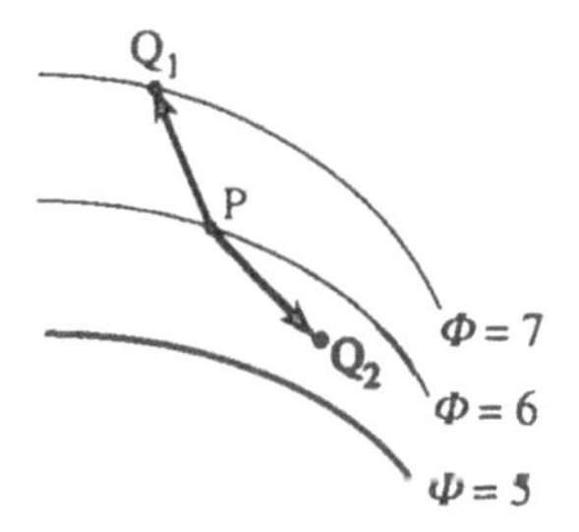

Fig 5.1
The change of field value depends on the direction of the displacement.

Vector fields are more complex than scalar fields. To describe the spatial variations of a vector field $\mathbf{F}$ we introduce two fields, one a scalar field called the *divergence* of $\mathbf{F}$ and the other a vector field called the *curl* of $\mathbf{F}$. These two fields describe different aspects of how $\mathbf{F}$ varies in space.

In order to calculate these new fields we have to know how to differentiate field functions. The basic concept here is the *directional derivative* which describes the spatial rate of change of a scalar field in a specified direction. This is just a generalisation of the familiar derivative df/dx from one dimension to three. We begin Section 5.1 by introducing the directional derivatives at a point and the three directional derivatives in directions parallel to the cartesian axes, known as *partial derivatives*. In Sections 5.2 to 5.4 we introduce the gradient, divergence and curl fields, show how these fields can be expressed in terms of partial derivatives and how the field values are calculated. Many physical phenomena depend on the spatial variations of scalar and vector fields and we shall discuss a selection of physical laws and processes that depend on the gradient of a scalar field or the divergence or curl of a vector field.

Because of the limited level and scope of this book, some of the important results are quoted and supported by plausibility arguments rather than derived formally. The emphasis is on an understanding of concepts, a mastery of calculation techniques and an appreciation of the relevance to scientific and engineering applications.

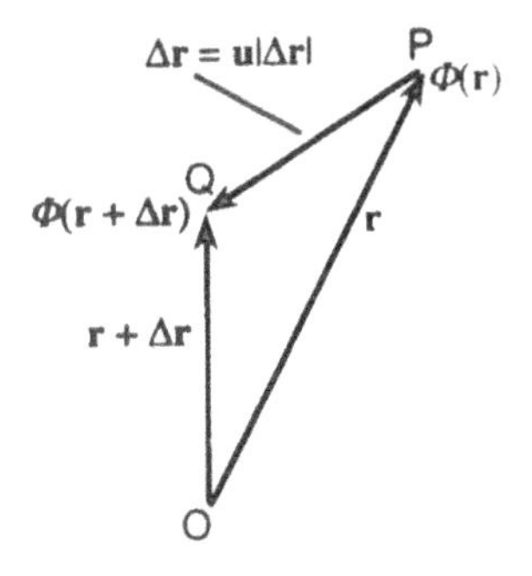

Fig 5.2
The field values $\Phi(\mathbf{r}+\Delta\mathbf{r})$ and $\Phi(\mathbf{r})$ at field points P and Q.

5.1 DIRECTIONAL DERIVATIVES AND PARTIAL DERIVATIVES

Let $\Phi(\mathbf{r})$ and $\Phi(\mathbf{r}+\Delta\mathbf{r})$ be scalar field values at field points P and Q which have position vectors $\mathbf{r}$ and $\mathbf{r}+\Delta\mathbf{r}$ (Fig 5.2). We shall specify the direction of the displacement $\Delta\mathbf{r} = \mathbf{PQ}$ by a unit vector $\mathbf{u}$, i.e. $\Delta\mathbf{r} = |\Delta\mathbf{r}|\mathbf{u}$. The spatial rate of change of Φ in the direction specified by $\mathbf{u}$ is denoted by $\Phi'_{\mathbf{u}}$ and defined to be the change in the field value, divided by the magnitude of the displacement, taken to the limit of small displacements. That is,

$$\Phi'_{\mathbf{u}}(\mathbf{r}) = \lim_{|\Delta\mathbf{r}|\to 0}\left[\frac{\Phi(\mathbf{r}+\Delta\mathbf{r})-\Phi(\mathbf{r})}{|\Delta\mathbf{r}|}\right] \tag{5.1}$$

We have assumed that the function Φ varies in space sufficiently smoothly that the limit (5.1) exists everywhere and for all directions, i.e. we have assumed that the function is **differentiable** and we shall continue with this assumption throughout the chapter.

$\Phi'_{\mathbf{u}}$ is called the **directional derivative** of Φ in the direction of $\mathbf{u}$ and the definition (5.1) is a generalisation to three dimensions of the definition of the derivative $f'(x)$ of a function $f(x)$ (Eq (3.13)). There is an infinite number of directional derivatives of Φ at any point P, one for each possible direction $\mathbf{u}$; this is why we need the label $\mathbf{u}$ on $\Phi'_{\mathbf{u}}$.

The directional derivatives at P have the following properties:

(i) $\Phi_{\mathbf{u}}' = -\Phi_{-\mathbf{u}}'$, i.e. if Φ increases in a particular direction then it decreases at the same rate in the opposite direction.

(ii) Let $\mathbf{n}$ be the unit vector at P perpendicular to the contour surface passing through P and pointing in the direction of increasing Φ (Fig 5.3). Then the largest and smallest directional derivatives at P are $\Phi_{\mathbf{n}}$ and $\Phi_{-\mathbf{n}}$ respectively, i.e. the "steepest" directions are at right-angles to the contours.

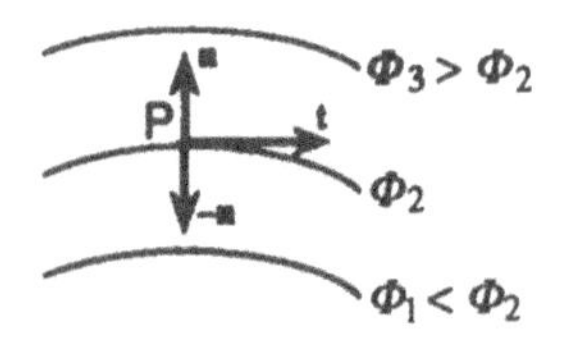

Fig 5.3
A section across contour surfaces. Unit vector **n** is at right-angles to the contour passing through P and points in the direction of increasing Φ. Unit vector **t** lies in the tangent plane of the contour.

(iii) Let $\mathbf{t}$ be any unit vector at P lying in the plane tangential to the contour surface passing through P. Then $\Phi'_{\mathbf{t}} = 0$. This property expresses the fact that Φ doesn't change when you move on a contour.

The definition (5.1) is not very convenient for calculations. We normally introduce a coordinate system and consider the directional derivatives in the directions of the unit base vectors. For example, the directional derivative in the x direction of a cartesian system (Fig 5.4) is found by putting $\Delta\mathbf{r} = \mathbf{i}\Delta x$ in

Eq (5.1) and expressing the position vectors **r** and **r** + Δ**r** as the cartesian triples (x,y,z) and $(x+\Delta x,y,z)$. Then Eq (5.1) takes the form

$$\Phi'_{\mathbf{i}}(x,y,z) = \lim_{\Delta x \to 0}\left[\frac{\Phi(x+\Delta x,y,z)-\Phi(x,y,z)}{\Delta x}\right] \tag{5.2}$$

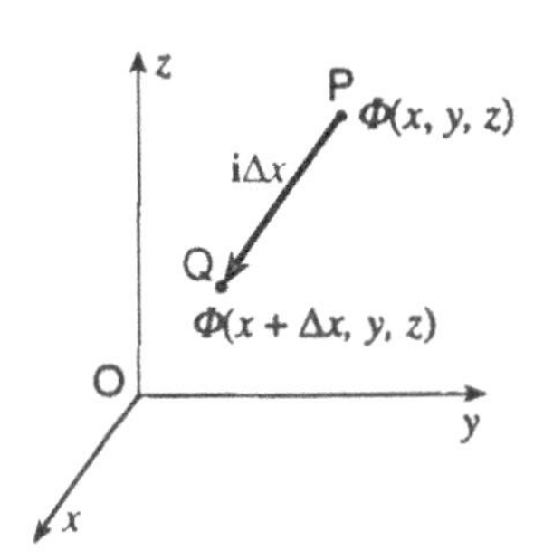

Fig 5.4
A displacement PQ in the x-direction.

This directional derivative is called the **partial derivative** of Φ with respect to x and is denoted by the symbol $\dfrac{\partial\Phi}{\partial x}(x,y,z)$ or simply $\dfrac{\partial\Phi}{\partial x}$. Thus

$$\Phi'_{\mathbf{i}}(x,y,z) = \frac{\partial\Phi}{\partial x}(x,y,z)$$

The curly dee is used to distinguish the partial derivative from the ordinary derivative df/dx of $f(x)$. Notice that the coordinates y and z do not change when a displacement $\mathbf{i}\Delta x$ is made, and so the symbols y and z in the definition (5.2) stay constant as Δx goes to zero ($\Delta x \to 0$). The definition (5.2) therefore has the same form as the definition of the derivative df/dx (Eq (3.13)). It follows that the partial derivative $\partial\Phi/dx$ can be calculated using the familiar rules of differentiation for functions $f(x)$ of a single variable x, provided we regard y and z in the expression for Φ as constants. For example, the partial derivative of the function

$$\Phi(x,y,z) = x^2y + 3y + xz^2 \tag{5.3}$$

with respect to x is

$$\frac{\partial\Phi}{\partial x}(x,y,z) = 2xy + z^2 \tag{5.4}$$

This is obtained by first differentiating the x^2y term with respect to x, regarding y as a constant number, to obtain $2xy$; then differentiating the term $3y$ to obtain 0 since y is treated as a constant; then finally differentiating xz^2 with respect to x treating z as a constant to give z^2.

The partial derivatives with respect to y and z are defined and calculated in a similar way. Thus differentiating Eq (5.3) partially with respect to y and z, we find $\partial\Phi/\partial y = x^2 + 3$ (treating x and z as constants) and $\partial\Phi/\partial z = 2xz$ (treating x and y as constants).

The partial derivatives are themselves scalar fields since they have scalar values at any field point $P(x,y,z)$. For example, the value of the partial derivative $\partial\Phi/\partial x$ (Eq (5.4)) at the field point (1,2,3) is found by substituting $x = 1$, $y = 2$ and $z = 3$ in Eq (5.4). Thus $\dfrac{\partial\Phi}{\partial x}(1,2,3) = 2\times1\times2 + 3^2 = 13$, and similarly $\dfrac{\partial\Phi}{\partial y}(1,2,3) = 1^2 + 3 = 4$ and $\dfrac{\partial\Phi}{\partial z}(1,2,3) = 2\times1\times3 = 6$. In practice we normally use the abbreviated notation $\partial\Phi/\partial x$ for both the field and the field value at a specified field point, relying on the context to make the meaning clear.

Since the partial derivatives are themselves scalar fields we can differentiate them to obtain partial derivatives of higher order. Thus we have the **second-order partial derivative** $\partial^2\Phi/\partial x^2$ obtained by partial differentiation with respect to x of the first-order partial derivative $\partial\Phi/\partial x$, i.e.

$$\frac{\partial^2\Phi}{\partial x^2} = \frac{\partial}{\partial x}\left(\frac{\partial\Phi}{\partial x}\right)$$

and similarly

$$\frac{\partial^2\Phi}{\partial y^2} = \frac{\partial}{\partial y}\left(\frac{\partial\Phi}{\partial y}\right) \quad \text{and} \quad \frac{\partial^2\Phi}{\partial z^2} = \frac{\partial}{\partial z}\left(\frac{\partial\Phi}{\partial z}\right)$$

There are also the **mixed second-order partial derivatives**, such as $\partial^2\Phi/\partial y\partial x = \partial(\partial\Phi/\partial x)/\partial y$, etc. The mixed partial derivatives always obey

$$\frac{\partial^2\Phi}{\partial x\partial y} = \frac{\partial^2\Phi}{\partial y\partial x} \tag{5.5}$$

and similarly for the other mixed partial derivatives.

Partial derivatives with respect to polar coordinates can be defined and evaluated in a similar way.

Summary of section 5.1

- The spatial rate of change of a scalar field Φ in the direction specified by a unit vector **u** is given by the **directional derivative** $\Phi'_{\mathbf{u}}$ (Eq (5.1)).

- The directional derivative of Φ in a direction parallel to the positive x-axis, $\Phi'_{\mathbf{i}}$, is the **partial derivative** with respect to x and is denoted by $\partial\Phi/\partial x$.

- The partial derivative $\partial\Phi/\partial x$ can be calculated by using the familiar rules for differentiating a function of x and regarding y and z as constants.

- The **second-order partial derivatives** are $\partial^2\Phi/\partial x^2 = \partial(\partial\Phi/\partial x)/\partial x$, $\partial^2\Phi/\partial y\partial x = \partial(\partial\Phi/\partial x)\partial y$, etc. The **mixed second-order partial derivatives** with respect to the same pair of variables are always equal (Eq (5.5))

Example 1.1 (*Objective 1*)

(a) Find the three first-order partial derivatives of the function

$$\Phi(x,y,z) = 2x - 3y + xy^2 - 4yz^2 + x^2yz$$

(b) Evaluate the partial derivative $\partial\Phi/\partial x$ at (i) the origin and (ii) the point $(1,-3,7)$.

Solution 1.1

(a) The first-order partial derivatives are $\partial\Phi/\partial x$, $\partial\Phi/\partial y$ and $\partial\Phi/\partial z$. To find $\partial\Phi/\partial x$ we regard y and z as constants and differentiate with respect to x. Thus

$$\frac{\partial\Phi}{\partial x} = 2 + y^2 + 2xyz$$

Regarding x and z as constants and differentiating Φ with respect to y, we obtain

$$\frac{\partial\Phi}{\partial y} = -3 + 2xy - 4z^2 + x^2z$$

Finally, differentiating Φ with respect to z, regarding x and y as constants, we find

$$\frac{\partial\Phi}{\partial y} = -8yz + x^2y$$

(b) (i) At the origin we put $x = 0$, $y = 0$ and $z = 0$ in the expression for $\partial\Phi/\partial x$ found in part (a), and obtain $\partial\Phi/\partial x = 2$.

(ii) At $(1,-3,7)$ we have $\partial\Phi/\partial x = 2 + (-3)^2 + 2\times1\times(-3)\times7 = -31$.

Example 1.2 (*Objective 1*) Find all the second-order partial derivatives of the two-dimensional scalar field function

$$h(x,y) = x^2y - xy^2$$

and evaluate them at the point $(1,6)$.

Solution 1.2 We must first find the two first-order partial derivatives. They are

$$\frac{\partial h}{\partial x} = 2xy - y^2 \quad \text{and} \quad \frac{\partial h}{\partial y} = x^2 - 2xy$$

From these, the four second-order partial derivatives are found:

$$\frac{\partial^2 h}{\partial x^2} = \frac{\partial}{\partial x}\left(\frac{\partial h}{\partial x}\right) = \frac{\partial}{\partial x}(2xy - y^2) = 2y$$

$$\frac{\partial^2 h}{\partial y^2} = \frac{\partial}{\partial y}\left(\frac{\partial h}{\partial y}\right) = \frac{\partial}{\partial y}(x^2 - 2xy) = -2x$$

$$\frac{\partial^2 h}{\partial y \partial x} = \frac{\partial}{\partial y}\left(\frac{\partial h}{\partial x}\right) = \frac{\partial}{\partial y}(2xy - y^2) = 2x - 2y$$

$$\frac{\partial^2 h}{\partial x \partial y} = \frac{\partial}{\partial x}\left(\frac{\partial h}{\partial y}\right) = \frac{\partial}{\partial x}(x^2 - 2xy) = 2x - 2y$$

Evaluating the second-order partial derivatives at the point (1,6) yields

$$\frac{\partial^2 h}{\partial x^2} = 12, \quad \frac{\partial^2 h}{\partial y^2} = -2, \quad \frac{\partial^2 h}{\partial y \partial x} = \frac{\partial^2 h}{\partial x \partial y} = 2 - 12 = -10$$

Example 1.3 (*Objective 1*) Determine the first-order partial derivatives with respect to spherical polar coordinates of the function

$$U(r,\theta,\phi) = \frac{k \sin\theta \cos\phi}{r^2} \qquad (r > 0,\ k \text{ is a constant})$$

Solution 1.3 We are asked to find $\partial U/\partial r$, $\partial U/\partial \theta$ and $\partial U/\partial \phi$. To find $\partial U/\partial r$ we differentiate U regarding θ and ϕ as constants. Thus we treat the factor $k\sin\theta\cos\phi$ as a constant and differentiate the $1/r^2$ factor with respect to r. This gives

$$\frac{\partial U}{\partial r} = \frac{-2k \sin\theta \cos\phi}{r^3}$$

We now differentiate U with respect to θ regarding r and ϕ as constants and obtain

$$\frac{\partial U}{\partial \theta} = \frac{k \cos\theta \cos\phi}{r^2}$$

Finally, regarding r and θ as constants we obtain

$$\frac{\partial U}{\partial \phi} = \frac{-k \sin\theta \sin\phi}{r^2}$$

Problem 1.1 (*Objective 1*) Find the partial derivatives (first order only) with respect to x, y and z of the following functions

$$p(x,y,z) = x^2 + y^2 + z^2 + 3xy$$

$$q(x,y,z) = \cos(5x)\sin(3y)$$

$$t(x,y,z) = x^2 + xy + y^2$$

$$c(x,y,z) = 1 - 6z$$

Problem 1.2 (*Objective 1*) The two mixed second-order partial derivatives of any scalar field with respect to the same pair of variables are equal. Demonstrate this for the function

$$\Phi(x,y,z) = z^2 x \exp(y) + z^2$$

Problem 1.3 (*Objective 1*) Consider the function

$$U(x,y) = \log_e(x^2 + y^2) \qquad (x,y) \neq (0,0)$$

Show that

$$\frac{\partial^2 U}{\partial x^2} + \frac{\partial^2 U}{\partial y^2} = 0$$

Problem 1.4 (*Objective 1*) Given the two-dimensional scalar field function $h(x,y) = \sin(xy) - x\cos(y)$, find all the second-order partial derivatives.

Problem 1.5 (*Objective 1*) Find the three first order partial derivatives of the function $\psi(\rho,\phi,z) = \exp(-\rho^2)\sin(2\phi)$.

Problem 1.6 (*Objective 1*) Show that the function

$$\phi(x,y,z) = (x^2 + y^2 + z^2)^{-1} \qquad (x,y,z) \neq (0,0,0)$$

satisfies $\dfrac{\partial^2 \phi}{\partial x^2} + \dfrac{\partial^2 \phi}{\partial y^2} + \dfrac{\partial^2 \phi}{\partial z^2} = 2\phi^2$.

5.2 GRADIENT OF A SCALAR FIELD

The spatial variations of a scalar field can be found from the three cartesian partial derivatives which are in fact the components of a vector called the *gradient vector*. We introduce the gradient vector in a familiar physical context where it describes the magnitude and direction of the steepest slope on a hillside, and indicate how this leads to an expression for the gradient vector in terms of partial derivatives. We then show how the gradient vector of a given scalar field can be calculated and how the slope of the field (i.e. the directional derivative) in any direction can be found from the gradient vector. Examples are given of the role played by the gradient vector in the expression of physical laws.

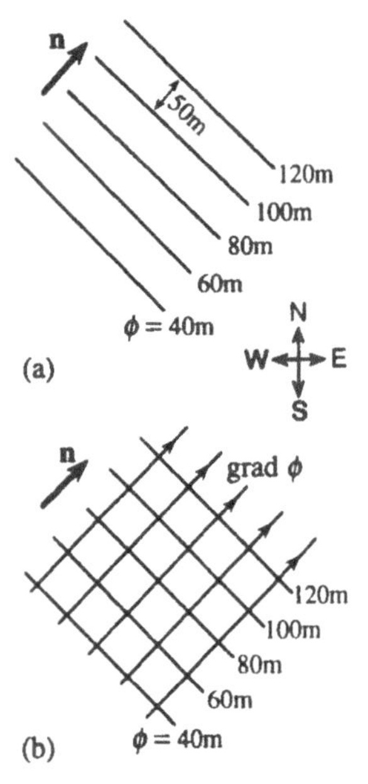

Fig 5.5
(a) A plane map showing contour lines on a hillside at intervals of 20 m vertical height. The horizontal distance between adjacent contours is 50 m.
(b) Field lines of the gradient vector cut the contours at right angles.

5.2.1 Introducing gradient

Consider a scalar field ϕ describing the land elevation in a hilly region. This two-dimensional field may be shown on a map by contour lines from which you can deduce various topographical features (Section 4.1). Of particular interest is the direction of steepest upward slope at a point P. This direction is specified by the unit vector $\mathbf{n}$ at right angles to the contour line passing through P and pointing in the direction of increasing ϕ. The magnitude of the steepest slope is the directional derivative $\phi'_{\mathbf{n}}$ which is just the increase in height per unit horizontal distance in the direction $\mathbf{n}$. The direction and magnitude of the steepest upward slope at any point define the gradient vector at that point. Since we can specify a gradient vector at each point we have a gradient vector field.

A simple case is illustrated in Fig 5.5a which shows contour lines of a scalar field ϕ representing a hillside that slopes uniformly upwards towards the north-east rising through 20 m vertical height for every 50 m horizontally. The gradient vector at any point is directed towards the north-east since this is the direction of steepest slope, and is of magnitude 20 m/50 m = 0.4. We can write this gradient vector as

$$\text{grad}\phi = 0.4\mathbf{n} \tag{5.6}$$

where $\mathbf{n}$ lies in the horizontal plane of the map and points towards the north-east. Fig 5.5b shows the field lines of this gradient vector field cutting the contours at right angles.

We may be interested in the slope along a path on the hillside in an arbitrary direction specified by a horizontal unit vector $\mathbf{u}$. This slope is the directional derivative $\phi_{\mathbf{u}}'$. Suppose $\mathbf{u}$ makes an angle α with the direction $\mathbf{n}$ of the gradient vector (Fig 5.6). The slope along a path PQ_2 in the direction of $\mathbf{u}$ is seen from Fig 5.6 to be

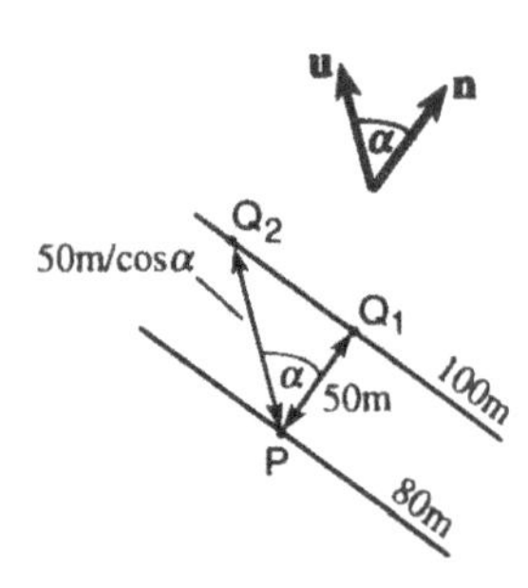

Fig 5.6
The two contours at 80 m and 100 m are separated by a horizontal distance of 50 m. $PQ_1 = 50$ m and $PQ_2 = 50$ m/cosα.

$$\phi'_{\mathbf{u}} = \frac{20\text{ m vertical height}}{\text{horizontal distance } PQ_2}$$

$$= \frac{20\text{ m}}{50\text{ m}/\cos\alpha} = 0.4\cos\alpha$$

But this is just the projection of the gradient vector (Eq (5.6)) onto $\mathbf{u}$, since $\mathbf{u}\,.\,\mathbf{n} = \cos\alpha$. Thus $\phi'_{\mathbf{u}} = \mathbf{u}\,.\,\text{grad}\phi = \mathbf{u}\,.\,(0.4\mathbf{n}) = 0.4\cos\alpha$.

We now introduce a cartesian coordinate system with x and y axes in the horizontal plane. The projections of the gradient vector onto the two cartesian unit vectors are, from the above, ϕ'_i and ϕ'_j. These projections are the cartesian components of the gradient vector. Remembering that ϕ'_i and ϕ'_j are the partial derivatives with respect to x and y respectively (Section 5.1) we can write $\text{grad}\phi = \mathbf{i}\partial\phi/\partial x + \mathbf{j}\partial\phi/\partial y$. This relationship is true for any two-dimensional scalar field ϕ and is generalised to three dimensions in an obvious way.

5.2.2 Calculating gradients

The **gradient vector** at any point P(x,y,z) in a scalar field Φ is defined by

$$\text{grad}\Phi(x,y,z) = \mathbf{i}\frac{\partial\Phi}{\partial x} + \mathbf{j}\frac{\partial\Phi}{\partial y} + \mathbf{k}\frac{\partial\Phi}{\partial z} \qquad (5.7)$$

We normally abbreviate $\text{grad}\Phi(x,y,z)$ simply as $\text{grad}\Phi$.

It follows that if A and C are constants,

$$\text{grad}(A\Phi) = A\,\text{grad}\Phi \qquad (A \text{ constant}) \qquad (5.8)$$

and

$$\text{grad}(\Phi + C) = \text{grad}\Phi \qquad (C \text{ constant}) \qquad (5.9)$$

The spatial rate of change of Φ in an arbitrary direction specified by a unit vector $\mathbf{u}$ is the directional derivative $\Phi'_\mathbf{u}$ and is given by the projection of $\text{grad}\Phi$ onto $\mathbf{u}$,

$$\Phi'_\mathbf{u} = \mathbf{u}\,.\,\text{grad}\Phi = |\text{grad}\Phi|\cos\alpha \qquad (5.10)$$

where α is the angle between $\mathbf{u}$ and the direction of $\text{grad}\Phi$. Eqs (5.7) and (5.10) enable us to calculate the gradient vector and any directional derivative once the partial derivatives are known.

As an example we calculate the gradient vector of the scalar field

$$U(x,y,z) = \frac{C}{2}(x^2 + y^2 + z^2) \qquad (C \text{ is a constant}) \qquad (5.11)$$

The partial derivatives are

$$\frac{\partial U}{\partial x} = Cx, \quad \frac{\partial U}{\partial y} = Cy, \quad \frac{\partial U}{\partial z} = Cz$$

Using Eq (5.7) we have

$$\text{grad}U(x,y,z) = C(x\mathbf{i} + y\mathbf{j} + z\mathbf{k}) \qquad (5.12)$$

Eq (5.12) is the gradient vector field. To find the gradient vector at a particular point, P(1,2,3) say, we simply substitute (1,2,3) for (x,y,z). Thus **grad**U at P(1,2,3) is the vector $C(\mathbf{i} + 2\mathbf{j} + 3\mathbf{k})$. Note that the substitution must be made in **grad**U, not in U, i.e. we substitute *after* the gradient field has been found from U.

Eq (5.7) is the cartesian form of the gradient vector. The polar forms are:

grad in cylindrical polar form

$$\mathbf{grad}\Phi = \mathbf{e}_\rho \frac{\partial \Phi}{\partial \rho} + \mathbf{e}_\phi \frac{1}{\rho}\frac{\partial \Phi}{\partial \phi} + \mathbf{e}_z \frac{\partial \Phi}{\partial z} \tag{5.13}$$

When the field Φ has cylindrical symmetry there is no ϕ or z variation and so Eq (5.13) reduces to

$$\mathbf{grad}\Phi = \mathbf{e}_\rho \frac{\partial \Phi}{\partial \rho} \qquad \text{(cylindrical symmetry)} \tag{5.14}$$

grad in spherical polar form

The cylindrical and spherical polar expressions for **grad**Φ are more complicated than the cartesian expression (Eq (5.7)) because the base vectors in polar coordinate systems vary with position while the cartesian base vectors are constants.

$$\mathbf{grad}\Phi = \mathbf{e}_r \frac{\partial \Phi}{\partial r} + \mathbf{e}_\theta \frac{1}{r}\frac{\partial \Phi}{\partial \theta} + \mathbf{e}_\phi \frac{1}{r\sin\theta}\frac{\partial \Phi}{\partial \phi} \tag{5.15}$$

When Φ is spherically symmetric there is no θ or ϕ variation and so

$$\mathbf{grad}\Phi = \mathbf{e}_r \frac{\partial \Phi}{\partial r} \qquad \text{(spherical symmetry)} \tag{5.16}$$

Eq (5.11) is an example of a spherically symmetric scalar field since it varies only with $r = (x^2 + y^2 + z^2)^{1/2}$, and has the very simple spherical polar form

$$U(r,\theta,\phi) = \frac{Cr^2}{2} \tag{5.17}$$

from which **grad**U is easily obtained using Eq (5.16)

$$\mathbf{grad}U = Cr\mathbf{e}_r \tag{5.18}$$

which is the spherical polar form of Eq (5.12).

5.2.3 Gradient and physical laws

Many physical laws relate a "flow rate" vector to the gradient of a scalar field which "drives" the flow. Consider, for example, the conduction of heat through the walls of a house. The spatial pattern of heat flow in the walls is represented by a vector field **h**. The direction of the vector **h** at any point shows the direction of heat flow and the magnitude of **h** is equal to the rate of flow of heat across unit area normal to the direction of flow (Wm^{-2}). The flow **h** is driven by the negative of the gradient of the temperature field T in the walls, i.e. heat flows from hot regions to cold regions, according to the law

$$\mathbf{h} = -\alpha \,\mathrm{grad}T \tag{5.19}$$

where the constant α is the thermal conductivity of the wall material.

The flow of electric current through certain electrical conductors follows a similar law,

$$\mathbf{J} = -\sigma \,\mathrm{grad}V \tag{5.20}$$

where the constant σ is the electrical conductivity, V is the electrostatic potential field in the conductor and **J** is the **current density vector**. The direction of **J** at any point in the conductor gives the direction of current flow and the magnitude of **J** is the electric current crossing unit area normal to the direction of flow at that point (Am^{-2}). We can make a connection with elementary circuit theory by applying Eq (5.20) to the case of a straight wire of length L and cross-sectional area A with its ends connected across the terminals of a battery of emf V_b. The electric potential in the wire falls uniformly along its length from the positive end to the negative end by an amount V_b over a distance L. Thus the magnitude of the gradient of the potential field in the wire is $|\mathrm{grad}V| = V_b/L$. The magnitude of the current density in the wire is $|\mathbf{J}| = I/A$ where I is the current. Equating the magnitudes of both sides of Eq (5.20) gives $|\mathbf{J}| = \sigma|\mathrm{grad}V|$, as $\sigma > 0$. Thus we have $I = V_b(A\sigma/L)$. But $L/\sigma A$ is equal to the resistance R of the wire and so $I = V_b/R$ which is Ohm's law applied to the wire. Note also that the direction of the current density vector in Eq (5.20) is opposite that of $\mathrm{grad}V$, i.e. the current flows "down the potential gradient" from the positive terminal to the negative terminal. Eq (5.20) is sometimes called the **differential form of Ohm's law**.

We now turn to a different kind of application of the gradient vector which is important in the study of conservative fields. A vector field **G** is a *conservative field* if it can be expressed as

$$\mathbf{G} = -\mathrm{grad}\phi \tag{5.21}$$

where ϕ is a scalar field called a **scalar potential** and the minus sign is a convention. Eq (5.21) is not a physical law but a defining statement of what is meant by a conservative field, i.e. a conservative field is one that can be expressed as a gradient field. Alternatively you can regard Eq (5.21) as the definition of a scalar potential field ϕ in terms of **G**. Of course ϕ is defined by this equation only to within an arbitrary constant scalar field C, since $\mathrm{grad}\phi = \mathrm{grad}(\phi + C)$.

In the case of a conservative force $\mathbf{F}(\mathbf{r})$ acting on a particle at **r** we have

$$\mathbf{F} = -\mathrm{grad}U \tag{5.22}$$

where the scalar potential U is the potential energy of the particle. Problems in mechanics can often be solved either by considering the forces involved and using Newton's laws of motion or by applying the law of conservation of energy. Eq (5.22) is the link between the two approaches.

When $\mathbf{F}$ is the electrostatic force acting on a charged particle we have $\mathbf{F} = q\mathbf{E}$ and $U = qV$, where q is the electric charge carried by the particle. $\mathbf{E}$ is the electrostatic field and V the electrostatic potential. Then Eq (5.22) gives

$$\mathbf{E} = -\text{grad}V \tag{5.23}$$

We can combine Eqs (5.20) and (5.23) to obtain

$$\mathbf{J} = \sigma\mathbf{E} \tag{5.24}$$

which is another way of writing Ohm's law. Conservative fields are discussed further in Sections 5.4.3, 6.4 and 6.6.

Summary of section 5.2

- There is a gradient vector field associated with any scalar field. The magnitude of the gradient vector at a point is the largest directional derivative of the scalar field at that point. The direction of the gradient vector is the direction in which this largest value occurs. The gradient vector field is everywhere at right angles to the contours of the scalar field.

- The cartesian components of the gradient field are the corresponding first-order partial derivatives of the scalar field Φ, i.e.

$$\text{grad}\Phi = \mathbf{i}\partial\Phi/\partial x + \mathbf{j}\partial\Phi/\partial y + \mathbf{k}\partial\Phi/\partial z \tag{5.7}$$

  The polar forms of gradΦ are given by Eqs (5.13) to (5.16).

- The directional derivative of Φ in any direction $\mathbf{u}$ is found from

$$\Phi'_{\mathbf{u}} = \mathbf{u} \,.\, \text{grad}\Phi = |\text{grad}\Phi|\cos\alpha \tag{5.10}$$

- Many physical laws describe a vector flow rate driven by the negative of the gradient of a scalar field (e.g. Eqs (5.19) and (5.20)).

- A conservative field is one that can be expressed as the gradient of a scalar field.

Example 2.1 (*Objective 2*) Determine the gradient vector fields corresponding to the following scalar fields:

(a) $f(x,y,z) = x^2y - y^2x - z^3$
(b) $\phi(x,y) = \sin(5x)\cos(5y)$
(c) $T(x,y,z) = \log_e(x^2 + y^2) \qquad (x^2 + y^2) \neq 0$
(d) $U(r,\theta,\phi) = \dfrac{k\cos\theta}{r^2}$ where k is a constant and $r \neq 0$.

Solution 2.1 Use Eq (5.7) for parts (a) to (c), to obtain:

(a) $\text{grad}\, f = (2xy - y^2)\mathbf{i} + (x^2 - 2yx)\mathbf{j} - 3z^2\mathbf{k}$
(b) $\text{grad}\, \phi = 5(\mathbf{i}\cos(5x)\cos(5y) - \mathbf{j}\sin(5x)\sin(5y))$
(c) $\text{grad}\, T = 2(x\mathbf{i} + y\mathbf{j})/(x^2 + y^2)$

For part (d) we use the spherical polar form given in Eq (5.15).

(d) $\text{grad}\, U = -\mathbf{e}_r \dfrac{2k\cos\theta}{r^3} - \mathbf{e}_\theta \dfrac{k\sin\theta}{r^3}$

Example 2.2 (*Objective 2*) Consider the scalar field

$$\phi(x,y) = \exp\left[-\frac{(x^2 + y^2)}{3}\right]$$

(a) Determine the components and the magnitude of the gradient vector at the point (0,1).

(b) Determine the directional derivative in the direction of the unit vector $\mathbf{e} = (\mathbf{i} + \mathbf{j})/\sqrt{2}$ at the point (0,1).

Solution 2.2

(a) We first find the components of gradϕ from Eq (5.7) and the magnitude of gradϕ at a general point (x,y). Thus

$$(\text{grad}\, \phi)_x = \frac{\partial\phi}{\partial x} = -\frac{2x}{3}\phi(x,y),$$

$$(\text{grad}\, \phi)_y = \frac{\partial\phi}{\partial y} = -\frac{2y}{3}\phi(x,y),$$

$$|\text{grad}\, \phi| = \frac{2}{3}\phi(x,y)(x^2 + y^2)^{1/2}$$

Evaluating these at the point (0,1) yields

$$(\mathrm{grad}\phi)_x = 0, \quad (\mathrm{grad}\phi)_y = -\frac{2}{3}\exp\left(-\frac{1}{3}\right) = -0.478, \quad |\mathrm{grad}\phi| = 0.478$$

(b) We have from part (a)

$$\mathrm{grad}\phi = -\frac{2}{3}\phi(x,y)(x\mathbf{i} + y\mathbf{j})$$

We now use Eq (5.10) to find the required directional derivative:

$$\phi'_{\mathbf{e}}(x,y) = \frac{(\mathbf{i}+\mathbf{j})}{\sqrt{2}} \cdot \mathrm{grad}\phi = -\frac{\sqrt{2}}{3}(x+y)\phi(x,y)$$

Evaluating this at (0,1) gives

$$\phi'_{\mathbf{e}}(0,1) = -\frac{\sqrt{2}}{3}\exp\left(-\frac{1}{3}\right) = -0.338$$

Example 2.3 (*Objective 2*) Find the y-component of grad f at points on the x-axis, for

$$f(x,y) = \sin(3y)\exp(-x^2)$$

Solution 2.3 The y-component of the gradient field is found from Eq (5.7):

$$(\mathrm{grad}\, f)_y = \frac{\partial f}{\partial y} = 3\cos(3y)\exp(-x^2)$$

The x-axis is the line $y = 0$, so we evaluate $(\mathrm{grad}\, f)_y$ at $(x,0)$ to obtain $3\exp(-x^2)$.

Example 2.4 (Objective 2)

(a) Find grad$V$ where V is the two dimensional field

$$V(\rho,\phi) = \log_e\left(\frac{\rho}{a}\right) \qquad (\rho > 0;\ a \text{ is a constant})$$

Give your answer in plane polar coordinates and in cartesian coordinates. Make a rough sketch showing one contour line of V and some field lines of grad$V$.

(b) Determine the directional derivatives of V in the directions of the unit vectors $\mathbf{u} = (\mathbf{i}+\mathbf{j})/\sqrt{2}$ and $\mathbf{v} = (-\mathbf{i}+\mathbf{j})/\sqrt{2}$ at the cartesian point (1,0).

(c) In which directions at the cartesian point (1,0) are the directional derivatives of V equal to zero?

Solution 2.4

(a) The scalar field V is given in plane polar coordinates and varies with ρ only. We can therefore use Eq (5.14) to give

$$\text{grad}V = \mathbf{e}_\rho \frac{\partial V}{\partial \rho} = \frac{\mathbf{e}_\rho}{\rho} \qquad (\rho > 0)$$

In cartesian coordinates we put $\mathbf{e}_\rho = (x\mathbf{i} + y\mathbf{j})/(x^2+y^2)^{1/2}$ and $\rho = (x^2+y^2)^{1/2}$ (Section 4.4) to obtain the cartesian form

$$\text{grad}V = \frac{x\mathbf{i}+y\mathbf{j}}{(x^2+y^2)} \qquad (x,y) \neq (0,0)$$

Since V depends on ρ only, the contour lines of V are circles centred on the origin. We have found that grad$V = \mathbf{e}_\rho/\rho$, and so the field lines of grad$V$ are directed outwards from the origin, cutting the contour lines of V at right angles (Fig 5.7).

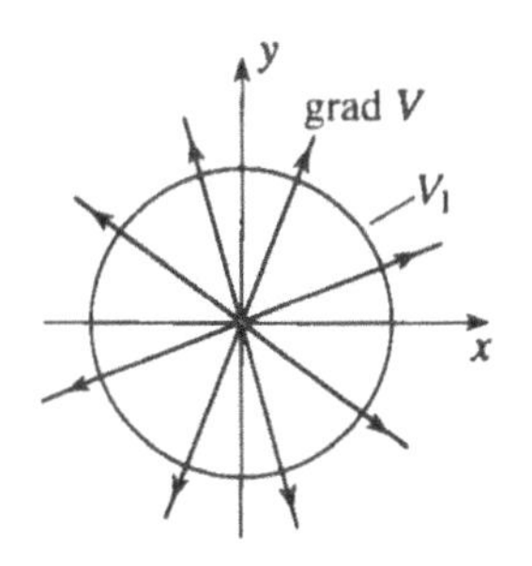

Fig 5.7
A circular contour line on which V has the constant value V_1, and field lines of grad$V$ directed outwards from the origin.

(b) Evaluating grad$V$ from part (a) at the cartesian point (1,0) gives grad$V = \mathbf{i}$. The directional derivatives of V at (1,0) in the directions specified by $\mathbf{u}$ and $\mathbf{v}$ are now found from Eq (5.10) to be

$$V'_{\mathbf{u}} = \mathbf{u}\,.\,\mathbf{i} = \frac{1}{\sqrt{2}} \quad \text{and} \quad V'_{\mathbf{v}} = \mathbf{v}\,.\,\mathbf{i} = -\frac{1}{\sqrt{2}}$$

(c) The directional derivatives at (1,0) are zero in directions specified by unit vectors $\mathbf{m}$ that satisfy the equation $\mathbf{m}\,.\,(\text{grad}V) = 0$, i.e. $\mathbf{m}\,.\,\mathbf{i} = 0$. This equation has solutions $\mathbf{m} = \mathbf{j}$ and $\mathbf{m} = -\mathbf{j}$. Hence the directional derivatives of V at (1,0) are zero in the positive and negative y directions.

Example 2.5 (*Objective 2*)

(a) The temperature on one face of a window pane is 20°C, the temperature on the other face is 15°C and the thickness of the glass is 5×10^{-3} m. Assuming that the temperature varies uniformly between the two faces, write down a function describing the temperature distribution in the glass and determine the temperature gradient field.

(b) Use Eq (5.19) to determine the rate of heat flow per square metre (Wm^{-2}) through the window pane in part (a) given that the thermal conductivity of the glass is 1.1 Wm^{-1} $°C^{-1}$.

Solution 2.5

(a) Choose a cartesian coordinate system with the x-y plane ($z = 0$) on the 20°C surface. The uniform temperature field is then independent of x and y, and depends linearly on z, i.e. it varies as $T = A + Bz$ where A and B are constants that can be found from the given surface temperatures. Thus at $z = 0$ we have $20 = A + B \times 0$, and at $z = 5 \times 10^{-3}$ we have $15 = A + B \times 0.005$. Hence $A = 20$°C and $B = -1000$ °C m^{-1}, and the temperature field can be written as

$$T(z) = 20 - 1000z \qquad (0 \le z \le 0.005)$$

where we have written $T(z)$ for $T(x,y,z)$. The temperature gradient field is found from Eq (5.7) to be $\mathbf{grad}T = \mathbf{k}\partial T/\partial z = -1000\mathbf{k}$ °C m^{-1}.

(b) We use Eq (5.19) to give $\mathbf{h} = -\alpha\,\mathbf{grad}T = 1.1 \times 1000\mathbf{k}$, from which $|\mathbf{h}| = 1100$, i.e. the rate of heat flow is 1.1 kWm^{-2}.

Problem 2.1 (*Objective 2*) Find the gradient vector fields for the following scalar fields:

(a) $f(x,y) = x^2y - y^2$
(b) $\phi(x,y,z) = x(\cos y + \sin z)$
(c) $h(\mathbf{r}) = 2|\mathbf{r}|$ where $\mathbf{r}$ is the position vector.
(d) $q(\rho,\phi,z) = \exp(-\rho/a)\cos(m\phi)$ where a and m are constants.

Problem 2.2 (*Objective 2*) Given the scalar field $f(x,y,z) = x^2 + yz + z^2$, find the gradient vector at the point (1,2,3) and determine the directional derivative at this point in the direction of the vector $(\mathbf{i} + \mathbf{j} + \mathbf{k})$. Confirm that this directional derivative is less than the magnitude of the gradient vector.

Problem 2.3 (*Objective 2*) Show that the gradient of $u(\mathbf{r}) = \exp(-|\mathbf{r}|^2)$ is directed radially inwards everywhere (except at the origin) and make a rough sketch showing some contour lines of u and some field lines of $\mathbf{grad}u$.

Problem 2.4 (*Objective 2*) A temperature distribution is given by $T(\rho,\phi,z) = (A - B\log(a/\rho))\exp(-z^2/c^2)$ ($\rho \ge a$) where A, B, a and c are positive constants. Determine the temperature gradient field and hence find the temperature gradient vector at a point a distance a perpendicularly out from the point $z = c$ on the z-axis.

Problem 2.5 (*Objective 2*) Find the *x*-component of gradϕ at points on the *y*-axis, where ϕ is the two-dimensional scalar field $\phi(x,y) = \sin(2x)\cos(y)$.

Problem 2.6 (*Objective 2*) Section 5.2.1 began by considering a hillside sloping uniformly upwards towards the north-east (Fig 5.5a). Introduce a cartesian system with *y*- and *x*- axes in the directions of north and east respectively and with the origin on the 100m contour. Deduce the land height function $\phi(x,y)$ that describes the hillside (give the rule only) and show that Eq (5.7) gives a gradient field of magnitude 0.4 directed towards the north-east.

Problem 2.7 (*Objective 2*) The potential energy of a particle in an isotropic simple harmonic oscillator potential well is $U(r,\theta,\phi) = (C/2)r^2$ where the force constant C is 0.2 Nm^{-1} and r is in metres. Determine the magnitude and direction of the force on the particle when it is at a distance of 3 mm from the origin.

Problem 2.8 (*Objective 2*) An electric charge Q is fixed at the origin O. The electrostatic potential at **r** is $= Q/4\pi\epsilon_0|\mathbf{r}|$ $(\mathbf{r} \neq \mathbf{0})$. Show that this is consistent with Coulomb's law which states that the electrostatic field at **r** is $Q\mathbf{r}/4\pi\epsilon_0|\mathbf{r}|^3$.

Problem 2.9 (*Objective 2*) The electrostatic potential field produced by a very small electric dipole is given by $V(\mathbf{r}) = \dfrac{\mathbf{p}\cdot\mathbf{r}}{4\pi\epsilon_0|\mathbf{r}|^3}$ $(\mathbf{r} \neq \mathbf{0})$ where **p** is the electric dipole moment of the dipole and **r** is the field point measured from the position of the dipole. Show that the electrostatic field produced by the dipole is given by Eq (4.28) with $C = 1/4\pi\epsilon_0$.

The expression given for the dipole potential V in Problem 2.9 is in fact an approximation that is valid when the distance $|\mathbf{r}|$ from the dipole is very large compared with the size of the dipole.

5.3 DIVERGENCE OF A VECTOR FIELD

We have seen that the spatial variations of a scalar field are described by the gradient vector field whose cartesian components are the three partial derivatives of the scalar field (Eq (5.7)). The spatial variations of a vector field **F** are more complex. There are nine partial derivatives, three for each scalar component field. From these we can construct two fields, one a scalar field called the *divergence* of **F**, the subject of this section, and the other a vector field called the *curl* of **F**, the subject of Section 5.4. These two fields play important roles in the description of many physical processes.

Section 5.3.1 introduces the concept of divergence in terms of the outward flux across a closed surface. Flux was introduced in Section 4.5 which should be studied before Section 5.3.1. Section 5.3.2 is concerned with calculating the divergence of given vector fields and Section 5.3.4 discusses the role played by divergence in physical laws.

5.3.1 Introducing divergence

We first review the concept of flux (Section 4.5). The flux of a vector field can be visualised by sketching field lines to show both the magnitude and direction of the field. In this picture the magnitude of the vector field is proportional to the density of field lines and the flux of the field across a surface is proportional to the total number of field lines crossing the surface. If the field lines are continuous lines everywhere in the domain of **F**, i.e. there are no sources or sinks of **F**, then the total number of lines entering the interior of any closed surface must be equal to the total number emerging from it (Fig 4.33) and the net outward flux is zero. Fields with this property are said to be *solenoidal*. For other vector fields the net outward flux is non-zero for some or for all closed surfaces. This *non-solenoidal* property is revealed in a field line picture by field lines terminating (sinks) or new ones beginning (sources) in the domain of the field (as in Fig 4.34). We now introduce the divergence of **F** as the outward flux of **F** per unit of enclosed volume, evaluated as a limit at a point.

Let Φ_0 be the outward flux of a vector field **F** across a closed surface S enclosing a volume V that includes the point P (Fig 5.8). Imagine now that the closed surface S is made to shrink, like a deflating balloon, eventually collapsing onto the point P. During this process both the outward flux Φ_0 and the enclosed volume V become smaller in magnitude and eventually zero, but the quotient Φ_0/V, the net outward flux per unit volume, tends to a definite limit which is independent of the shape of the surface S and the details of how it is made to collapse. The limit depends only on the field **F** and the position of the point P and is called the divergence of **F** at P, abbreviated as div**F**. Thus

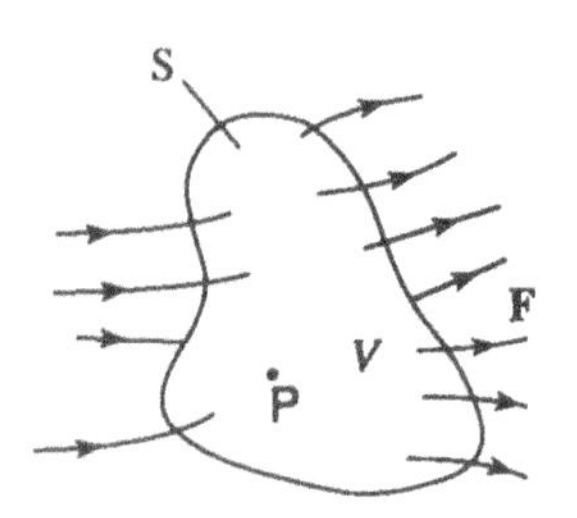

Fig 5.8
Field lines of **F** crossing a closed surface S enclosing a volume V that includes the point P.

$$\text{div}\mathbf{F} = \lim_{V\to 0}(\Phi_0/V) \tag{5.25}$$

Note that divergence is a scalar quantity associated with a point P in the domain of **F** and is therefore a scalar field with the same domain as **F**.

The limit (5.25) is not convenient for calculating the divergence. Instead we use an expression for divergence in terms of the partial derivatives of the components of **F** (Eq (5.26) below). We now indicate how this expression follows from Eq (5.25).

Consider a point P with coordinates (x_1,y_1,z_1) at the centre of a closed cube of edge length a (Fig 5.9). The net outward flux of the arbitrary vector field

$$\mathbf{F}(x,y,z) = F_x(x,y,z)\mathbf{i} + F_y(x,y,z)\mathbf{j} + F_z(x,y,z)\mathbf{k}$$

across the closed surface of the cube is the sum of the outward fluxes across the six faces. Consider the two opposite faces in the planes $x = x_1 + a/2$ and $x = x_1 - a/2$. The unit outward normals on these faces are $\mathbf{n} = \mathbf{i}$ and $\mathbf{n} = -\mathbf{i}$ respectively (Fig 5.9).

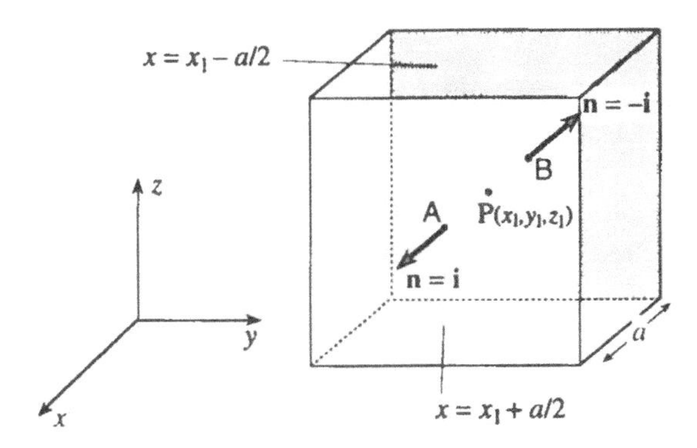

Fig 5.9
$P(x_1,y_1,z_1)$ is at the centre of the cube. A and B are the centre points of the cube faces in the planes $x = x_1 + a/2$ and $x = x_1 - a/2$; the unit outward normals on these faces are $\mathbf{n} = \mathbf{i}$ and $\mathbf{n} = -\mathbf{i}$ respectively.

The outward normal components of the field (i.e. $\mathbf{n} \cdot \mathbf{F}$) at the centre points A and B of these faces are respectively

$$\mathbf{i} \cdot \mathbf{F}(x_1 + a/2, y_1, z_1) = F_x(x_1 + a/2, y_1, z_1)$$

and

$$-\mathbf{i} \cdot \mathbf{F}(x_1 - a/2, y_1, z_1) = -F_x(x_1 - a/2, y_1, z_1)$$

We now make the assumption that the cube is sufficiently small that these values apply everywhere on the respective faces (i.e. we are neglecting the variations of F_x with y and z over the faces). With this assumption we can use Eq (4.29) of Chapter 4 to write the net outward flux across this pair of faces as

$$\Phi_o = a^2[F_x(x_1 + a/2, y_1, z_1) - F_x(x_1 - a/2, y_1, z_1)]$$

where a^2 is the area of each face. Following Eq (5.25) we divide this net outward flux by the volume a^3 and take the limit as the cube shrinks onto P (i.e. as $a \to 0$). The contribution of this pair of faces to the divergence of $\mathbf{F}$ is therefore

$$\lim_{a\to 0}\left[\frac{F_x(x+a/2,y_1,z_1) - F_x(x_1 - a/2,y_1,z_1)}{a}\right] = \frac{\partial F_x}{\partial x}$$

where we have recognised the limit as the partial derivative $\partial F_x/\partial x$ at $P(x_1,y_1,z_1)$, see Eq (5.2). Thus by considering the net outward flux across the two opposite faces we have a contribution to the divergence at P of $\partial F_x/\partial x$. Similarly we can show that the other two pairs of faces give contributions $\partial F_y/\partial y$ and $\partial F_z/\partial z$.

5.3.2 Calculating divergence

The argument of the preceding section leads to the following cartesian expression for the **divergence of a vector field F** at a point $P(x,y,z)$

We normally abbreviate $\operatorname{div}\mathbf{F}(x,y,z)$ simply as $\operatorname{div}\mathbf{F}$.

$$\operatorname{div}\mathbf{F}(x,y,z) = \frac{\partial F_x}{\partial x} + \frac{\partial F_y}{\partial y} + \frac{\partial F_z}{\partial z} \tag{5.26}$$

Eq (5.26) is the cartesian form of Eq (5.25). Either expression may be taken as the definition of divergence, but Eq (5.26) is more useful because we know how to calculate partial derivatives. As an example we now use Eq (5.26) to calculate the divergence of the vector field

$$\mathbf{F}(x,y,z) = xy\mathbf{i} + yz\mathbf{j} - z^2\mathbf{k} \tag{5.27}$$

The components of **F** are

$$F_x = xy, \quad F_y = yz, \quad F_z = -z^2$$

From these we obtain the partial derivatives

$$\frac{\partial F_x}{\partial x} = y, \quad \frac{\partial F_y}{\partial y} = z, \quad \frac{\partial F_z}{\partial z} = -2z$$

and so Eq (5.26) gives

$$\operatorname{div}\mathbf{F}(x,y,z) = y - z \tag{5.28}$$

The divergence, Eq (5.28), is a scalar field with the same domain as **F**. We can easily evaluate it at particular points. For example, the value of div**F** at the origin is $\operatorname{div}\mathbf{F}(0,0,0) = 0 - 0 = 0$ and the value at the point (1,2,3) is $\operatorname{div}\mathbf{F}(1,2,3) = 2 - 3 = -1$. We normally use the abbreviated notation div**F** to denote both the field and particular field values, relying on the context to make the meaning clear.

Eq (5.26) is the cartesian expression for divergence. The equivalent polar expressions are:

divergence in cylindrical polar coordinates

$$\operatorname{div}\mathbf{F} = \frac{\partial F_\rho}{\partial \rho} + \frac{1}{\rho}F_\rho + \frac{1}{\rho}\frac{\partial F_\phi}{\partial \phi} + \frac{\partial F_z}{\partial z} \tag{5.29}$$

divergence in spherical polar coordinates

$$\operatorname{div}\mathbf{F} = \frac{\partial F_r}{\partial r} + \frac{2}{r}F_r + \frac{1}{r}\frac{\partial F_\theta}{\partial \theta} + \frac{\cot\theta}{r}F_\theta + \frac{1}{r\sin\theta}\frac{\partial F_\phi}{\partial \phi} \tag{5.30}$$

These expressions look quite complicated, but for fields with cylindrical or spherical symmetry some of the terms vanish. Consider the spherically symmetric inverse square law field given in spherical polar coordinates by

$$\mathbf{E}(r,\theta,\phi) = \frac{C}{r^2}\mathbf{e}_r \qquad (r > 0;\ C \text{ is a constant}) \qquad (5.31)$$

The components of **E** are $E_r = C/r^2$, $E_\theta = E_\phi = 0$, and so the last three terms of Eq (5.30) are zero and we obtain div**E** $= -2C/r^3 + (2/r)(C/r^2) = 0$. Thus div**E** = 0 everywhere in the domain of **E**.

5.3.3 Divergence and physical law

In many applications the flux of a vector field across a surface describes the rate of flow of mass, energy or some other entity across the surface. For example, the pattern of heat flow in a conducting material is described by a vector field **h** (Wm^{-2}) defined in Section 5.2.3. The flux of **h** across a surface is the rate at which heat flows across the surface (in watts) and so div**h** (Wm^{-3}) is the rate of outward flow of heat per unit of enclosed volume, evaluated as a limit at a point (as in Eq (5.25)). When conditions are steady (i.e. temperatures are independent of time) and when there are no sources of heat (such as electrical heating) inside the material, the spatial pattern of heat flow obeys the physical law

$$\mathrm{div}\,\mathbf{h} = 0 \qquad \text{(steady conditions; no sources)} \qquad (5.32)$$

This equation states that the net outflow of heat per unit volume evaluated at any point is zero. In simple terms this means that the heat flowing into any small region of the material is balanced by an equal amount of heat flowing out of it.

When there are sources of heat in the conducting material the heat flow is governed by

$$\mathrm{div}\,\mathbf{h} = S \qquad \text{(steady conditions with heat sources)} \qquad (5.33)$$

where S is a scalar source field representing the rate at which heat is generated per unit volume of material (Wm^{-3}). Eq (5.33) governs the steady pattern of heat flow in the bar of an electric fire in which heat is generated electrically, or in a nuclear reactor fuel rod in which heat is generated by nuclear fission.

It is often important to consider heat flow under non-steady conditions. Suppose an electric bar heater is switched off at time $t = 0$. Then for $t > 0$ there are no sources of heat in the bar but there is an outward flow of heat from any small region as the bar cools. The time-dependent pattern of this heat flow is governed by

$$\mathrm{div}\,\mathbf{h} = -c\rho_{\mathrm{m}}\frac{\partial T}{\partial t} \qquad \text{(non-steady conditions; no sources)} \qquad (5.34)$$

where c is the specific heat and ρ_m the mass density of the material, and $-\partial T/\partial t$ is the rate at which the temperature falls at any point. The right-hand side of Eq (5.34) describes the rate at which heat is lost per unit volume of the material. Thus the equation equates the outward flow of heat per unit volume per second (div**h**) to the rate at which heat is lost per unit volume by conduction, and therefore expresses the law of conservation of energy. Eq (5.34) is known as the **continuity equation** for heat conduction.

When an electric bar heater is switched on there is a transient period when the bar is heating up. The non-steady heat flow in the bar is then described by

$$\text{div}\mathbf{h} = -c\rho_m \frac{\partial T}{\partial t} + S \quad \text{(non-steady conditions with sources)} \qquad (5.35)$$

where S is the electrical power dissipated as heat per unit volume in the bar, as in Eq (5.33). Eqs (5.32) to (5.34) are clearly special cases of Eq (5.35).

Equations similar to those above describe the flow of gases and liquids and the flow of electric charge. For example, there is a continuity equation for electric charge which states that

$$\text{div}\mathbf{J} = -\frac{\partial \rho_c}{\partial t} \qquad \text{(always true)} \qquad (5.36)$$

where **J** is the current density vector (defined in Section 5.2.3) and ρ_c is the electric charge density. This equation equates the outward current flow per unit volume evaluated at a point to the rate at which the electric charge per unit volume is decreasing at that point, a requirement of the law of conservation of electric charge. There is no charge source term here analogous to the heat source field S in Eq (5.35), since this would represent the creation of electric charge which would violate the law of conservation of electric charge.

A scalar field equation like Eq (5.37) is satisfied at each field point. For example, if the electric charge density ρ_c is zero at a point P then div**E** is zero at P. This does not of course mean that **E** is zero at P since div**E** involves only the partial derivatives of **E**, not **E** itself.

There are other examples where equations similar to those above apply but there is no physical flow of any kind. Perhaps the most familiar examples are in electromagnetism where the divergence of the electrostatic field **E** at a point is equal to the electric charge density ρ_c at that point divided by ϵ_0, i.e.

$$\text{div}\mathbf{E} = \frac{\rho_c}{\epsilon_0} \qquad (5.37)$$

A pictorial expression of Eq (5.37) is found in field line diagrams where lines of **E** begin on positive charges (sources of **E**) and terminate on negative charges (sinks of **E**). Eq (5.37) is known as the **differential form of Gauss's law**.

There is a Gauss's law for magnetic fields which states that

$$\text{div}\mathbf{B} = 0 \qquad \text{(always true)} \qquad (5.38)$$

This law expresses the fact that magnetic fields do not have scalar sources; i.e. magnetic monopoles, once thought to be the magnetic analogues of electric charges, do not exist. This is why we can't produce an isolated north or south magnetic pole by, say, cutting the end off a magnet – we just get two magnets. Of course magnetic fields do have physical sources, such as magnets (i.e. magnetic dipoles) and electric currents, but these are vector sources (to be explained in Section 5.4) and do not give rise to any divergence of **B**. The law div**B** = 0, signifying the non-existence of magnetic monopole sources, is always true and consequently magnetic field lines are always closed loops.

Summary of section 5.3

- The **divergence of a vector field** is the outward flux of the field per unit of enclosed volume, evaluated as a limit at a point (Eq (5.25)).
- The divergence of a cartesian vector field **F** can be calculated from

$$\text{div}\mathbf{F} = \partial F_x/\partial x + \partial F_y/\partial y + \partial F_z/\partial z \qquad (5.26)$$

- The polar forms of div**F** are given by Eqs (5.29) and (5.30).
- Many physical laws relate the divergence of a vector field to a scalar source field.

Example 3.1 (*Objective 3*) Use Eqs (5.26), (5.29) or (5.30) to calculate the divergence of the following vector fields:

(a) $\mathbf{f}(x,y,z) = y\mathbf{e}_x + x\mathbf{e}_y + \mathbf{e}_z$

(b) $\mathbf{G}(x,y,z) = xy^2\mathbf{e}_x + x^2y\mathbf{e}_y + z^3\mathbf{e}_z$

(c) $\mathbf{h}(\rho,\phi,z) = \dfrac{1}{\rho}\mathbf{e}_\phi$ $\qquad (\rho \neq 0)$

(d) $\mathbf{Q}(r,\theta,\phi) = \mathbf{e}_r\dfrac{2k}{r^3}\cos\theta + \mathbf{e}_\theta\dfrac{k}{r^3}\sin\theta$ $\qquad$ ($r \neq 0$ and k is a constant)

(e) $\mathbf{R}(\mathbf{r}) = \mathbf{r}$, where $\mathbf{r}$ is the position vector.

(f) $\mathbf{T}(r,\theta,\phi) = \mathbf{e}_\theta$

(g) $\mathbf{v}(x,y,z) = \alpha y(d-y)\mathbf{i} + \mathbf{k}$ $\qquad$ ($0 \leq y \leq d$; α and d are constants)

Solution 3.1

(a) The components of **f** are $f_x = y$, $f_y = x$ and $f_z = 1$ and so Eq (5.26) gives div**f** = 0.

(b) Similarly, using Eq (5.26), we have

$$\begin{aligned}\text{div}\mathbf{G} &= \frac{\partial(xy^2)}{\partial x}+\frac{\partial(x^2y)}{\partial y}+\frac{\partial(z^3)}{\partial z}\\ &= y^2+x^2+3z^2.\end{aligned}$$

(c) The cylindrical polar components are $h_\rho = 0$, $h_\phi = 1/\rho$ and $h_z = 0$ and so Eq (5.29) yields $\text{div}\mathbf{h} = 0$.

(d) Here the spherical polar components are $Q_r = \frac{2k}{r^3}\cos\theta$, $Q_\theta = \frac{k}{r^3}\sin\theta$ and, $Q_\phi = 0$ and so Eq (5.30) gives

$$\text{div}\mathbf{Q} = -\frac{6k}{r^4}\cos\theta+\left(\frac{2}{r}\right)\left(\frac{2k}{r^3}\right)\cos\theta+\left(\frac{1}{r}\right)\frac{k}{r^3}\cos\theta+\frac{1}{r}\cot\theta\frac{k}{r^3}\sin\theta=0$$

(e) In cartesian form, $\mathbf{R}(x,y,z) = x\mathbf{i} + y\mathbf{j} + z\mathbf{k}$ and so Eq (5.26) gives $\text{div}\mathbf{R} = 1+1+1 = 3$. (Alternatively, in spherical polar form, $\mathbf{R}(r,\theta,\phi) = r\mathbf{e}_r$ and the first two terms of Eq (5.30) give $\text{div}\mathbf{R} = 1 + 2 = 3$.)

$\mathbf{e}_\theta$ is not a constant vector; its direction depends on the angles θ and ϕ. You should not be surprised therefore to find that $\text{div}\mathbf{e}_\theta$ is non-zero.

(f) $T_\theta = 1$ and $T_r = T_\phi = 0$, and the fourth term in Eq (5.30) gives the only contribution: $\text{div}\mathbf{T} = (\cot\theta)/r$.

(g) The components are $v_x = \alpha y(d - y)$, $v_y = 0$ and $v_z = 1$. Hence $\partial v_x/\partial x = \partial v_y/\partial y = \partial v_z/\partial z = 0$, and so $\text{div}\mathbf{v} = 0$.

Example 3.2 (*Objective 3*) Given the vector field

$$\mathbf{F}(x,y,z) = (x\mathbf{i} + y\mathbf{j})(x^2+y^2)$$

find the divergence at the point (1,2,0) and on the line (1,0,z).

Solution 3.2 We have $F_x = x(x^2+y^2)$, $F_y = y(x^2+y^2)$ and $F_z = 0$. Differentiating, we find $\partial F_x/\partial x = 3x^2+y^2$, $\partial F_y/\partial y = x^2+3y^2$ and $\partial F_z/\partial z = 0$. Thus

$$\text{div}\mathbf{F} = 4(x^2+y^2)$$

At the point (1,2,0), $\text{div}\mathbf{F} = 4(1^2+2^2) = 20$. On the line (1,0,$z$), $\text{div}\mathbf{F} = 4$.

(An alternative approach is to recognise that the field has cylindrical symmetry and write it as $\mathbf{F}(\rho,\phi,z) = \rho^3\mathbf{e}_\rho$. The divergence is then found from the first two terms of Eq (5.29) with $F_\rho = \rho^3$. Thus $\text{div}\mathbf{F} = 3\rho^2 + (1/\rho)\rho^3 = 4\rho^2 = 4(x^2+y^2)$, as before.)

Example 3.3 (*Objective 3*) Show that the velocity field of a solid cylinder spinning with constant angular velocity ω about its axis has a divergence of zero everywhere.

Solution 3.3 The velocity at a point in the cylinder with position vector **r** measured from a reference point O on the axis is $\mathbf{v}(\mathbf{r}) = \boldsymbol{\omega} \times \mathbf{r}$. This leads directly to the spherical polar form $\mathbf{v}(r,\theta,\phi) = \omega r(\sin\theta)\mathbf{e}_\phi$ where the z-axis is along the axis of rotation (Fig 5.10). However it is better to take full advantage of the cylindrical symmetry and use cylindrical polar coordinates with $\rho = r\sin\theta$. Thus $\mathbf{v}(\rho,\phi,z) = \omega\rho\mathbf{e}_\phi$. The components are $v_\rho = 0$, $v_\phi = \omega\rho$ and $v_z = 0$, and so Eq (5.29) gives $\text{div}\mathbf{v} = 0$.

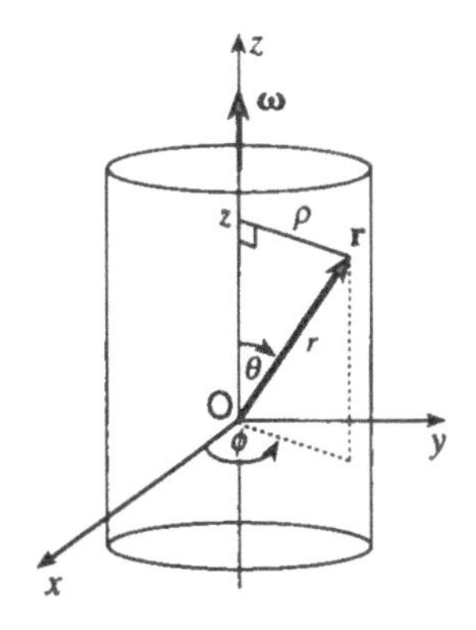

Fig 5.10
Spherical and cylindrical polar coordinates of a point of position vector **r** in a solid cylinder.

Example 3.4 (*Objective 3*) Consider a point light source of power W (in watts) located at the origin. When the source is radiating into a transparent medium (such as air) the rate of flow of light energy across unit area normal to the radial direction is described by a flow vector **P** obeying an inverse square law,

$$\mathbf{P}(r) = \frac{W}{4\pi r^2}\mathbf{e}_r \qquad (r > 0)$$

When the source is surrounded by a uniform absorbing medium (a street lamp in fog for example) the flow vector is found to be

$$\mathbf{P}_\text{m}(r) = \frac{W}{4\pi r^2}\exp(-\alpha r)\mathbf{e}_r \qquad (r > 0)$$

where α is a positive constant called the absorption coefficient of the medium. Find $\text{div}\mathbf{P}$ and $\text{div}\mathbf{P}_\text{m}$ and interpret the results in terms of sources and sinks of the flow vector.

Solution 3.4 The divergence of any spherically symmetric vector field such as **P** or $\mathbf{P}_\text{m}$ is given by the first two terms of Eq (5.30). Thus

$$\text{div}\mathbf{P} = \frac{W}{4\pi}\left[\frac{\partial}{\partial r}\left(\frac{1}{r^2}\right) + \frac{2}{r}\left(\frac{1}{r^2}\right)\right] = 0$$

$$\text{div}\mathbf{P}_\text{m} = \frac{W}{4\pi}\left[\frac{\partial}{\partial r}\left(\frac{\exp(-\alpha r)}{r^2}\right) + \frac{2}{r}\left(\frac{\exp(-\alpha r)}{r^2}\right)\right]$$

$$= \frac{W}{4\pi}\left(-\frac{\alpha}{r^2}\right)\exp(-\alpha r) = -\alpha\,|\mathbf{P}_\text{m}(r)|$$

Note that $\mathbf{P}_m$ is in Wm^{-2} and so $\mathrm{div}\mathbf{P}_m$ is in Wm^{-3} since a derivative symbol such as $\partial/\partial x$ carries units of m^{-1} when x is in metres.

When the light source is radiating into a transparent medium there are no sources or sinks in the domain of the flow vector (the light source itself is at $r = 0$ outside the domain) and so $\mathrm{div}\mathbf{P} = 0$ everywhere. An absorbing medium (such as fog) is a sink of light. This is indicated by the negative value found for $\mathrm{div}\mathbf{P}_m$. $|\mathrm{div}\mathbf{P}_m|$ is equal to the light power absorbed per unit volume of the medium.

Example 3.5 (*Objective 3*) Verify that the vector field

$$\mathbf{F}(r,\theta,\phi) = \frac{\mathbf{e}_r}{r^n} \qquad (r > 0;\ n \text{ is an integer})$$

is divergence-free (solenoidal) only when $n = 2$.

Solution 3.5 Use Eq (5.30) with $F_r = 1/r^n$ and $F_\theta = F_\phi = 0$, and we find

$$\mathrm{div}\mathbf{F} = -\frac{n}{r^{n+1}} + \left(\frac{2}{r}\right)\left(\frac{1}{r^n}\right) = \frac{2-n}{r^{n+1}}$$

which is zero only for $n = 2$.

Problem 3.1 (*Objective 3*) Find the divergence of the following vector fields:

(a) $\mathbf{F}(x,y,z) = yz\mathbf{i} + zx\mathbf{j} + xy\mathbf{k}$
(b) $\mathbf{G}(x,y,z) = \mathbf{i}\cos x + \mathbf{j}\sin y$
(c) $\mathbf{H}(x,y,z) = \mathbf{i}\cos y + \mathbf{j}\sin x$
(d) $\mathbf{C}(x,y,z) = \mathbf{i} + \mathbf{j} + \mathbf{k}$
(e) $\mathbf{V}(\rho,\phi,z) = (1/\rho)\mathbf{e}_\phi \qquad (\rho \neq 0)$
(f) $\mathbf{W}(r,\theta,\phi) = r\exp(-r)\mathbf{e}_r$

Problem 3.2 (*Objective 3*) $\mathbf{v}$ is the cartesian vector field defined by

$$\mathbf{v}(x,y,z) = ((x^2 + 2xy), y^2, xy)$$

Find $\mathrm{div}\mathbf{v}$ at the origin and at the point on the positive x-axis at unit distance from the origin. What is the value of $\mathrm{div}\mathbf{v}$ on the z-axis?

Problem 3.3 (*Objective 3*) Find the divergence of the field

$$\mathbf{E}(x,y,z) = x\exp(-z^2)\mathbf{i}$$

at points with coordinates $(x,y,-3)$ and say where these points lie.

Problem 3.4 (*Objective 3*) An electric charge Q is distributed uniformly in a spherical volume of radius R. The electric field in the sphere due to this charge distribution is found to be (Example 5.3 of Chapter 4)

$$\mathbf{E}(r,\theta,\phi) = \frac{Qr\mathbf{e}_r}{4\pi\epsilon_0 R^3} \qquad (r < R)$$

Find div**E** and confirm that it satisfies Gauss's law (Eq (5.37)).

Problem 3.5 (*Objective 3*) Use Gauss's law (Eq (5.37)) to show that an electric field of the form

$$\mathbf{E}(x,y,z) = [Az(1 - z/2d) + B]\mathbf{k} \qquad (0 < z < d),$$

where A, B and d are non-zero constants, cannot exist unless there is electric charge in the domain.

Problem 3.6 (*Objective 3*) Find the integral value n for which the vector field

$$\mathbf{F}(\rho,\phi,z) = \mathbf{e}_\rho/\rho^n \qquad (\rho > 0)$$

is solenoidal.

5.4 CURL OF A VECTOR FIELD

There are nine partial derivatives of a three-dimensional vector field. They describe the spatial rates of change of the three scalar component fields in directions parallel to the three coordinate axes. We can display them as a three by three array where each row gives the partial derivatives of one component of the field:

$$\begin{array}{ccc} \partial F_x/\partial x & \partial F_x/\partial y & \partial F_x/\partial z \\ \partial F_y/\partial x & \partial F_y/\partial y & \partial F_y/\partial z \\ \partial F_z/\partial x & \partial F_z/\partial y & \partial F_z/\partial z \end{array}$$

We have seen (Eq (5.26)) that the divergence of the vector field is the scalar obtained by adding the three partial derivatives along the main diagonal of the array (top left to bottom right). Loosely speaking, these three partial derivatives describe the way the field is changing along the field direction. We now introduce a vector field called the curl of **F**, denoted by curl**F**, which is constructed from the other six elements of the array and the cartesian unit vectors. You can think of these six partial derivatives as "sideways" derivatives describing how the field changes in directions perpendicular to the field direction.

Curl is introduced in Section 5.4.1 as a development of the idea of circulation around a closed loop introduced in Section 4.5 which should be studied before Section 5.4.1. Section 5.4.2 is concerned with calculating the curls of vector fields in cartesian and polar coordinates. Section 5.4.3 gives a physical interpretation of curl as a measure of the local rotation in a vector field, and then gives examples of physical laws expressed in terms of curl.

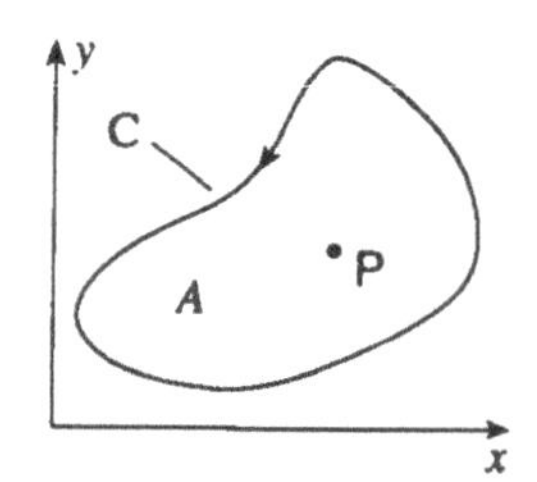

Fig 5.11
A closed loop C of area A in the x-y plane enclosing a point P.

5.4.1 Introducing curl

Let P be a point in the domain of a two-dimensional vector field $\mathbf{f}(x,y)$ and let C be a loop in the x-y plane enclosing an area A which includes the point P (Fig 5.11). The anticlockwise circulation W_0 of the field $\mathbf{f}$ around the loop can be found by the method described in Section 4.5 (or more generally by methods to be described in Chapter 6). Now suppose the loop is made to shrink, like a tightening noose, until it eventually collapses onto the point P. During this process the circulation W_0 and the enclosed area A both become smaller in magnitude and eventually zero, but the quotient W_0/A, the circulation per unit area, tends to a limit which is independent of the shape of the loop and the details of how it is made to shrink; it depends only on the field $\mathbf{f}$ and the position of the point P. The magnitude of this limit is the magnitude of a vector called the curl of $\mathbf{f}$ at P or curl$\mathbf{f}$. The direction of the vector curl$\mathbf{f}$ is at right-angles to the x-y plane (the plane of the loop), and in the positive z-direction when the limit of W_0/A is positive. Thus

$$\text{curl}\mathbf{f}(x,y) = [\lim_{A\to 0}(W_0/A)]\mathbf{k} \tag{5.39}$$

This picture of curl will be extended to the case of three-dimensional vector fields in Section 6.4.

The limit (5.39) is not useful for calculating the curl of a vector field. Instead we use an expression for curl in terms of partial derivatives (Eq (5.40) below). We now indicate how this expression is obtained from the limit (5.39).

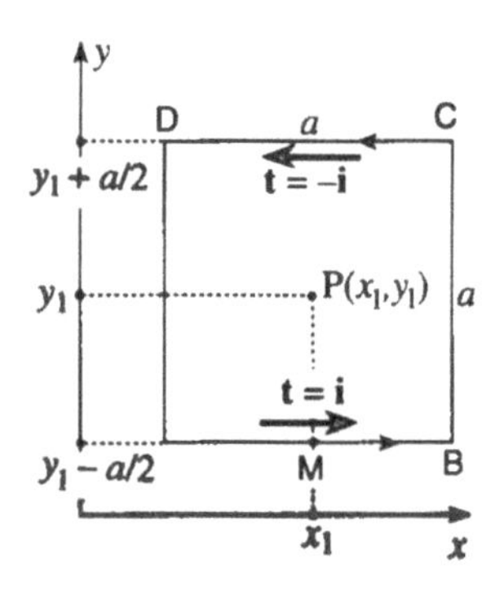

Fig 5.12
The point $P(x_1,y_1)$ at the centre of a square loop of side length a.

Let the point $P(x_1,y_1)$ be at the centre of a closed square loop ABCD in the x-y plane with sides of length a (Fig 5.12). The anticlockwise circulation of a general two-dimensional vector field

$$\mathbf{f}(x,y) = \mathbf{i}\,f_x(x,y) + \mathbf{j}\,f_y(x,y)$$

is the sum of contributions along the four sides. Consider the side AB. The unit tangent vector on this side for traversal from A to B is $\mathbf{t} = \mathbf{i}$ and the tangential component of $\mathbf{f}$ at the midpoint M of AB is $\mathbf{i}\cdot\mathbf{f} = f_x(x_1, y_1 - a/2)$. We now make the assumption that the loop is sufficiently small that this value applies along the whole length of AB (i.e. we are ignoring the variation of f_x on AB). With this assumption we use Eq (4.30) to obtain the contribution of the path segment AB to the anticlockwise circulation: $W_{AB} = af_x(x_1, y_1 - a/2)$. Now consider side CD. The unit tangent vector here is $\mathbf{t} = -\mathbf{i}$ and, by a similar argument, the contribution to the circulation is $W_{CD} = -af_x(x_1, y_1 + a/2)$. The net contribution of sides AB and CD is therefore $af_x(x_1,y_1 - a/2) - af_x(x_1, y_1 + a/2)$. Following

Eq (5.39), we now divide this by the area a^2 of the loop and take the limit as the loop shrinks onto P. Thus we have the following contribution to the curl

$$-\lim_{a\to 0}\left[\frac{f_x(x_1,y_1+a/2)-f_x(x_1,y_1-a/2)}{a}\right]\mathbf{k}=-\frac{\partial f_x}{\partial y}\mathbf{k}$$

where we have recognised the limit as the definition of the partial derivative (Eq (5.2)). The contribution of the other pair of sides BC and DA is found by a similar argument to be $(\partial f_y/\partial x)\mathbf{k}$. We conclude that curl**f** at P is $(\partial f_y/\partial x - \partial f_x/\partial y)\mathbf{k}$.

5.4.2 Calculating curl

The argument of the preceding section gives the following expression for the **curl of a two-dimensional vector field** in the *x-y* plane:

$$\mathrm{curl}\mathbf{f}(x,y)=\left(\frac{\partial f_y}{\partial x}-\frac{\partial f_x}{\partial y}\right)\mathbf{k} \qquad (5.40)$$

Eq (5.40) is equivalent to Eq (5.39) but is more convenient in calculations. Consider for example the field

$$\mathbf{u}(x,y)=\omega(-y\mathbf{i}+x\mathbf{j}) \qquad (5.41)$$

Eq (5.41) is the velocity field on the surface of a disc in the *x-y* plane rotating uniformly with angular speed ω about the *z*-axis.

where ω is a constant. The components of **u** are $u_x = -\omega y$ and $u_y = \omega x$ and the relevant partial derivatives are $\partial u_y/\partial x = \omega$ and $\partial u_x/\partial y = -\omega$, and so Eq (5.40) gives

$$\mathrm{curl}\mathbf{u}=2\omega\mathbf{k} \qquad (5.42)$$

We now quote the expression for the **curl of a three-dimensional vector field F**(x,y,z)

$$\mathrm{curl}\mathbf{F}(x,y,z)=\left(\frac{\partial F_z}{\partial y}-\frac{\partial F_y}{\partial z}\right)\mathbf{i}+\left(\frac{\partial F_x}{\partial z}-\frac{\partial F_z}{\partial x}\right)\mathbf{j}+\left(\frac{\partial F_y}{\partial x}-\frac{\partial F_x}{\partial y}\right)\mathbf{k} \qquad (5.43a)$$

Fortunately this expression is very much simpler than it appears to be at first sight. You should spend some time looking at it carefully and noting the cyclic symmetries (Fig 5.13). Note first of all that the third term, the *z*-component of curl, is equal to the right-hand side of Eq (5.40), as expected; it involves only the partial derivatives of F_y and F_x with respect to *x* and *y* respectively. The *x*-component, $(\mathrm{curl}\mathbf{F})_x = \partial F_z/\partial y - \partial F_y/\partial z$, has the same form as the *z*-component but with the cyclic replacements $x \to y$ and $y \to z$; and similarly, the *y*-component of curl is obtained from the *x*-component by the replacements $y \to z$

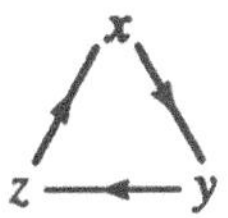

Fig 5.13
Cyclic replacements.

and $z \to x$. In fact the vector curl**F** has the same form and symmetry as the familiar vector product Eq (2.28a), a point we shall take up in Section 5.5.2.

Eq (5.43a) is often written as a cartesian triple

$$\text{curl}\mathbf{F} = \left(\frac{\partial F_z}{\partial y} - \frac{\partial F_y}{\partial z}, \ \frac{\partial F_x}{\partial z} - \frac{\partial F_z}{\partial x}, \ \frac{\partial F_y}{\partial x} - \frac{\partial F_x}{\partial y} \right) \qquad (5.43b)$$

or as a determinant

$$\text{curl}\mathbf{F} = \begin{vmatrix} \mathbf{i} & \mathbf{j} & \mathbf{k} \\ \dfrac{\partial}{\partial x} & \dfrac{\partial}{\partial y} & \dfrac{\partial}{\partial z} \\ F_x & F_y & F_z \end{vmatrix} \qquad (5.43c)$$

The determinant form exhibits the cyclic symmetries nicely and is easily remembered.

We have assumed that **F** is differentiable everywhere in its domain and so the vector curl**F** exists at each point. curl**F** is a therefore a vector field with the same domain as **F**. (Now is a good time to study Example 4.1 (a) to (e) and Problems 4.1 and 4.2.)

The expressions for curl**F** in polar coordinates are:

curl in cylindrical polar coordinates

$$\text{curl}\mathbf{F} = \mathbf{e}_\rho \left[\frac{1}{\rho}\frac{\partial F_z}{\partial \phi} - \frac{\partial F_\phi}{\partial z} \right] + \mathbf{e}_\phi \left[\frac{\partial F_\rho}{\partial z} - \frac{\partial F_z}{\partial \rho} \right]$$

$$+ \mathbf{e}_z \left[\frac{\partial F_\phi}{\partial \rho} + \frac{F_\phi}{\rho} - \frac{1}{\rho}\frac{\partial F_\rho}{\partial \phi} \right] \qquad (5.44)$$

The curl of a two-dimensional field in plane polar coordinates is given by the last term (z-component only) of Eq (5.44).

curl in spherical polar coordinates

$$\text{curl}\mathbf{F} = \mathbf{e}_r \left[\frac{1}{r}\frac{\partial F_\phi}{\partial \theta} + \frac{\cot\theta}{r} F_\phi - \frac{1}{r\sin\theta}\frac{\partial F_\theta}{\partial \phi} \right]$$

$$+ \mathbf{e}_\theta \left[\frac{1}{r\sin\theta}\frac{\partial F_r}{\partial \phi} - \frac{\partial F_\phi}{\partial r} - \frac{F_\phi}{r} \right] + \mathbf{e}_\phi \left[\frac{\partial F_\theta}{\partial r} + \frac{F_\theta}{r} - \frac{1}{r}\frac{\partial F_r}{\partial \theta} \right] \qquad (5.45)$$

These rather complicated expressions are often surprisingly easy to use when the field **F** has the appropriate symmetry since some of the terms are then zero. For example, the field **u** (Eq (5.41)) has circular symmetry and can be written in plane polar form as $\mathbf{u}(\rho,\phi) = \mathbf{e}_\phi\omega\rho$. The components are $u_\rho = 0$, $u_\phi = \omega\rho$ and $u_z = 0$. The only non-zero partial derivative is therefore $\partial u_\phi/\partial\rho = \omega$. Thus all terms in Eq (5.44) are zero except for the first two terms of the *z*-component. These yield $\mathrm{curl}\mathbf{u} = \mathbf{e}_z[\partial u_\phi/\partial\rho + u_\phi/\rho] = \mathbf{e}_z 2\omega$, as found before.

5.4.3 Curl and physical law

We first discuss an interpretation of the curl of a vector field as a measure of the local rotation in the field. Consider the velocity field $\mathbf{v}(x,y)$ produced on the surface of a canal in the region between one bank ($y = 0$) and the near side of a passing ship ($y = d$). The ship moves parallel to the bank in the *x*-direction (Fig 5.14) dragging the water along to produce a surface velocity shear-like field that can often be described by

$$\mathbf{v}(x,y) = \mathbf{i}Cy \qquad (0 < y < d;\ C \text{ is a constant}) \qquad (5.46)$$

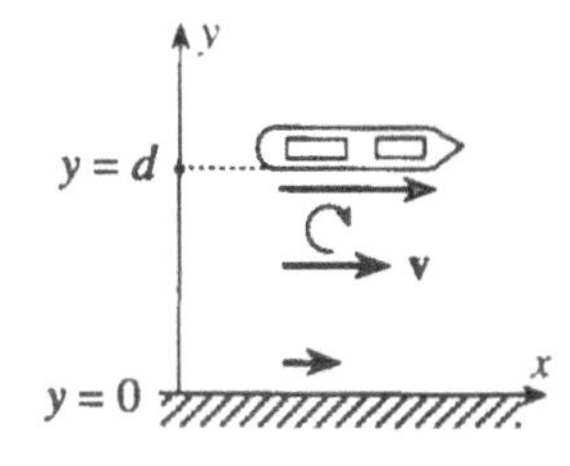

Fig 5.14
The three arrows depict the velocity field **v** between the canal bank ($y = 0$) and the near side of a passing ship ($y = d$).

The shear constant $C > 0$ describes the rate of increase of surface speed with distance out from the bank. The components of this field are $v_x = Cy$ and $v_y = 0$, and so the curl is found from Eq (5.40) to be

$$\mathrm{curl}\mathbf{v} = -C\,\mathbf{k}$$

Standing on the bank with the ship moving towards your right, you would observe that any small floating object between the bank and the ship rotates clockwise about a vertical *z*-axis as it floats along parallel to the bank. This is because its far side is dragged along by the water faster than its near side. The actual rate of rotation ω depends on the physical properties of the object but is proportional to the shear constant *C*, while the clockwise sense of the rotation in the *x*-*y* plane is specified by the unit vector $-\mathbf{k}$, according to the right-hand rule. Thus the angular velocity of the floating object is proportional to $-C\mathbf{k} = \mathrm{curl}\mathbf{v}$. In practice we can explore the curl of a fluid velocity field by measuring the rate of rotation of a floating probe of standard size and shape placed at various points in the fluid.

In fluid mechanics the curl of a velocity field is called the **vorticity**, a term that expresses the interpretation of curl as a measure of the local rotation in the fluid.

Another example is a disc spinning with angular velocity $\boldsymbol{\omega} = \omega\mathbf{k}$. Any small element of the disc undergoes a local rotational motion with angular velocity $\boldsymbol{\omega}$ as it travels along a circular path around the centre of the disc. The surface velocity field is $\mathbf{u}(x,y) = \omega(-y\mathbf{i} + x\mathbf{j})$ which has $\mathrm{curl}\mathbf{u} = 2\omega\mathbf{k} = 2\boldsymbol{\omega}$ (Eq (5.42)) so again we find a proportionality between the local rotation and the curl of the velocity.

A familiar example of translation along a circular path without rotation can be found in a fair ground where the chairs and passengers on the Ferris wheel keep upright as the wheel rotates in a vertical plane. Rotation of the chairs would occur if the bearings on which they are hung were to accidentally lock.

It should be emphasised that the curl refers to the local rotational motion and not the translational or bulk motion. In fact fluid velocity fields exist in which a floating object would travel along a circular path without rotating, indicating that the velocity field has zero curl. Such a field occurs in the outer region of a vortex, see Problem 4.5.

In a velocity field there is actual movement of material. In other kinds of vector field such as electric and magnetic fields there is no actual movement of material but the term local rotation is often used figuratively to refer to the curl of the field. In fact an alternative notation that is sometimes used for curl$\mathbf{F}$ is rot$\mathbf{F}$, and any vector field that has zero curl is said to be **irrotational**.

We now consider some physical laws that are expressed in terms of curl. In Section 5.3.3 you saw several examples of physical laws that equate the divergence of a vector field to a scalar source field (5.37). In a similar manner there are physical laws that relate the curl of a vector field to another vector field which can sometimes be interpreted as a vector source field. An example is the differential form of **Ampère's circuital law** for magnetic fields produced by steady currents

$$\text{curl}\mathbf{B} = \mu_0 \mathbf{J} \qquad \text{(static fields)} \qquad (5.47)$$

A vector field equation like Eq (5.47) is satisfied at every field point. Thus, for example, if $\mathbf{J} = \mathbf{0}$ at a point P then curl$\mathbf{B} = \mathbf{0}$ at P. This does not of course mean that $\mathbf{B}$ is zero at P since curl$\mathbf{B}$ involves only the partial derivatives of $\mathbf{B}$. A case in point is in the space outside a current-carrying wire where $\mathbf{J}$ and curl$\mathbf{B}$ are both zero but $\mathbf{B}$ is non-zero.

The current density vector $\mathbf{J}$ is the source of the magnetic field. When the fields vary in time, Ampère's circuital law is modified by the addition of a second vector source term called the **displacement current density** which is proportional to the rate of change with time of the electric field,

$$\text{curl}\mathbf{B} = \mu_0 \mathbf{J} + \mu_0 \epsilon_0 \frac{\partial \mathbf{E}}{\partial t} \qquad (5.48)$$

There is a similar law which states that a changing magnetic field is a vector source of an electric field,

$$\text{curl}\mathbf{E} = -\frac{\partial \mathbf{B}}{\partial t} \qquad (5.49)$$

This is in fact **Faraday's law** of electromagnetic induction in differential form.

The four equations connecting the electromagnetic fields $\mathbf{E}$ and $\mathbf{B}$, Eqs (5.37), (5.38), (5.48) and (5.49), govern all electromagnetic phenomena (in the absence of dielectric and magnetic materials) and are known collectively as **Maxwell's equations**.

As a final application of curl we return to the topic of conservative fields. We have seen (Section 5.2.3) that any conservative field $\mathbf{G}$ can be expressed as the gradient of a scalar field, $\mathbf{G} = -\text{grad}\phi$. We shall show in Example 4.6 that the curl of any gradient vector field is always zero, i.e. curl(gradϕ) = $\mathbf{0}$ everywhere for all fields ϕ. It follows that any conservative field is irrotational, i.e. it has zero curl everywhere in its domain. This result provides us with a useful test for conservative fields

$$\text{curl}\mathbf{G} = \mathbf{0} \qquad \text{(the curl test for a conservative field)} \qquad (5.50)$$

We discuss conservative fields further in Section 6.4.

Summary of section 5.4

- The curl of a vector field is itself a vector field. The magnitude of the curl of a two-dimensional vector field **f** at a point is the magnitude of the circulation per unit enclosed area at that point (Eq (5.39)).

- The **curl of a three-dimensional vector field** $\mathbf{F}(x,y,z)$ can be expressed in terms of the partial derivatives

$$\text{curl}\mathbf{F} = \left(\frac{\partial F_z}{\partial y} - \frac{\partial F_y}{\partial z}\right)\mathbf{i} + \left(\frac{\partial F_x}{\partial z} - \frac{\partial F_z}{\partial x}\right)\mathbf{j} + \left(\frac{\partial F_y}{\partial x} - \frac{\partial F_x}{\partial y}\right)\mathbf{k} \qquad (5.43a)$$

  This can be written as a cartesian triple (Eq (5.43b)) or as a determinant (Eq (5.43c)). The curl of a two-dimensional field **f** in the x-y plane is $\text{curl}\mathbf{f} = (\partial f_y/\partial x - \partial f_x/\partial y)\mathbf{k}$. The polar forms of curl are given by Eqs (5.44) and (5.45).

- The vector field curl**F** describes the local rotation in **F**. Some physical laws equate the curl of a vector field **F** to another vector field which can be interpreted as a vector source of **F**.

- If curl**G** = **0** the vector field **G** is **irrotational** and conservative.

Example 4.1 (*Objective 4*) Determine the curl of each of the following vector fields:

(a) $\mathbf{f}(x,y,z) = x^2y\mathbf{i} + y^2z\mathbf{j} + z^2x\mathbf{k}$
(b) $\mathbf{g}(x,y,z) = x(\mathbf{i} + \mathbf{j}) - 3\mathbf{k}$
(c) $\mathbf{h}(x,y,z) = \mathbf{k}(x^2 + y^2)^{-1}$ $\quad (x,y) \neq (0,0)$
(d) $\mathbf{p}(x,y) = (x - y,\ x + y)$
(e) $\mathbf{t}(x,y,z) = (\sin(xy),\ \cos(xy),\ 0)$
(f) $\mathbf{y}(\rho,\phi,z) = \mathbf{e}_\rho \log_e(\rho/a)$ $\quad (\rho > 0,\ a$ is a constant)
(g) $\mathbf{q}(r,\theta,\phi) = \mathbf{e}_r(\cos\theta)/r^3$ $\quad (r > 0)$

Solution 4.1 For parts (a) to (e) use any of the three versions of Eq (5.43).

(a) The components of **f** are $f_x = x^2y$, $f_y = y^2z$ and $f_z = z^2x$. Hence

$$\text{curl}\mathbf{f} = \left(\frac{\partial(z^2x)}{\partial y} - \frac{\partial(y^2z)}{\partial z}\right)\mathbf{i} + \left(\frac{\partial(x^2y)}{\partial z} - \frac{\partial(z^2x)}{\partial x}\right)\mathbf{j}$$

$$+\left(\frac{\partial(y^2z)}{\partial x} - \frac{\partial(x^2y)}{\partial y}\right)\mathbf{k}$$

$$= -y^2\mathbf{i} - z^2\mathbf{j} - x^2\mathbf{k}$$

(b) The components are $g_x = x$, $g_y = x$ and $g_z = -3$. The only relevant non-zero partial derivative is $\partial g_y/\partial x = 1$ and this gives $\text{curl}\mathbf{g} = \mathbf{k}$.

(c) The components are $h_x = h_y = 0$ and $h_z = 1/x^2 + y^2$, and the non-zero partial derivatives are $\partial h_z/\partial x = -2x/(x^2+y^2)^2$ and $\partial h_z/\partial y = -2y/(x^2+y^2)^2$. Thus
$\text{curl}\mathbf{h} = -2y\mathbf{i}/(x^2+y^2)^2 + 2x\mathbf{j}/(x^2+y^2)^2 = -2(y\mathbf{i} - x\mathbf{j})/(x^2+y^2)^2$.

(d) $p_x = x - y$, $p_y = x + y$, $\partial p_y/\partial x = 1$, $\partial p_x/\partial y = -1$ and so $\text{curl}\mathbf{p} = 2\mathbf{k}$.

(e) The components are $t_x = \sin(xy)$, $t_y = \cos(xy)$ and $t_z = 0$. The relevant non-zero partial derivatives are $\partial t_x/\partial y = x\cos(xy)$ and $\partial t_y/\partial x = -y\sin(xy)$, and so $\text{curl}\mathbf{t} = -(y\sin(xy) + x\cos(xy))\mathbf{k}$.

(f) The cylindrical polar components are $y_\rho = \log_e(\rho/a)$, $y_\phi = y_z = 0$, and so the only non-vanishing partial derivative is $\partial y_\rho/\partial\rho$ which does not appear in the appropriate expression for curl, Eq (5.44), and so $\text{curl}\mathbf{y} = 0$.

(g) The spherical polar components are $q_r = (\cos\theta)/r^3$, $q_\theta = q_\phi = 0$. The only non-zero partial derivative of q appearing in the appropriate expression for curl, Eq (5.45), is $\partial q_r/\partial\theta = (-\sin\theta)/r^3$ and so $\text{curl}\mathbf{q} = \mathbf{e}_\phi(-1/r)(-\sin\theta)/r^3 = \mathbf{e}_\phi\sin(\theta)/r^4$.

Example 4.2 (*Objective 4*) Given

$$\mathbf{F}(x,y,z) = x(y-z)\mathbf{i} + 3x^2\mathbf{j} + yz\mathbf{k}$$

determine $\text{curl}\mathbf{F}$ at (a) the origin, (b) the point (1,2,3) and (c) the point 5 units from the origin on the positive z-axis.

Solution 4.2 Use Eq (5.43a). The components of $\mathbf{F}$ are $F_x = x(y-z)$, $F_y = 3x^2$ and $F_z = yz$, and so

$$\text{curl}\mathbf{F} = z\mathbf{i} - x\mathbf{j} + 5x\mathbf{k}$$

Now evaluate curl**F** at the specified points:

(a) At the origin, curl**F** = **0**
(b) At (1,2,3), curl**F** = 3**i** − **j** + 5**k**
(c) At (0,0,5), curl**F** = 5**i**.

Example 4.3 (*Objective 4*) Given the vector field

$$\mathbf{a}(x,y,z) = (\sin(\lambda y), \cos(\lambda x), 0)$$

where λ is a positive constant, determine $(\text{curl}\mathbf{a})_x$, $(\text{curl}\mathbf{a})_y$, $(\text{curl}\mathbf{a})_z$ and $|\text{curl}\mathbf{a}|$. What is the magnitude of curl**a** at the origin?

Solution 4.3 The given field has components $a_x = \sin(\lambda y)$, $a_y = \cos(\lambda x)$ and $a_z = 0$. The components of curl**a** are found from Eq (5.43b) to be

$$(\text{curl}\mathbf{a})_x = 0$$

$$(\text{curl}\mathbf{a})_y = 0$$

$$(\text{curl}\mathbf{a})_z = \frac{\partial}{\partial x}(\cos(\lambda x)) - \frac{\partial}{\partial y}(\sin(\lambda y)) = -\lambda(\sin(\lambda x) + \cos(\lambda y))$$

The magnitude of curl**a** at a general field point is

$$|\text{curl}\mathbf{a}| = \lambda\,|\sin(\lambda x) + \cos(\lambda y)|$$

At the origin, $|\text{curl}\mathbf{a}| = \lambda\,|\sin(0) + \cos(0)| = \lambda$.

Example 4.4 (*Objective 4*) A long straight cylindrical wire of radius a with its axis lying on the z-axis carries a steady electric current I in the positive z-direction. The magnetic field produced by this current inside and outside the wire is given by

$$\mathbf{B}(\rho,\phi,z) = \frac{\mu_0 I \rho \mathbf{e}_\phi}{2\pi a^2} \qquad (\rho \le a)$$

$$\mathbf{B}(\rho,\phi,z) = \frac{\mu_0 I \mathbf{e}_\phi}{2\pi\rho} \qquad (\rho > a)$$

Confirm that these fields satisfy the differential form of Ampère's law, i.e. $\text{curl}\mathbf{B} = \mu_0\mathbf{J}$, where **J** is the current density vector introduced in Section 5.2.3.

Solution 4.4 Use Eq (5.44) for the curl in cylindrical polar coordinates.

Inside the wire ($\rho \le a$) we have $B_\rho = 0$, $B_\phi = (\mu_0 I\rho/2\pi a^2)$ and $B_z = 0$, and so

$$\text{curl}\mathbf{B} = \mathbf{e}_z\left[\left(\frac{\mu_0 I}{2\pi a^2}\right) + \frac{1}{\rho}\left(\frac{\mu_0 I\rho}{2\pi a^2}\right)\right] = \frac{\mathbf{e}_z \mu_0 I}{\pi a^2}$$

But $\mathbf{e}_z I/\pi a^2 = \mathbf{J}$. Hence curl$\mathbf{B} = \mu_0\mathbf{J}$ ($\rho \le a$).

Outside the wire ($\rho > a$) we have $B_\rho = 0$, $B_\phi = (\mu_0 I/2\pi\rho)$ and $B_z = 0$, and so

$$\text{curl}\mathbf{B} = \mathbf{e}_z\left[\left(-\frac{\mu_0 I}{2\pi\rho^2}\right) + \frac{1}{\rho}\left(\frac{\mu_0 I}{2\pi\rho}\right)\right] = 0$$

Thus we have curl$\mathbf{B} = \mathbf{0}$ ($\rho > a$) which satisfies Ampère's law since $\mathbf{J} = \mathbf{0}$ outside the wire.

Example 4.5 (*Objective 4*) The surface velocity on a river of width d is given by

$$\mathbf{v}(x,y) = Ay(y-d)\mathbf{i} \qquad (0 \le y \le d)$$

where A is a positive constant and y is the distance out from one bank. Determine curl**v** and hence state where you would expect an object to float downstream without rotating. Is **v** a conservative field?

Solution 4.5 Use Eq (5.43a) with $v_x = Ay(y-d)$ and $v_y = 0$ to obtain

$$\text{curl}\mathbf{v} = \mathbf{k}A(-2y + d)$$

This gives curl$\mathbf{v} = \mathbf{0}$ at $y = d/2$. Thus the field has zero curl in midstream and so this is where an object would float without rotating. **v** is not conservative because curl**v** is not zero everywhere.

Example 4.6 (*Objective 4*) Show that curl(gradϕ) = **0** for any scalar field ϕ.

Solution 4.6 Introduce a cartesian system, and let **G** = gradϕ. Since the x-axis may be chosen to be in any direction it is sufficient to show that the x component of curl(gradϕ) is zero. Using Eq (5.43a), we find $(\text{curl}\mathbf{G})_x = \partial G_z/\partial y - \partial G_y/\partial z$ where $G_z = (\text{grad}\phi)_z = \partial\phi/\partial z$ and $G_y = \partial\phi/\partial y$ (Eq (5.7)). Therefore $(\text{curl}\mathbf{G})_x = \dfrac{\partial^2\phi}{\partial y\partial z} - \dfrac{\partial^2\phi}{\partial z\partial y} = 0$, since the two mixed second-order partial derivatives are equal.

Problem 4.1 (*Objective 4*) Given $\mathbf{F}(x,y,z) = (xy, x^2, xz)$, find curl**F** and evaluate $(\text{curl}\mathbf{F})_z$ at the point (1,–1,3). In what region is curl**F** = **0**?

Problem 4.2 (*Objective 4*) Find curl**f** where $\mathbf{f}(x,y,z) = xyz\mathbf{i}$. Determine $|\text{curl}\mathbf{f}|$ at a general field point and at the point (3,–2,5).

Problem 4.3 (*Objective 4*) Consider the vector field $\mathbf{q}(\mathbf{r}) = \mathbf{p} \cdot \mathbf{r}/|\mathbf{r}|^3$ $(\mathbf{r} \neq \mathbf{0})$ where **p** is a constant vector and **r** is the position vector. Express **q** in spherical polar coordinates and determine curl**q**.

Problem 4.4 (*Objective 4*) Find the value(s) of n for which curl**F** = **0** where the vector field $\mathbf{F}(\rho,\phi,z)$ is given by (a) $\mathbf{F} = \mathbf{e}_\rho/\rho^n$, (b) $\mathbf{F} = \mathbf{e}_\phi/\rho^n$ and (c) $\mathbf{F} = \mathbf{e}_z/\rho^n$, for $\rho > 0$ in each case.

Problem 4.5 (*Objective 4*) Vortices are often seen in the flow of fluids. A small-scale example is the swirling funnel above the plughole of an emptying bathtub; a tornado is a large-scale example. We can often distinguish two regions: the vortex itself where the velocity increases with distance ρ from the centre and an outer region where the velocity falls off with distance. The following velocity function is often used to describe the vortex region and the outer region:

$$\mathbf{v}(\rho,\phi) = \omega\rho\mathbf{e}_\phi \qquad (\rho \le a)$$

$$\mathbf{v}(\rho,\phi) = \frac{\omega a^2 \mathbf{e}_\phi}{\rho} \qquad (\rho > a)$$

where a is a constant. Sketch the flow lines (i.e. field lines) and calculate curl**v** in the two regions. State how a small floating object would move in the outer region.

Problem 4.6 (*Objective 4*) Find the curl of the vector field

$$\mathbf{v}(x,y,z) = (-yz^2, xz^2, 0)$$

and describe how curl**v** varies along the positive z-axis.

Problem 4.7 (*Objective 4*) Starting from Eqs (5.26) and (5.43a), show that div(curl**F**) = 0 for any vector field **F**.

5.5 THE VECTOR DIFFERENTIAL OPERATOR "del"

We now introduce the vector differential operator known as "del" or "nabla". This operator provides us with a unifying insight into the structure of differential vector calculus as well as an efficient alternative notation for grad, div and curl. We begin by explaining what is meant by a *vector differential operator*.

5.5.1 Introducing differential operators

The simplest example of a differential operator is the derivative symbol $\mathrm{d}/\mathrm{d}x$ which acquires meaning only when it is given a function $f(x)$ on its right-hand side to act or operate on. Then we have $(\mathrm{d}/\mathrm{d}x)f(x) = f'(x)$, the derivative; for example $(\mathrm{d}/\mathrm{d}x)x^2 = \mathrm{d}(x^2)/\mathrm{d}x = 2x$. Note that the function $f(x)$ to be operated on is written on the immediate right of the operator, i.e. we write $(\mathrm{d}/\mathrm{d}x)f(x)$, not $f(x)\mathrm{d}/\mathrm{d}x$ which is another operator since it ends with $\mathrm{d}/\mathrm{d}x$.

The operator $\mathrm{d}/\mathrm{d}x$ is an example of a *scalar differential operator*. When $\mathrm{d}/\mathrm{d}x$ operates on a scalar function f it produces another scalar function f'. When it operates on a vector function $\mathbf{a}(x)$ the result is another vector function $\mathbf{b}(x) = \mathrm{d}\mathbf{a}/\mathrm{d}x$. The velocity of a particle for example is defined by the operator $\mathrm{d}/\mathrm{d}t$ acting on the position vector $\mathbf{r}$ of a particle: $\mathbf{v}(t) = \mathrm{d}\mathbf{r}/\mathrm{d}t$.

Another simple example of a scalar differential operator is the partial derivative operator $\partial/\partial x$ which acts on a scalar field function to produce another scalar field function, for example $(\partial/\partial x)(x^2 + xy + z^2) = \partial(x^2 + xy + z^2)/\partial x = 2x + y$.

An operator such as $\mathbf{i}\partial/\partial x$ is an example of a **vector differential operator** since it has both vector and operator characteristics. Note that we can write it as $(\partial/\partial x)\mathbf{i}$ since $\mathbf{i}$ is a constant vector independent of x and can therefore be moved through the $\partial/\partial x$ symbol. The action of this vector operator on a scalar field function is to give a vector field function; for example, $(\mathbf{i}\partial/\partial x)(x^2 + xy + z^2) = \mathbf{i}(2x + y)$.

Consider now how $\mathbf{i}\partial/\partial x$ might operate on a vector function such as $\mathbf{V}(x,y) = x^2y\mathbf{i} + xy^2\mathbf{j}$. Operating with the $\partial/\partial x$ part is straightforward; we simply take the partial derivative with respect to x of the two components of $\mathbf{V}$. But the vector $\mathbf{i}$ in the operator $\mathbf{i}\partial/\partial x$ can act on a vector field in one of two ways, either as a scalar product or as a vector product. Operating as a scalar product gives $(\mathbf{i}\partial/\partial x)\,.\,\mathbf{V} = (\partial/\partial x)\mathbf{i}\,.\,\mathbf{V} = (\partial/\partial x)(x^2y) = 2xy$ which is a scalar field. Operating as a vector product on the same field $\mathbf{V}$ gives $(\mathbf{i}\partial/\partial x) \times \mathbf{V} = (\partial/\partial x)\mathbf{i} \times \mathbf{V} = (\partial/\partial x)(xy^2\mathbf{k}) = y^2\mathbf{k}$ which is a vector field.

Example 5.1 and Problem 5.1 give further practice in using these simple differential operators.

5.5.2 The "del" operator

The vector differential operator

$$\nabla = \mathbf{i}\frac{\partial}{\partial x} + \mathbf{j}\frac{\partial}{\partial y} + \mathbf{k}\frac{\partial}{\partial z} \qquad (5.51)$$

is called **del** or **nabla** and is denoted by the upside-down Greek delta symbol ∇ printed in bold type. Being a vector differential operator, del acts on a scalar field U to produce a vector field,

$$\nabla U = \left(\mathbf{i}\frac{\partial}{\partial x} + \mathbf{j}\frac{\partial}{\partial y} + \mathbf{k}\frac{\partial}{\partial z}\right)U = \mathbf{i}\frac{\partial U}{\partial x} + \mathbf{j}\frac{\partial U}{\partial y} + \mathbf{k}\frac{\partial U}{\partial z} \qquad (5.52)$$

You should recognise the vector expression on the right-hand side (refer to Eq (5.7)); it is grad$U$. Thus we have the identity

$$\nabla U = \text{grad}U \qquad (5.53)$$

∇U is another way of writing grad$U$.

The unifying role of the del operator begins to emerge when we allow it to act on a vector function $\mathbf{F}(x,y,z) = \mathbf{i}F_x(x,y,z) + \mathbf{j}F_y(x,y,z) + \mathbf{k}F_z(x,y,z)$. When ∇ acts on $\mathbf{F}$ as a scalar product we obtain

$$\nabla . \mathbf{F} = \left(\mathbf{i}\frac{\partial}{\partial x} + \mathbf{j}\frac{\partial}{\partial y} + \mathbf{k}\frac{\partial}{\partial z}\right) . (\mathbf{i}F_x + \mathbf{j}F_y + \mathbf{k}F_z)$$

$$= \left(\frac{\partial}{\partial x}\right)F_x + \left(\frac{\partial}{\partial y}\right)F_y + \left(\frac{\partial}{\partial z}\right)F_z$$

$$= \frac{\partial F_x}{\partial x} + \frac{\partial F_y}{\partial y} + \frac{\partial F_z}{\partial z} \qquad (5.54)$$

which is div$\mathbf{F}$ (Eq (5.26)). Thus we have the identity

$$\nabla . \mathbf{F} = \text{div}\mathbf{F} \qquad (5.55)$$

Now see what happens when ∇ acts on $\mathbf{F}$ as a vector product – you can probably guess the result. Using the vector product rule (Section 2.2.5), we obtain

$$\nabla \times \mathbf{F} = \left(\mathbf{i}\frac{\partial}{\partial x} + \mathbf{j}\frac{\partial}{\partial y} + \mathbf{k}\frac{\partial}{\partial z}\right) \times (\mathbf{i}F_x + \mathbf{j}F_y + \mathbf{k}F_z)$$

$$= \mathbf{i}\left(\frac{\partial F_z}{\partial y} - \frac{\partial F_y}{\partial z}\right) + \mathbf{j}\left(\frac{\partial F_x}{\partial z} - \frac{\partial F_z}{\partial x}\right) + \mathbf{k}\left(\frac{\partial F_y}{\partial x} - \frac{\partial F_x}{\partial y}\right)$$

$$= \text{curl}\mathbf{F}$$

Thus we have the identity

$$\nabla \times \mathbf{F} = \text{curl}\mathbf{F}$$

We conclude that del not only provides us with an alternative notation for grad, div and curl,

$$\nabla U = \text{grad}U \qquad \nabla \cdot \mathbf{F} = \text{div}\mathbf{F} \qquad \text{and} \qquad \nabla \times \mathbf{F} = \text{curl}\mathbf{F} \tag{5.56}$$

but also reveals an underlying mathematical structure: grad is the operator ∇ acting like scalar multiplication on a scalar field, while div and curl are, respectively, the operators $\nabla.$ and $\nabla\times$ acting on a vector field.

5.5.3 The Laplacian operator

Consider a gradient field $\mathbf{G} = \text{grad}\phi$, where ϕ is any scalar field. The divergence of $\mathbf{G}$ is $\text{div}\mathbf{G} = \text{div}(\text{grad}\phi)$. Using Eqs (5.56) we can write this as $\nabla \cdot (\nabla\phi)$, and using Eq (5.51) we can write it out as

$$\begin{aligned}\text{div}(\text{grad}\phi) &= \nabla \cdot (\nabla\phi)\\ &= \left(\mathbf{i}\frac{\partial}{\partial x} + \mathbf{j}\frac{\partial}{\partial y} + \mathbf{k}\frac{\partial}{\partial z}\right) \cdot \left(\mathbf{i}\frac{\partial\phi}{\partial x} + \mathbf{j}\frac{\partial\phi}{\partial y} + \mathbf{k}\frac{\partial\phi}{\partial z}\right)\\ &= \frac{\partial^2\phi}{\partial x^2} + \frac{\partial^2\phi}{\partial y^2} + \frac{\partial^2\phi}{\partial z^2}\end{aligned} \tag{5.57}$$

This same expression is obtained by working out $(\nabla \cdot \nabla)\phi$, where $\nabla \cdot \nabla$ is the operator obtained by formally taking the scalar product of ∇ with itself. The operator $\nabla \cdot \nabla$ is commonly written as ∇^2 "del squared" and has the form

$$\begin{aligned}\nabla^2 = \nabla \cdot \nabla &= \left(\mathbf{i}\frac{\partial}{\partial x} + \mathbf{j}\frac{\partial}{\partial y} + \mathbf{k}\frac{\partial}{\partial z}\right) \cdot \left(\mathbf{i}\frac{\partial}{\partial x} + \mathbf{j}\frac{\partial}{\partial y} + \mathbf{k}\frac{\partial}{\partial z}\right)\\ &= \frac{\partial^2}{\partial x^2} + \frac{\partial^2}{\partial y^2} + \frac{\partial^2}{\partial z^2}\end{aligned} \tag{5.58}$$

Thus we have the identity

$$\text{div}(\text{grad}\phi) = (\nabla \cdot \nabla)\phi = \nabla^2\phi \tag{5.59}$$

The operator ∇^2 is called the **Laplacian operator**. We often refer to it as "del squared" or "nabla squared". ∇^2 can act on any scalar field ϕ, as in Eq (5.59), or on a vector field (Example 5.3b).

Section 5.3.3 discussed physical laws of the form div**F** = S (Eq (5.33)). When **F** is a conservative field we can write $\mathbf{F} = -\nabla\phi$ and hence $\nabla^2\phi = -S$. For example, Gauss's law of electrostatics (Eq (5.37)) can be expressed as

$$\nabla^2 V = \frac{\partial^2 V}{\partial x^2} + \frac{\partial^2 V}{\partial y^2} + \frac{\partial^2 V}{\partial z^2} = -\frac{\rho_c}{\epsilon_0} \tag{5.60}$$

where the scalar field V is the electrostatic potential, i.e. $\mathbf{E} = -\nabla V$.

Any equation of the form $\nabla^2\phi = -S$ where S is a scalar field, such as Eq (5.60), is known as **Poisson's equation**. When the source term $S = 0$, Poisson's equation takes the form $\nabla^2\phi = 0$ which is known as **Laplace's equation**.

Poisson's and Laplace's equations are examples of partial differential equations. They can be solved, i.e. integrated, to find the unknown scalar field ϕ, provided certain information about the values of ϕ (or its derivatives) on the boundaries of the domain are given. Such problems, known as *boundary value problems*, are outside the scope of this book.

5.5.4 Vector field identities

The operators grad, div and curl often appear in compositions of two or more, such as div(grad$U$); they can also operate on products of field functions, as in div(ϕ**F**). Sometimes it is possible to simplify such expressions using vector field identities. You have seen two examples (Example 4.6 and Problem 4.7):

$$\text{curl}(\text{grad}U) = \mathbf{0} \tag{5.61}$$

$$\text{div}(\text{curl}\mathbf{F}) = 0 \tag{5.62}$$

In terms of del, these identities are, respectively,

$$\nabla \times (\nabla U) = \mathbf{0} \tag{5.63}$$

$$\nabla . (\nabla \times \mathbf{F}) = 0 \tag{5.64}$$

Some other examples of vector field identities are

$$\nabla \times (\nabla \times \mathbf{F}) = \nabla(\nabla . \mathbf{F}) - \nabla^2\mathbf{F} \tag{5.65}$$

$$\nabla . (\phi\mathbf{F}) = \phi\nabla . \mathbf{F} + \mathbf{F} . \nabla\phi \tag{5.66}$$

$$\nabla \times (\phi\mathbf{F}) = \phi\nabla \times \mathbf{F} - \mathbf{F} \times (\nabla\phi) \tag{5.67}$$

These identities can be verified by expressing ∇ and the fields **F** and ϕ in component form using Eq (5.51) for ∇, as in Example 4.6. An alternative approach is to manipulate ∇ as a vector satisfying the ordinary vector identities (Ch 2) while respecting its differential operator character and making use of the rules of differential calculus (Example 5.5).

Summary of section 5.5

- The vector differential operator **del** or **nabla** is defined by

$$\nabla = \mathbf{i}\partial/\partial x + \mathbf{j}\partial/\partial y + \mathbf{k}\partial/\partial z \tag{5.51}$$

  Then grad$U = \nabla U$, div$\mathbf{F} = \nabla \cdot \mathbf{F}$ and curl$\mathbf{F} = \nabla \times \mathbf{F}$. (5.56)

- The scalar differential operator

$$\nabla^2 = \nabla \cdot \nabla = \partial^2/\partial x^2 + \partial^2/\partial y^2 + \partial^2/\partial z^2 \tag{5.58}$$

  is called the **Laplacian operator** or "del squared".

- An equation of the form $\nabla^2\phi = -S$ is known as **Poisson's equation**. When $S = 0$ we have $\nabla^2\phi = 0$ which is known as **Laplace's equation**.

- Vector field identities, such as Eqs (5.61) to (5.67), are useful for simplifying scalar and vector field expressions.

Example 5.1 (*Objective 5*) Consider the vector differential operator $\mathbf{i}(\mathrm{d}/\mathrm{d}x)$ which acts on functions $f(x)$ of a single variable.

(a) Find the action of this operator on the scalar function $f(x) = x^3 - x - 5$.

(b) Demonstrate the two ways in which the operator can act on the vector function $\mathbf{g}(x) = \mathbf{i}\cos x + \mathbf{j}\sin x$.

(c) Find

$$\left(\mathbf{i}\frac{\mathrm{d}}{\mathrm{d}x}\right)\cdot\left(\mathbf{i}\frac{\mathrm{d}}{\mathrm{d}x}\right)f(x) \quad \text{and} \quad \left(\mathbf{i}\frac{\mathrm{d}}{\mathrm{d}x}\right)\times\left(\mathbf{i}\frac{\mathrm{d}}{\mathrm{d}x}\right)f(x)$$

where f is the function in part (a).

Solution 5.1

(a) $\left(\mathbf{i}\frac{\mathrm{d}}{\mathrm{d}x}\right)f(x) = \mathbf{i}\frac{\mathrm{d}f}{\mathrm{d}x} = \mathbf{i}(3x^2 - 1)$, a vector function.

(b) Operating as a scalar product we have the scalar function

$$\left(\mathbf{i}\frac{\mathrm{d}}{\mathrm{d}x}\right)\cdot\mathbf{g}(x) = \mathbf{i}\cdot\mathbf{i}\frac{\mathrm{d}}{\mathrm{d}x}(\cos x) + \mathbf{i}\cdot\mathbf{j}\frac{\mathrm{d}}{\mathrm{d}x}(\sin x) = -\sin x$$

Operating as a vector product we have the vector function

$$\left(\mathbf{i}\frac{d}{dx}\right)\times \mathbf{g}(x) = \mathbf{i}\times\mathbf{i}\frac{d}{dx}(\cos x) + \mathbf{i}\times\mathbf{j}\frac{d}{dx}(\sin x) = \mathbf{k}\cos x$$

(c) Using the answer to part (a), we have

$$\left(\mathbf{i}\frac{d}{dx}\right).\left(\mathbf{i}\frac{d}{dx}\right)f(x) = \left(\mathbf{i}\frac{d}{dx}\right).\mathbf{i}(3x^2-1) = 6x$$

and

$$\left(\mathbf{i}\frac{d}{dx}\right)\times\left(\mathbf{i}\frac{d}{dx}\right)f(x) = \left(\mathbf{i}\frac{d}{dx}\right)\times\mathbf{i}(3x^2-1) = 0$$

since $\mathbf{i}\times\mathbf{i} = 0$. Alternatively, note that since the unit vector **i** is independent of x we can move it through the differentiation symbols and so write

$$\left(\mathbf{i}\frac{d}{dx}\right)\times\left(\mathbf{i}\frac{d}{dx}\right)f(x) = \mathbf{i}\times\mathbf{i}\frac{d^2 f}{dx^2} = 0$$

Example 5.2 (*Objective 6*) Given the fields

$$h(x,y,z) = x^2 - z^2 \quad \text{and} \quad \mathbf{f}(x,y,z) = y\mathbf{i} - x\mathbf{j} + 5\mathbf{k}$$

determine where possible: ∇h, $\nabla . \mathbf{f}$, $\nabla \times \mathbf{f}$, $\nabla^2 h$, $\nabla . (h\mathbf{f})$, $\nabla . h$ and $\nabla\mathbf{f}$.

Solution 5.2

$$\nabla h = \text{grad}h = 2x\mathbf{i} - 2z\mathbf{k},$$
$$\nabla . \mathbf{f} = \text{div}\mathbf{f} = 0,$$
$$\nabla \times \mathbf{f} = \text{curl}\mathbf{f} = \mathbf{k}((-1) - 1) = -2\mathbf{k}$$

We can recognise $\nabla^2 h$ as $\text{div}(\text{grad}h) = \text{div}(2x\mathbf{i} - 2z\mathbf{k}) = 2 + (-2) = 0$. (Alternatively use Eq (5.57) to obtain the same result.)

$$\nabla . (h\mathbf{f}) = \text{div}[(x^2 - z^2)(y\mathbf{i} - x\mathbf{j} + 5\mathbf{k})] = 2xy - 10z$$

(Alternatively use the identity (5.66). This gives

$$\nabla . (h\mathbf{f}) = h\text{div}\mathbf{f} + \mathbf{f} . \text{grad}h = 0 + (y\mathbf{i} - x\mathbf{j} + 5\mathbf{k}) . (2x\mathbf{i} - 2z\mathbf{k}) = 2xy - 10z.)$$

$\nabla . h$ is meaningless because the scalar product, indicated by the dot, can be formed only with two vector objects. $\nabla\mathbf{f}$ is meaningless as a product because two adjacent vector objects must be separated by a dot or a cross.

Example 5.3 (*Objective 6*)

(a) Find $\nabla^2\phi$ where

$$\phi(x,y,z) = \frac{1}{(x^2+y^2+z^2)^{1/2}} \qquad (x,y,z) \neq (0,0,0)$$

(b) The Laplacian operator can act on a vector function by operating on each scalar component. Find $\nabla^2\mathbf{F}$ where

$$\mathbf{F}(x,y,z) = x^2y\mathbf{i} + xy^2\mathbf{j} + z^3\mathbf{k}$$

Solution 5.3

(a) $\nabla^2\phi = \partial^2\phi/\partial x^2 + \partial^2\phi/\partial y^2 + \partial^2\phi/\partial z^2$. We first find

$$\frac{\partial\phi}{\partial x} = \left(-\frac{1}{2}\right)\frac{2x}{(x^2+y^2+z^2)^{3/2}}$$

Then

$$\frac{\partial^2\phi}{\partial x^2} = \frac{(x^2+y^2+z^2)^{3/2}(-1) - (-x)(3/2)(2x)(x^2+y^2+z^2)^{1/2}}{(x^2+y^2+z^2)^3}$$

This simplifies to

$$\frac{\partial^2\phi}{\partial x^2} = \frac{-(x^2+y^2+z^2)^{3/2} + 3x^2(x^2+y^2+z^2)^{1/2}}{(x^2+y^2+z^2)^3}$$

Now notice that the field ϕ and the operator ∇^2 are unchanged when any two of the three cartesian variables are interchanged. It follows that we can write down the expression for $\partial^2\phi/\partial y^2$ simply by interchanging x and y in the above expression for $\partial^2\phi/\partial x^2$, and similarly for $\partial^2\phi/\partial z^2$. The denominator and the first term in the numerator of $\partial^2\phi/\partial x^2$ are unchanged by any interchange, while the factor $3x^2$ in the second term becomes $3y^2$ and $3z^2$ in the corresponding expressions for $\partial^2\phi/\partial y^2$ and $\partial^2\phi/\partial z^2$. When we put the three second order partial derivatives together to form ∇^2 we obtain

$$\frac{-3(x^2+y^2+z^2)^{3/2} + (3x^2+3y^2+3z^2)(x^2+y^2+z^2)^{1/2}}{(x^2+y^2+z^2)^3} = 0$$

Hence $\nabla^2\phi = 0$. (See Problem 5.4 for a quicker solution.)

(b) $\nabla^2\mathbf{F} = \nabla^2(x^2y)\mathbf{i} + \nabla^2(xy^2)\mathbf{j} + \nabla^2(z^3)\mathbf{k} = 2y\mathbf{i} + 2x\mathbf{j} + 6z\mathbf{k}$

Example 5.4 (*Objective 6*) Find curl(curl**F**) at the point (3,–2,5) where **F** is the vector field function given in part (b) of Example 5.3.

Solution 5.4 Recognise the identity given in Eq (5.65) as

$$\text{curl}(\text{curl}\mathbf{F}) = \text{grad}(\text{div}\mathbf{F}) - \nabla^2\mathbf{F}$$

and note that $\nabla^2\mathbf{F}$ was found in Solution 5.3b. We need div**F** $= 2xy + 2xy + 3z^2$ and grad(div**F**) $= \mathbf{i}4y + \mathbf{j}4x + \mathbf{k}6z$. Hence curl(curl**F**) $= 2y\mathbf{i} + 2x\mathbf{j}$. Evaluating this at (3,–2,5) gives the vector $-4\mathbf{i} + 6\mathbf{j}$. (We could alternatively have worked out curl(curl**F**) directly.)

Example 5.5 (*Objective 6*) Derive the identity given in Eq (5.65) by exploiting the vector nature of ∇ and respecting its differential operator properties.

Solution 5.5 $\nabla \times (\nabla \times \mathbf{F})$ has the form of a vector triple product. We therefore apply the triple product expansion (Eq (2.37)) to give $\nabla \times (\nabla \times \mathbf{F}) = \nabla(\nabla \cdot \mathbf{F}) - (\nabla \cdot \nabla)\mathbf{F}$, where we have positioned the del operator so that each term is a vector. (Note that it would be meaningless to write $\nabla \times (\nabla \times \mathbf{F}) = (\nabla \cdot \mathbf{F})\nabla - (\nabla \cdot \nabla)\mathbf{F}$ since the term $(\nabla \cdot \mathbf{F})\nabla$ is a differential operator, not a vector; for the same reason it would be wrong to write this term as $\nabla(\mathbf{F} \cdot \nabla)$.)

Example 5.6 (*Objective 6*) Find the vector field $(\mathbf{F} \cdot \nabla)\mathbf{G}$ where

$$\mathbf{F} = x^2y\mathbf{i} + y^3\mathbf{j} \quad \text{and} \quad \mathbf{G} = \mathbf{i} - x\mathbf{j}$$

Solution 5.6 ∇ cannot operate directly on **G** here since there is no dot or cross separating them. We must therefore first evaluate the scalar differential operator $\mathbf{F} \cdot \nabla$, as indicated by the brackets, which can then operate on **G**. Thus

$$\mathbf{F} \cdot \nabla = (x^2y\mathbf{i} + y^3\mathbf{j}) \cdot \left(\mathbf{i}\frac{\partial}{\partial x} + \mathbf{j}\frac{\partial}{\partial y} + \mathbf{k}\frac{\partial}{\partial z}\right) = x^2y\frac{\partial}{\partial x} + y^3\frac{\partial}{\partial y}$$

and

$$(\mathbf{F} \cdot \nabla)\mathbf{G} = \left(x^2y\frac{\partial}{\partial x} + y^3\frac{\partial}{\partial y}\right)(\mathbf{i} - x\mathbf{j}) = -x^2y\mathbf{j}$$

Example 5.7 (*Objective 6*) A long straight cylindrical rod is heated by passing a steady electric current along it. When the axis of the rod is on the z-axis, the steady temperature field in the rod is given by

$$T(\rho,\phi,z) = 30 - (2 \times 10^5)\rho^2 \qquad (\rho \le \text{radius of rod})$$

where ρ is in metres and T in °C.

(a) Show that T obeys Poisson's equation, $\nabla^2 T = -\Psi$, and hence specify the scalar field Ψ. Use the cylindrical polar form of ∇^2 given below,

$$\nabla^2 = \frac{\partial^2}{\partial \rho^2} + \frac{1}{\rho}\frac{\partial}{\partial \rho} + \frac{1}{\rho^2}\frac{\partial^2}{\partial \phi^2} + \frac{\partial^2}{\partial z^2} \qquad (5.68)$$

(b) Find the heat flow rate vector given by $\mathbf{h} = -\alpha \mathrm{grad}T$ where the thermal conductivity of the material is $\alpha = 100\ \mathrm{Wm^{-1}\,°C^{-1}}$.

(c) Find $\nabla \cdot \mathbf{h}$. Hence show that $\psi = S/\alpha$ where S is the heat source density in the rod.

(d) Make a rough sketch of the field lines of **h** to indicate the magnitude as well as the direction of **h**.

Solution 5.7

(a) T is a cylindrically symmetric function (i.e. varies with ρ only). The last two terms of $\nabla^2 T$ (Eq (5.68)) are therefore zero. We find $\partial T/\partial \rho = -4 \times 10^5 \rho$ and $\partial^2 T/\partial \rho^2 = -4 \times 10^5$, and so

$$\nabla^2 T = \partial^2 T/\partial \rho^2 + (1/\rho)\partial T/\partial \rho = -8 \times 10^5$$

Thus T satisfies Poisson's equation with $\psi = 8 \times 10^5\ \mathrm{°Cm^{-2}}$.

(b) Use Eq (5.13) to obtain

$$\mathrm{grad}T = \mathbf{e}_\rho \partial T/\partial \rho = \mathbf{e}_\rho(-4 \times 10^5 \rho) \qquad \text{in °Cm}^{-1}$$

and so

$$\mathbf{h} = -\alpha \mathrm{grad}T = \mathbf{e}_\rho 4 \times 10^7 \rho \qquad \text{in Wm}^{-2}.$$

(c) Eq (5.29) gives $\nabla \cdot \mathbf{h}$ or $\mathrm{div}\mathbf{h} = \partial h_\rho \partial \rho + (1/\rho)h_\rho = 8 \times 10^7\ \mathrm{Wm^{-3}}$. Referring to Section 5.3.3 and the law $\mathrm{div}\mathbf{h} = S$, we conclude that the electrical heating of the rod produces a heat source density $S = 8 \times 10^7\ \mathrm{Wm^{-3}}$ throughout the rod. We can write $\mathrm{div}\mathbf{h} = \mathrm{div}(-\alpha \mathrm{grad}T) = -\alpha \nabla^2 T$ which gives $\nabla^2 T = (-8 \times 10^7\ \mathrm{Wm^{-3}})/\alpha = -8 \times 10^5\ \mathrm{°Cm^{-2}}$, in agreement with the answer to part (a). Thus $\psi = S/\alpha$.

(d) The field lines are sketched in Fig 5.15. The density of lines in the rod increases with radial distance to indicate the increasing magnitude of **h**. The lines have beginning points throughout the rod indicating the positive divergence of **h** due to the presence of the heat source density (8×10^7 Wm^{-3}) generated by the current.

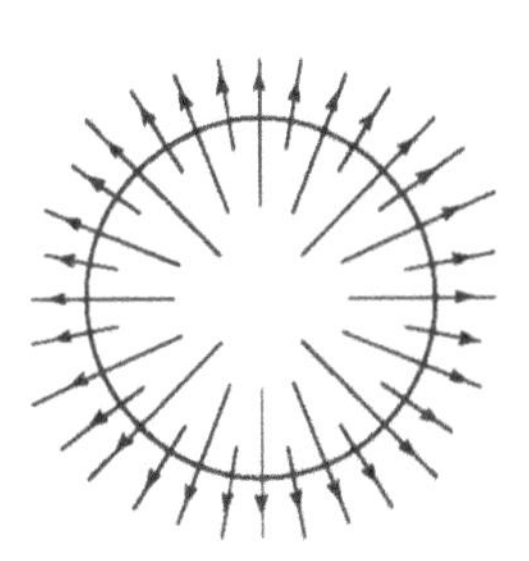

Fig 5.15
The field lines of the heat flow vector are directed outwards from the axis of the rod. They have beginning points (sources) throughout the rod and become more closely spaced with increasing radial distance.

Problem 5.1 (*Objective 5*) Consider the differential operators $\mathbf{i}\partial/\partial x$ and $\mathbf{j}\partial/\partial y$, and the functions $f(x,y) = x^3y - 2y$ and $\mathbf{g}(x,y) = xy\mathbf{i} + y^2\mathbf{j}$. Evaluate each of the following in terms of x and y, and in each case state whether your answer is a scalar function, a vector function, a scalar differential operator or a vector differential operator.

(a) $(\mathbf{i}\partial/\partial x)f(x,y)$
(b) $f(x,y)(\mathbf{i}\partial/\partial x)$
(c) $f(x,y)(\mathbf{i}\partial/\partial x)f(x,y)$
(d) $(\mathbf{j}\partial/\partial y) \cdot \mathbf{g}(x,y)$
(e) $(\mathbf{j}\partial/\partial y) \times \mathbf{g}(x,y)$
(f) $\mathbf{g}(x,y) \times (\mathbf{j}\partial/\partial y)$
(g) $(\mathbf{i}\partial/\partial x) \times (\mathbf{j}\partial/\partial y)f(x,y)$
(h) $(\mathbf{j}\partial/\partial y) \times (\mathbf{i}\partial/\partial x)f(x,y)$
(i) $(\mathbf{i}\partial/\partial x) \cdot (\mathbf{i}\partial/\partial x)$ also written as $(\mathbf{i}\partial/\partial x)^2$
(j) $(\mathbf{i}\partial/\partial x)^2 f(x)$

Problem 5.2 (*Objective 6*) Given the fields

$$\phi(x,y,z) = x^2 + y^2 \quad \text{and} \quad \mathbf{h}(x,y,z) = x^2\mathbf{i} - y^2\mathbf{j} - xy\mathbf{k}$$

determine where possible $\nabla\phi$, $\nabla\mathbf{h}$, $\nabla \cdot \phi$, $\nabla \cdot \mathbf{h}$, $\nabla \times \mathbf{h}$, $\nabla^2\phi$ and $\nabla^2\mathbf{h}$.

Problem 5.3 (*Objective 6*)

(a) Show that $\nabla \times \nabla = \mathbf{0}$.

(b) Show that div(curl**F**) = 0 for any vector field **F**.

Problem 5.4 (*Objective 6*) The spherical polar form of ∇^2 is

$$\nabla^2 = \frac{\partial^2}{\partial r^2} + \frac{2}{r}\frac{\partial}{\partial r} + \frac{1}{r^2}\frac{\partial^2}{\partial \theta^2} + \frac{\cot\theta}{r^2}\frac{\partial}{\partial \theta} + \frac{1}{r^2\sin^2\theta}\frac{\partial^2}{\partial \phi^2} \qquad (5.69)$$

Find $\nabla^2\phi$ where ϕ is the scalar field in Example 5.3.

Problem 5.5 (*Objective 6*) Find the scalar field $(\mathbf{F} \cdot \nabla)\phi$ where

$$\mathbf{F} = xy\mathbf{i} + y^2\mathbf{j} - z^2k \quad \text{and} \quad \phi = x^2 - yx - 5z^2$$

Find also $\mathbf{F} \cdot (\nabla\phi)$.

Problem 5.6 (*Objective 6*) A cylindrical wire of radius a with its axis on the z-axis is enclosed in a sleeve of electrical insulation of outer radius b. When the wire is heated by a steady current the temperature in the insulation is given by

$$T(x,y,z) = 50 - 30\log_e\left[\frac{(x^2+y^2)^{1/2}}{a}\right]\frac{1}{\log_e\left(\dfrac{b}{a}\right)} \qquad (a^2 < (x^2+y^2) < b^2)$$

(a) Show that this temperature field satisfies Laplace's equation.

(b) Find the heat flow rate vector $\mathbf{h} = -\alpha\,\mathbf{grad}T$ in the insulation where α is a constant, and show directly that $\text{div}\mathbf{h} = 0$. Sketch roughly the field lines of $\mathbf{h}$.

6

Integrating fields

After you have studied this chapter you should be able to

Objectives

- Express simple line integrals as definite integrals in cartesian or polar variables, and evaluate them (*Objective 1*).
- Evaluate line integrals along given parameterised curves (*Objective 2*)
- Define a conservative field in terms of line integrals and do calculations involving work, potential and potential energy; evaluate line integrals using scalar potential functions (*Objective 3*).
- Express the flux of a vector field across a surface as a surface integral (*Objective 4*).
- Express surface integrals as double integrals in cartesian or polar coordinates, and evaluate them (*Objective 5*).
- Use Stokes's theorem to transform line integrals into surface integrals and vice versa (*Objective 6*).
- Express volume integrals as triple integrals in cartesian or polar coordinates, and evaluate them (*Objective 7*).
- Use Gauss's theorem (i.e. the divergence theorem) to transform surface integrals into volume integrals and vice-versa (*Objective 8*).

The integral calculus of fields is concerned with summing scalar and vector field values along curves, on surfaces and throughout volumes, the sums being in the form of line integrals, surface integrals and volume integrals. These integrals are generalisations to three dimensions of the ordinary definite integral of a single variable, which sums function values $f(x)$ along the x-axis. This chapter shows how line integrals, surface integrals and volume integrals are defined as limits of sums and how they are evaluated, and illustrates their relevance to physical problems.

We shall introduce two important vector integral theorems known as Stokes's theorem and Gauss's theorem and indicate informally how they are derived. Armed with these theorems we shall look again at several physical applications discussed in previous chapters including the important topics of conservative vector fields and scalar potentials.

The basic strategy for evaluating line integrals, surface integrals and volume integrals is to express them in terms of ordinary definite integrals. It is essential therefore that you can evaluate definite integrals. We begin with a review of the definite integral.

6.1 DEFINITE INTEGRALS – A REVIEW

Let f be a continuous scalar function of a single scalar variable x. To see how the definite integral of f is defined, consider an interval of the x-axis between fixed points $x = a$ and $x = b$ (Fig 6.1a). Divide this interval into N small segments of lengths $\Delta x_1, \Delta x_2, \ldots, \Delta x_n, \ldots, \Delta x_N$, where the subscripts 1, 2, ..., n, ..., N label the segments. Let $f(x_n)$ be the function value at any chosen point x_n in the nth segment; in fact it is common to consider x_n to be at the left-hand end of the segment, the right-hand end then being at $x_n + \Delta x_n$ (Fig 6.1b). Now consider the sum

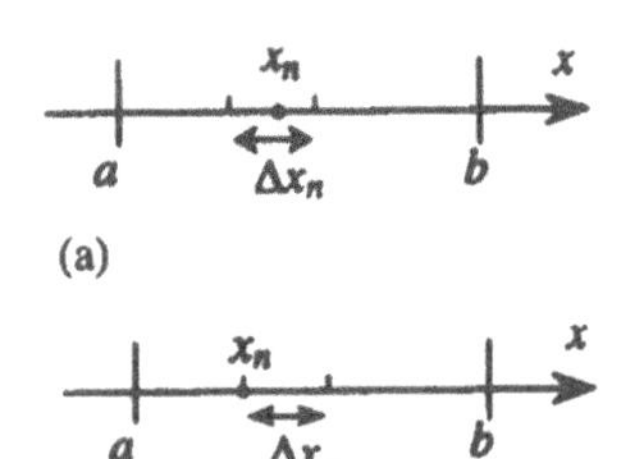

Fig 6.1
(a) x_n is any point on the nth segment of length Δx_n. (b) x_n is chosen to be the left-hand endpoint of the segment.

$$\sum_{n=1}^{N} f(x_n)\Delta x_n = f(x_1)\Delta x_1 + f(x_2)\Delta x_2 + \ldots f(x_n)\Delta x_n + \ldots f(x_N)\Delta x_N \qquad (6.1)$$

When N is very large and all segments are very short ($N \to \infty$; $\Delta x_n \to 0$), the sum (6.1) approaches a definite limit which is independent of the details of how the interval is divided and of where the points x_n are located in the segments. This limit is called the **definite integral** of f from a to b and is denoted by $\int_a^b f(x)\mathrm{d}x$. Thus

$$\lim_{N\to\infty}\left[\sum_{n=1}^{N} f(x_n)\Delta x_n\right] = \int_a^b f(x)\mathrm{d}x \qquad (6.2)$$

The definite integral is just a number (i.e. a scalar).

When the graph of f is above the x-axis everywhere between a and b, the definite integral is positive and equal to the area enclosed by the graph of f, the x-axis and the vertical lines $x = a$ and $x = b$. When the graph is below the x-axis the definite integral is negative and its magnitude is equal to the enclosed area (Fig 6.2). The values $x = a$ and $x = b$ are called respectively the **lower limit** and **upper limit** of the integral. The function $f(x)$ inside the integral, i.e. sitting between $\int_a^b$ and $\mathrm{d}x$, is called the **integrand**. The symbol $\mathrm{d}x$ is part of the integration symbol and indicates the **variable of integration** (x here). Although the symbol $\mathrm{d}x$ is not a physical quantity it is often useful to regard it as an infinitesimal segment of the x-axis carrying the physical dimensions of x. This ties in with the fact that the definite integral is an area when $f(x)$ and x are both lengths.

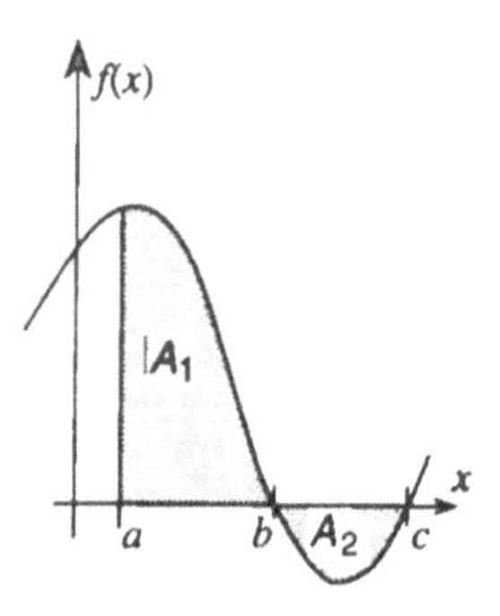

Fig 6.2
Area $A_1 = \int_a^b f(x)\mathrm{d}x$;
$A_2 = -\int_b^c f(x)\mathrm{d}x$.

Eq (6.2) defines the definite integral as the limit of a sum, and this is the form in which definite integrals often arise in scientific and engineering problems. However, the limit is not normally used to evaluate definite integrals. Instead we make use of a theorem from calculus theory called the **fundamental theorem of calculus** which relates integration to differentiation. The theorem states that

$$\int_a^b f(x)\mathrm{d}x = [F(x)]_a^b = F(b) - F(a) \qquad (6.3)$$

where $F(x)$ is any function such that $\mathrm{d}F/\mathrm{d}x = f(x)$, i.e. f is the derivative of F. The symbol $[F(x)]_a^b$ is standard notation meaning substitute b for x to obtain $F(b)$ and subtract from this the value of $F(a)$ obtained by substituting a for x. The function F is called a **primitive of** f or an **indefinite integral** of f. A table of indefinite integrals of common functions is given in Appendix B.

Consider as an example, the evaluation of the definite integral of the function $f(x) = x^2$ with lower limit $x = 2$ and upper limit $x = 5$. Using Eq (6.3) we write

$$\int_2^5 x^2\mathrm{d}x = \left[\frac{x^3}{3}\right]_2^5 = \frac{5^3}{3} - \frac{2^3}{3} = 39 \qquad (6.4)$$

Note that x, the variable of integration, does not appear in the answer which is just a number. Any other symbol can be used in place of x. Thus if the function were written as $f(y) = y^2$ we would simply replace x by y in Eq (6.4) and obtain the same answer, $\int_2^5 y^2\mathrm{d}y = 39$.

A definite integral has the following properties which follow from Eq (6.3):

(i) Any constant factor C in the integrand can be taken outside the integral sign,

$$\int_a^b Cf(x)\mathrm{d}x = C\int_a^b f(x)\mathrm{d}x \qquad (6.5)$$

(ii) The definite integral of the sum of two functions is the sum of the definite integrals

$$\int_a^b [f(x) + g(x)]\mathrm{d}x = \int_a^b f(x)\mathrm{d}x + \int_a^b g(x)\mathrm{d}x \qquad (6.6)$$

(iii) A definite integral changes sign when the path of integration is reversed, i.e. when the upper and lower limits are interchanged,

$$\int_a^b f(x)\mathrm{d}x = -\int_b^a f(x)\mathrm{d}x \qquad (6.7)$$

Summary of section 6.1

- The **definite integral** is defined as the limit of a sum (Eq (6.2)) and is evaluated by using the **fundamental theorem of calculus** which states that

$$\int_a^b f(x)\mathrm{d}x = [F(x)]_a^b = F(b) - F(a) \qquad (6.3)$$

Here F is a **primitive** (or **indefinite integral**) of f, i.e. $\mathrm{d}F/\mathrm{d}x = f$. The number a is the **lower limit** and b is the **upper limit**.

- For any definite integral

$$\int_a^b Cf(x)\mathrm{d}x = C\int_a^b f(x)\mathrm{d}x \qquad (C \text{ is any constant}) \qquad (6.5)$$

$$\int_a^b [f(x)+g(x)]\mathrm{d}x = \int_a^b f(x)\mathrm{d}x + \int_a^b g(x)\mathrm{d}x \qquad (6.6)$$

$$\int_a^b f(x)\mathrm{d}x = -\int_b^a f(x)\mathrm{d}x \qquad (6.7)$$

Example 1.1 (Review of definite integrals) Evaluate the following definite integrals:

(a) $\int_1^3 (x^3+1)\mathrm{d}x$ (b) $\int_{-3}^5 \mathrm{d}x$

(c) $\int_0^\pi \sin x\,\mathrm{d}x$ (d) $\int_0^4 (-5x^2)\mathrm{d}x$

Solution 1.1

(a) The integrand is $x^3 + 1$. A primitive (i.e. an indefinite integral) of $x^3 + 1$ is $\dfrac{x^4}{4}+x$. Hence Eq (6.3) gives

$$\int_1^3 (x^3+1)\,\mathrm{d}x = \left[\frac{x^4}{4}+x\right]_1^3 = \left(\frac{3^4}{4}+3\right)-\left(\frac{1^4}{4}+1\right) = \frac{93}{4}-\frac{5}{4} = 22$$

(b) The integrand is 1 (not shown explicitly) and so a primitive is x and Eq (6.3) gives

$$\int_{-3}^5 \mathrm{d}x = [x]_{-3}^5 = 5-(-3) = 8$$

(c) The integrand is $\sin x$ and a primitive of $\sin x$ is $-\cos x$, and so Eq (6.3) gives

$$\int_0^\pi \sin x\,\mathrm{d}x = [-\cos x]_0^\pi = (-\cos\pi)-(-\cos 0) = 1+1 = 2$$

(d) Take the constant factor –5 outside the integral sign. Thus

$$\int_0^4 (-5x^2)\mathrm{d}x = -5\int_0^4 x^2\mathrm{d}x = -5\left[\frac{x^3}{3}\right]_0^4 = -5\left[\left(\frac{4^3}{3}\right)-\frac{0^3}{3}\right] = -\frac{320}{3}$$

Problem 1.1 (review of definite integrals) Evaluate the following definite integrals:

(a) $\int_{-1}^{1} x^3\mathrm{d}x$ (b) $\int_{-1}^{1} x^4\mathrm{d}x$

(c) $\int_0^{\pi}(-3\cos x)\mathrm{d}x$ (d) $\int_1^2 \left(x-\frac{7}{x}\right)\mathrm{d}x$

Problem 1.2 (review of definite integrals) Given that $\int_c^d g(x)\mathrm{d}x = I$, where c, d and I are constants, state the value of

(a) $\int_d^c g(x)\mathrm{d}x$ (b) $\int_c^d C\,g(x)\mathrm{d}x$ where C is a constant

(c) $\int_c^d g(y)\mathrm{d}y$ (d) $\int_c^d \pi g(t)\mathrm{d}t$

6.2 LINE INTEGRALS

A line integral is a generalisation of the ordinary definite integral to two and three dimensions, the path of integration being along any curve in space. Some line integrals have scalar values while others are vectors. We shall deal mainly with the former which we refer to as *scalar line integrals*. Section 6.2.1 defines the scalar line integral as the limit of a sum. Section 6.2.2 shows how simple line integrals can be expressed as definite integrals and evaluated. A general technique for evaluating line integrals is given in Section 6.3.

6.2.1 Defining the scalar line integral

One way in which scalar line integrals arise is in the description of energy exchange between a force field and a particle. Consider a vector field $\mathbf{F}(\mathbf{r})$ describing the force acting on a particle at $\mathbf{r}$ and suppose the particle undergoes a displacement $\Delta\mathbf{r}$ along a straight line on which the force remains constant. Then an amount of energy $\Delta W = \mathbf{F}(\mathbf{r}) \cdot \Delta\mathbf{r}$ is transferred from the field to the particle or to some other system to which the particle is connected; we say that the field does work ΔW. Now suppose a particle moves from a point A to a point

B along a curved path labelled C with the force varying along the path. We can use the following procedure to calculate the work done.

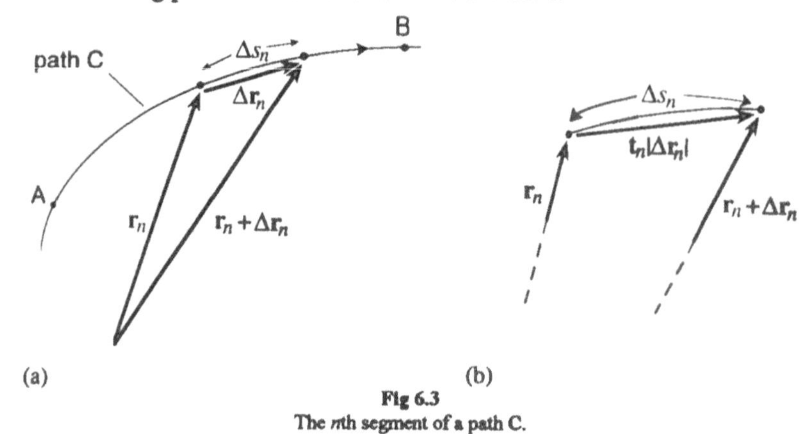

Fig 6.3
The *n*th segment of a path C.

Divide the path C into N small segments of lengths $\Delta s_1, \Delta s_2, ..., \Delta s_n, ...,$ Δs_N, where the symbol s is the distance of the particle measured along the path from a fixed reference point on the path. The nth segment is shown in Fig 6.3a with the position vectors of its endpoints, $\mathbf{r}_n$ and $\mathbf{r}_n + \Delta\mathbf{r}_n$. If the segments are short enough and nearly straight, the force may be almost constant on any segment. Then it is a good approximation to take the force in the nth segment to have the constant value $\mathbf{F}(\mathbf{r}_n)$, and the work done when the particle moves along the segment from $\mathbf{r}_n$ to $\mathbf{r}_n + \Delta\mathbf{r}_n$ to be $\Delta W_n \approx \mathbf{F}(\mathbf{r}_n) \,.\, \Delta\mathbf{r}_n$. An approximate value for the total work done along the whole path from A to B is then given by summing $\mathbf{F}(\mathbf{r}_n) \,.\, \Delta\mathbf{r}_n$ over all segments n. When N is very large and the lengths of the segments very small ($N \to \infty$; $\Delta s_n \to 0$), this sum approaches a limit which is independent of the manner in which the subdivisions are made, and is exactly equal to the work done along the path C. This limit is a number (scalar) called the **scalar line integral** of **F** along the curve or path C and is denoted by the integral symbol $\int_C$. Thus we have

$$W = \lim_{N\to\infty}\left[\sum_{n=1}^{N}\mathbf{F}(\mathbf{r}_n).\Delta\mathbf{r}_n\right] = \int_C \mathbf{F}(\mathbf{r}).\,d\mathbf{r} \qquad (6.8a)$$

The path C is a directed curve with a beginning point A and an endpoint B, and is called the **path of integration**. If the direction of the path is reversed (from B to A) then the value of the line integral changes sign. The vector symbol **dr** is referred to as the **vector line element**. Strictly speaking **dr** is part of the line integral symbol and is not a physical quantity. However it is convenient to think of **dr** as an infinitesimal directed segment of the curve C.

The scalar line integral can be expressed another way by writing $\Delta\mathbf{r}_n$ in Eq (6.8a) as $\Delta\mathbf{r}_n = \mathbf{t}_n|\Delta\mathbf{r}_n|$, where $\mathbf{t}_n$ is a unit vector (Fig 6.3b). We then have $\mathbf{F}(\mathbf{r}_n) \,.\, \Delta\mathbf{r}_n = \mathbf{F}(\mathbf{r}_n) \,.\, \mathbf{t}_n|\Delta\mathbf{r}_n|$. As N becomes large, $|\Delta\mathbf{r}_n|$ becomes the same as the path segment Δs_n, the unit vector $\mathbf{t}_n$ becomes the unit tangent vector $\mathbf{t}$ pointing in the positive direction along the path (from A to B) and $\mathbf{F}(\mathbf{r}_n) \cdot \mathbf{t}_n$ becomes F_t, the component of **F** tangential to the path. Thus we can express the line integral as

$$\lim_{N \to \infty} \left[\sum_{n=1}^{N} \mathbf{F}(\mathbf{r}_n) \cdot \mathbf{t}_n |\Delta \mathbf{r}_n| \right] = \int_{\mathrm{C}} F_t(\mathbf{r})\, \mathrm{d}s \tag{6.8b}$$

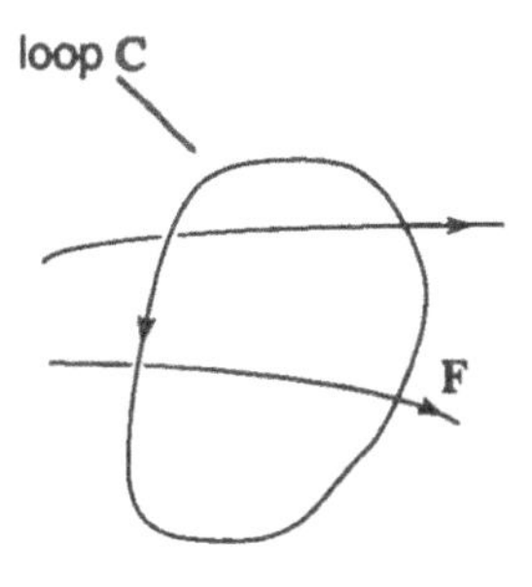

Fig 6.4
The path C is a closed loop.

The scalar symbol d*s*, equivalent to $|\mathrm{d}\mathbf{r}|$, is the scalar line element. Eqs (6.8a) and (6.8b) are different ways of writing the same scalar line integral.

When the path of integration is a plane closed loop (Fig 6.4) the scalar line integral around the loop, starting from any point on it and returning to the same point, is called the **circulation** of **F** around the loop, and is denoted by the special integral symbol $\oint$. Thus

$$\oint \mathbf{F}(\mathbf{r}) \cdot \mathrm{d}\mathbf{r} = \text{circulation of } \mathbf{F} \text{ around a loop} \tag{6.9}$$

When a positive direction normal to the plane of the loop is defined (such as the positive *z*-axis for a loop in the *x*-*y* plane), the traversal of the loop is taken to be in the anticlockwise sense (by the right-hand rule), i.e. the direction of traversal is such that the enclosed area lies to the left as the loop is traversed.

We now make a connection with Section 4.5.2. Consider a special case where the tangential component of the field is constant along the entire path. We can then put $F_t(\mathbf{r})$ equal to a constant F_t which can then be taken outside the summation symbol in Eq (6.8b) to give the line integral

$$\int_{\mathrm{C}} F_t(\mathbf{r})\mathrm{d}s = F_t \int_{\mathrm{C}} \mathrm{d}s = F_t L \qquad (F_t \text{ constant}) \tag{6.10}$$

where $L = \int_{\mathrm{C}} \mathrm{d}s$ is the total length of the curve C. Cases of this kind can be found in the Examples and Problems in Section 4.5.2.

Scalar line integrals having the form of Eqs (6.8a) and (6.8b) can be defined for any kind of vector field, not just force fields; only the interpretation in terms of work is specific to force fields. Furthermore, as the form of Eq (6.8b) suggests, a scalar line integral can be defined for any scalar field $\rho(\mathbf{r})$ by replacing $F_t(\mathbf{r})$ in Eq (6.8b) by $\rho(\mathbf{r})$. For example, the total mass of a wire bent into the shape of a curve C is given by

$$M = \int_{\mathrm{C}} \rho(\mathbf{r})\mathrm{d}s \tag{6.11}$$

where $\rho(\mathbf{r})$ is the mass density per unit length of the wire (kg m^{-1}) at the point on the wire whose position vector is **r**, and it is assumed that the path is traversed in the direction that makes the line integral positive. $\rho(\mathbf{r})\mathrm{d}s$ can be interpreted as the mass of the scalar line element d*s*.

Vector line integrals also occur in many applications, especially in the theory of magnetism. Consider a wire lying along a curve C carrying an electric current I in the positive direction along C. Then if the wire is situated in a region where there is a magnetic field $\mathbf{B}(\mathbf{r})$, the magnetic force acting on the wire is given by

$$\mathbf{F} = \int_C I\mathrm{d}\mathbf{r} \times \mathbf{B}(\mathbf{r}) = -I\int_C \mathbf{B}(\mathbf{r}) \times \mathrm{d}\mathbf{r} \tag{6.12}$$

Another example of a vector line integral is the Biot-Savart law (defined in Example 2.3) giving the magnetic field produced by a current-carrying wire.

6.2.2 Evaluating simple line integrals

A line integral is evaluated by introducing a coordinate system and expressing the line integral as one or more ordinary definite integrals which can then be evaluated by the techniques in Section 6.1.

In cartesian coordinates a three-dimensional vector field $\mathbf{F}(\mathbf{r})$ is expressed as

$$\mathbf{F}(\mathbf{r}) = \mathbf{i}F_x(x,y,z) + \mathbf{j}F_y(x,y,z) + \mathbf{k}F_z(x,y,z) \tag{6.13}$$

and the vector line element $\mathrm{d}\mathbf{r}$ is

$$\mathrm{d}\mathbf{r} = \mathbf{i}\mathrm{d}x + \mathbf{j}\mathrm{d}y + \mathbf{k}\mathrm{d}z \tag{6.14}$$

Using Eqs (6.13) and (6.14) in the line integral Eq (6.8a) and evaluating the scalar products, we obtain the sum of three cartesian integrals

$$\int_C \mathbf{F}(\mathbf{r})\,.\,\mathrm{d}\mathbf{r} = \int_C [\mathbf{i}F_x(x,y,z) + \mathbf{j}F_y(x,y,z) + \mathbf{k}F_z(x,y,z)]\,.\,(\mathbf{i}\mathrm{d}x + \mathbf{j}\mathrm{d}y + \mathbf{k}\mathrm{d}z)$$

$$= \int_C F_x(x,y,z)\mathrm{d}x + \int_C F_y(x,y,z)\mathrm{d}y + \int_C F_z(x,y,z)\mathrm{d}z \tag{6.15}$$

For a path confined to the x-y plane, the third integral on the right-hand side of Eq (6.15) is zero since the line element has no component in the z direction, i.e. $\mathrm{d}\mathbf{r} = \mathbf{i}\mathrm{d}x + \mathbf{j}\mathrm{d}y$. Similarly, we find that the scalar line integral of a two-dimensional field $\mathbf{f}(x,y)$ along a curve in the x-y plane is

$$\int_C \mathbf{f}(\mathbf{r}).\mathrm{d}\mathbf{r} = \int_C f_x(x,y)\mathrm{d}x + \int_C f_y(x,y)\mathrm{d}y \qquad \text{(C in } x\text{-}y \text{ plane)} \tag{6.16}$$

A systematic way of evaluating the integrals in Eqs (6.15) and (6.16) is described in Section 6.3. We conclude this section by describing an informal and intuitive approach which is often adequate for simple two-dimensional line integrals.

Simple examples – an informal approach

We shall evaluate some line integrals of the vector field

$$\mathbf{f}(x,y) = -y\mathbf{i} + x\mathbf{j} \tag{6.17}$$

You may recognise **f**, apart from an overall scaling, as the velocity field on the surface of a rotating disc (Eq (5.41)).

The cartesian components of this field are $f_x(x,y) = -y$ and $f_y(x,y) = x$ and so the line integral of **f** along any path C, Eq (6.16), is

$$\int_C \mathbf{f}(\mathbf{r})\,.\,\mathrm{d}\mathbf{r} = \int_C (-y)\mathrm{d}x + \int_C x\mathrm{d}y \tag{6.18}$$

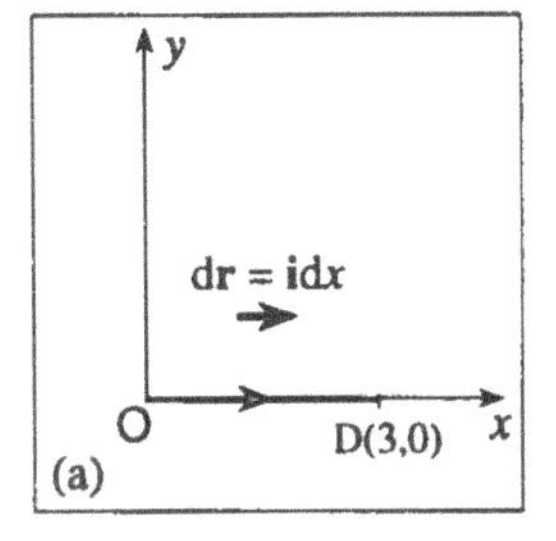

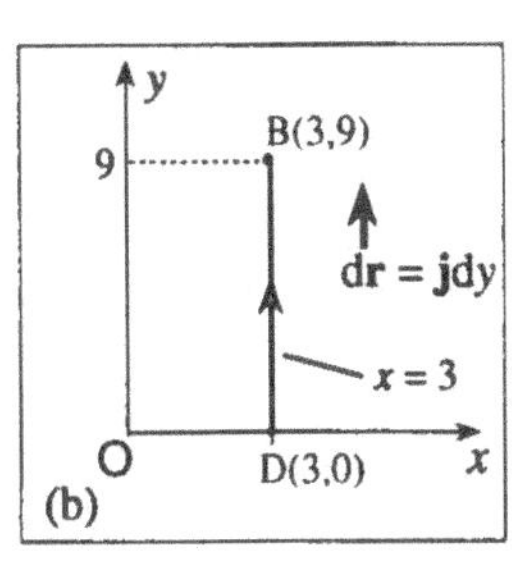

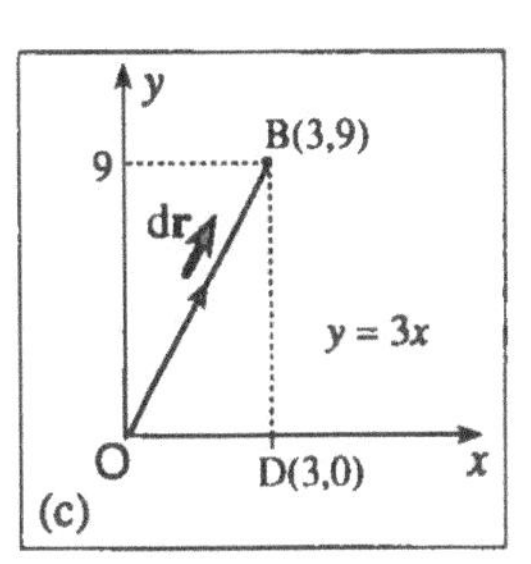

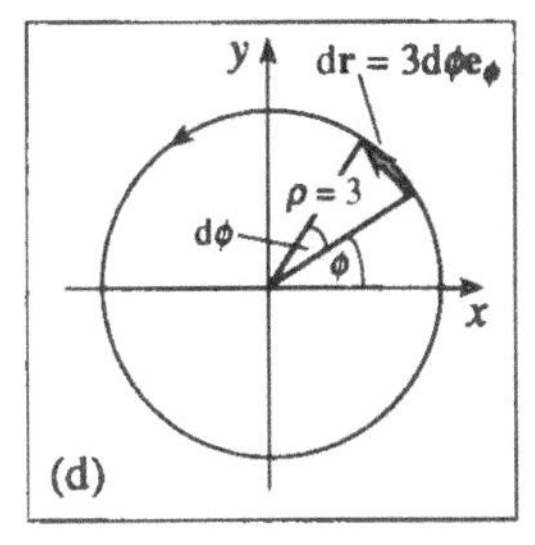

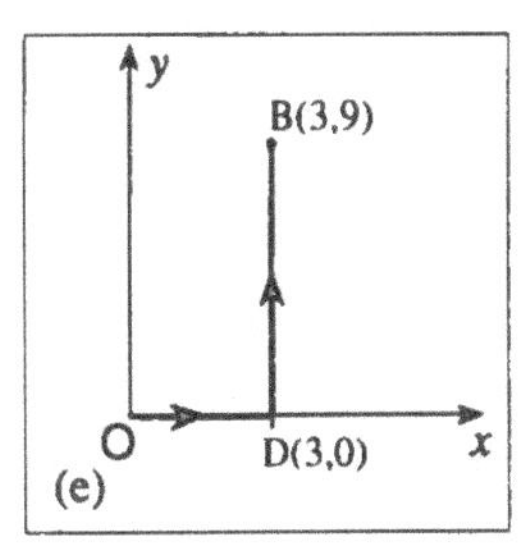

Fig 6.5
(a) The path OD on the x-axis. (b) The path DB parallel to the y-axis. (c) The straight-line path OB. (d) A circular path in the x-y plane. (e) The path ODB.

We now evaluate the line integral of **f** along four different paths:

(i) *Along the segment OD of the x-axis* (Fig 6.5a). We have $y = 0$ for any point on the x-axis and so the first integral in Eq (6.18) is zero. The second integral is also zero because the vector line element on the x-axis is simply $\mathbf{dr} = \mathbf{i}\mathrm{d}x$, i.e. we put $\mathrm{d}y = 0$ along the x-axis. Thus we have the result $\int_{\mathrm{OD}} \mathbf{f}(\mathbf{r}).\mathrm{d}\mathbf{r} = 0$.

We could have anticipated the result in (i) from the fact that **f** points in the y-direction everywhere on the x-axis, $\mathbf{f}(x,0) = x\mathbf{j}$, and so the tangential component of **f** is zero everywhere on the path OD.

(ii) *Along the straight path DB parallel to the y-axis* (Fig 6.5b). The first integral in Eq (6.18) vanishes for this path because the vector line element is $\mathbf{dr} = \mathbf{j}\mathrm{d}y$, i.e. we put $\mathrm{d}x = 0$ along any path parallel to the y-axis. Now consider the second integral in Eq (6.18). The path DB is a segment of the straight line $x = 3$, and so we can put $x = 3$ in the integrand. The integration variable is y and this varies from $y = 0$ at the beginning point D of the path to $y = 9$ at the end point B. Thus Eq (6.18) yields

$$\int_{DB} \mathbf{f}(\mathbf{r}) \cdot d\mathbf{r} = \int_{DB} 3\,dy = \int_0^9 3\,dy = [3y]_0^9 = 27$$

(iii) *Along the path OB* (Fig 6.5c). This path lies on the straight line described by the equation $y = 3x$. We can use this equation to express each integrand in the two integrals on the right-hand side of Eq (6.18) in terms of the corresponding variable of integration, the variable x in the first integral and y in the second. Thus we write the integrand $(-y)$ in the first integral as $(-3x)$ and the integrand x in the second integral as $y/3$. We next put in the lower and upper limits of each variable of integration corresponding to the beginning point (0,0) of the path and end point (3,9). The two integrals now become ordinary definite integrals and the line integral can be evaluated:

We could have anticipated the zero result in (iii) by noticing that the vector field $\mathbf{f}$ has the plane polar form $\mathbf{f}(\rho,\phi) = \rho\mathbf{e}_\phi$ and is therefore at right-angles to all radial lines. It follows that the tangential component f_t is zero everywhere on the straight line path OB.

$$\int_{OB} \mathbf{f}(\mathbf{r}) \cdot d\mathbf{r} = \int_0^3 (-3x)\,dx + \int_0^9 \left(\frac{y}{3}\right) dy$$

$$= \left[-\frac{3x^2}{2}\right]_0^3 + \left[\frac{y^2}{6}\right]_0^9 = -\frac{27}{2} + \frac{81}{6} = 0$$

(iv) *One complete anticlockwise circuit of the circular path of radius 3* (Fig 6.5d). Symmetry considerations suggest we use plane polar coordinates in which the vector field $\mathbf{f}$ has the form $\mathbf{f}(\rho,\phi) = \rho\mathbf{e}_\phi$. The coordinate ρ has the constant value $\rho = 3$ on the circular path while ϕ varies through its range from 0 to 2π, and so the function value on the path is $\mathbf{f}(3,\phi) = 3\mathbf{e}_\phi$. The vector line element on this circle is simply the arc element $d\mathbf{r} = 3d\phi\mathbf{e}_\phi$ and so the line integral, or circulation of $\mathbf{f}$ around the circle, is

We could have obtained the result 18π by using Eq (6.10) since the tangential component of $\mathbf{f}$ on the circle is $\mathbf{e}_\phi \cdot \mathbf{f} = \mathbf{e}_\phi \cdot \rho\mathbf{e}_\phi = \rho = 3$ which is constant everywhere on the circle, and so the circulation is simply $f_t L = 3\times2\pi\times3 = 18\pi$.

$$\oint \mathbf{f}(\mathbf{r}) \cdot d\mathbf{r} = \int_0^{2\pi} \mathbf{f}(3,\phi) \cdot 3d\phi\,\mathbf{e}_\phi = \int_0^{2\pi} 3\mathbf{e}_\phi \cdot 3\,d\phi\,\mathbf{e}_\phi$$

$$= 9\int_0^{2\pi} d\phi = 9\,[\phi]_0^{2\pi} = 18\pi$$

Several lessons can be learned from these examples. First, when the path of integration is along a segment of one of the coordinate axes, or along a line parallel to a coordinate axis, the line integral reduces immediately to a single definite integral, as in (i) and (ii) above.

Secondly, you can sometimes save yourself the effort of evaluating line integrals by spotting cases where the tangential component f_t is zero everywhere on the path, or where f_t is constant on the path, as in (iii) and (iv) above, when Eq (6.10) applies.

Finally, examples (i) to (iii) above illustrate an important property of the vector field $\mathbf{f}$. In (iii) we found that the line integral of $\mathbf{f}$ along the direct path

OB is zero. But we can get from O to B along the path ODB (Fig 6.5e). The line integral along this composite path is found by adding the results in (i) and (ii) above to give 0 + 27 = 27. Thus the scalar line integral of **f** between the two fixed points O and B depends on the actual path connecting the points! We shall discuss the path-dependence of line integrals in Section 6.3.

Summary of section 6.2

- The **scalar line integral** of a vector field **F** along a path C is the limit of the sum of the tangential components of **F** along C; it is denoted by $\int_C \mathbf{F}(\mathbf{r}) \cdot d\mathbf{r}$ or $\int_C F_t(\mathbf{r})ds$. C is a directed curve called the **path of integration**. The symbol d**r** is called the **vector line element**.

- When $F_t(\mathbf{r})$ has the constant value F_t everywhere on the path, the line integral is simply F_tL where L is the path length.

- When the path of integration is closed, the line integral for one complete traversal of the path in the anticlockwise sense is called the **circulation** of **F** and is denoted by using the symbol $\oint$.

- The cartesian form of the line integral is

$$\int_C \mathbf{F}(\mathbf{r}) \cdot d\mathbf{r} = \int_C F_x(x,y,z)dx + \int_C F_y(x,y,z)dy + \int_C F_z(x,y,z)dz \qquad (6.15)$$

  For paths confined to the x-y plane we put $dz = 0$ (Eq (6.16)) and for a path parallel to the x-axis (or y-axis) we also put $dy = 0$ (or $dx = 0$).

- **Vector line integrals** also exist – see Eq (6.12) and the Biot-Savart law (Example 2.3).

Example 2.1 (*Objective 1*)

(a) Find the scalar line integral of the vector field $\mathbf{F}(x,y) = x^2y\mathbf{i} + y^2x\mathbf{j}$ along each directed segment of the square of side length 3 shown in Fig 6.6.

(b) What is the anticlockwise circulation of **F** around the square?

(c) What is the line integral of **F** along (i) the path OAB and (ii) the path OCB?

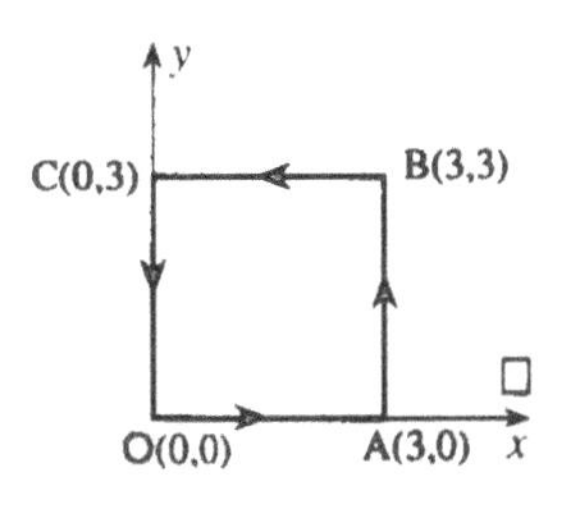

Fig 6.6
Paths along the sides of a square of side length 3.

Solution 2.1 The components of **F** are $F_x = x^2y$ and $F_y = y^2x$. Using Eq (6.16), the line integral of **F** along a path C is

$$\int_C \mathbf{F}(\mathbf{r})\,.\,d\mathbf{r} = \int_C x^2 y\,dx + \int_C y^2 x\,dy$$

(a) Along OA we have $y = 0$, and so both integrands are zero. Hence $\int_{OA} \mathbf{F}(\mathbf{r})\,.\,d\mathbf{r} = 0$.

Along the path AB we put $dx = 0$ and $x = 3$, and so we have $\int_{AB} \mathbf{F}(\mathbf{r})\,.\,d\mathbf{r} = 0 + \int_0^3 y^2 \times 3\,dy = \left[y^3\right]_0^3 = 27$.

Along the path BC we put $dy = 0$ and $y = 3$, and so we have $\int_{BC} \mathbf{F}(\mathbf{r})\,.\,d\mathbf{r} = \int_3^0 x^2 \times 3\,dx + 0 = \left[x^3\right]_3^0 = -27$.

Along CO we put $dx = 0$ and $x = 0$, which gives $\int_{CO} \mathbf{F}(\mathbf{r})\,.\,d\mathbf{r} = 0$.

(b) The circulation of **F** around the loop is the sum of the line integrals along the four sides of the square, calculated in part (a). Thus the circulation is $\oint \mathbf{F}(\mathbf{r})\,.\,d\mathbf{r} = 0 + 27 + (-27) + 0 = 0$.

(c) (i) The line integral along OAB is the sum of the line integrals along OA and AB, that is $0 + 27 = 27$.

(ii) The path OCB is the reverse of the composite path BCO along which the line integral is $-27 + 0$. Thus the line integral along the reverse path OCB has the value 27.

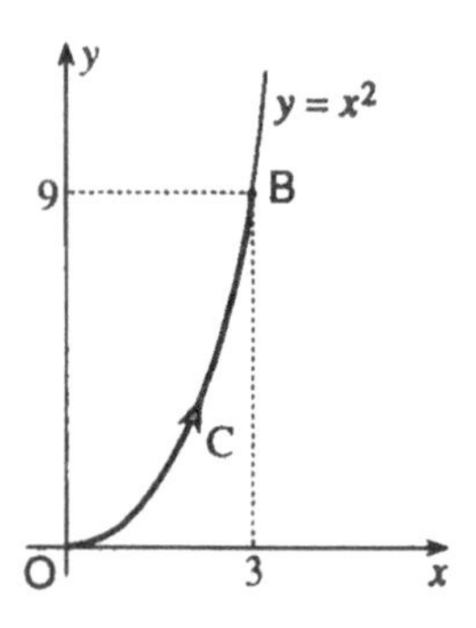

Fig 6.7
The path OB on the parabola $y = x^2$.

Example 2.2 (*Objective 1*) Determine the scalar line integral of the vector field $\mathbf{f}(x,y) = -y\mathbf{i} + x\mathbf{j}$ along the path C consisting of the segment OB of the parabola described by the equation $y = x^2$ and shown in Fig 6.7.

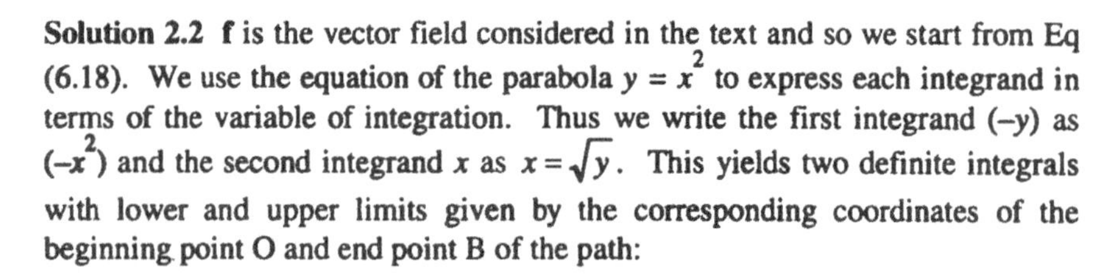

Solution 2.2 **f** is the vector field considered in the text and so we start from Eq (6.18). We use the equation of the parabola $y = x^2$ to express each integrand in terms of the variable of integration. Thus we write the first integrand $(-y)$ as $(-x^2)$ and the second integrand x as $x = \sqrt{y}$. This yields two definite integrals with lower and upper limits given by the corresponding coordinates of the beginning point O and end point B of the path:

$$\int_{OB} \mathbf{f}(\mathbf{r})\,.\,d\mathbf{r} = \int_0^3 (-x^2)\,dx + \int_0^9 y^{1/2}\,dy = \left[-\frac{x^3}{3}\right]_0^3 + \left[\frac{2}{3}y^{3/2}\right]_0^9 = -9 + 18 = 9$$

Example 2.3 (*Objective 1*) The Biot-Savart law gives the magnetic field vector **B** at a point Q due to a steady electric current I flowing in a circuit C. If we specify points on C by their position vectors **r** measured from the point Q, the law can be expressed as the vector line integral

$$\mathbf{B}_Q = \lambda \int_C \frac{\mathbf{r} \times d\mathbf{r}}{|\mathbf{r}|^3} \tag{6.19}$$

where $\lambda = \mu_0 I/4\pi$, and the direction of the current determines the positive direction around C.

Determine **B** at the centre of a circular current loop of radius R.

Solution 2.3 We use the Biot-Savart law with the origin point Q at the centre of the loop. Using plane polar coordinates in the plane of the loop (Fig 6.8) we have $\mathbf{r} = R\mathbf{e}_\rho$ and $d\mathbf{r} = \mathbf{e}_\phi R d\phi$. Thus Eq (6.19) becomes

$$\mathbf{B} = \lambda \oint_C \frac{R\mathbf{e}_\rho \times (\mathbf{e}_\phi R d\phi)}{R^3} = \frac{\lambda \mathbf{e}_z}{R} \int_0^{2\pi} d\phi = \frac{2\pi\lambda \mathbf{e}_z}{R}$$

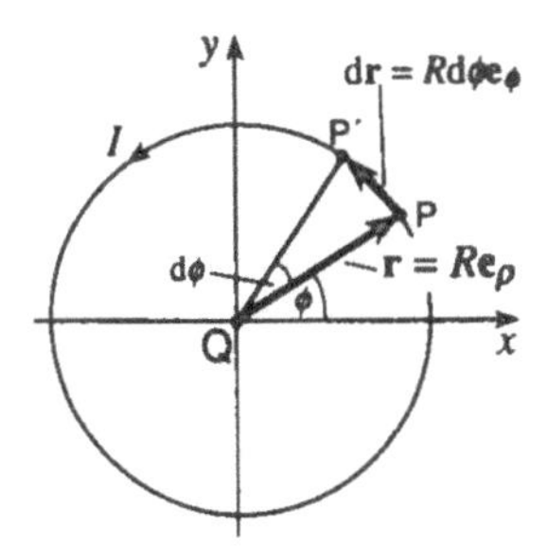

Fig 6.8
A circular wire carrying a current I with its centre at an origin Q.

We have assumed that the current flows anticlockwise around the loop so that the lower and upper limits of ϕ are 0 and 2π respectively. Thus we have shown that the magnetic field at the centre of the loop has magnitude $2\pi\lambda/R = \mu_0 I/2R$, and is directed normal to the plane of the loop in the positive z-direction.

Problem 2.1 (*Objective 1*) Determine the scalar line integral of the vector field $\mathbf{q}(x,y,z) = (x^2 - yz)\mathbf{i} + (y^2 + 3)\mathbf{j} - z^2\mathbf{k}$ along the y-axis from $y = 1$ to $y = 5$.

Problem 2.2 (*Objective 1*)

(a) Evaluate the scalar line integral of the two-dimensional vector field $\mathbf{h}(\mathbf{r}) = \mu\mathbf{r}$ where μ is a constant and $\mathbf{r}$ is in the x-y plane, along each of the following paths (Fig 6.9):

(i) the path OA on the x-axis,
(ii) the path AB on the parabola $y = 2 - 2x^2$,
(iii) the path BO on the y-axis.

(b) What is the circulation of **h** around the closed path comprised of the three segments above?

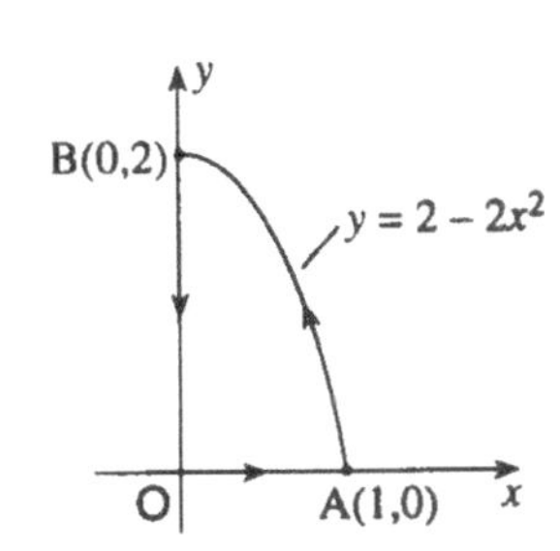

Fig 6.9
The paths of integration for Problem 2.2.

Problem 2.3 (*Objective 1*) Find the scalar line integral of the vector field $\mathbf{p}(x,y) = \mathbf{i}\sin(2y)$ along the line $y = -4x$ from the origin to the point $(\pi/8, -\pi/2)$.

Problem 2.4 (*Objective 1*) Find the circulation of the vector field $\mathbf{G}(x,y,z) = (-y\mathbf{i} + x\mathbf{j})/(x^2 + y^2)$ around a circular path in the x-y plane of radius a centred at the origin.

Problem 2.5 (*Objective 1. This problem involves evaluating vector line integrals, Eq (6.12).*) A straight segment of wire of length L carries an electric current I and lies in the magnetic field $\mathbf{B}(\rho,\phi,z) = \mu_0 I_1 \mathbf{e}_\phi/(2\pi\rho)$ $(\rho > 0)$ produced

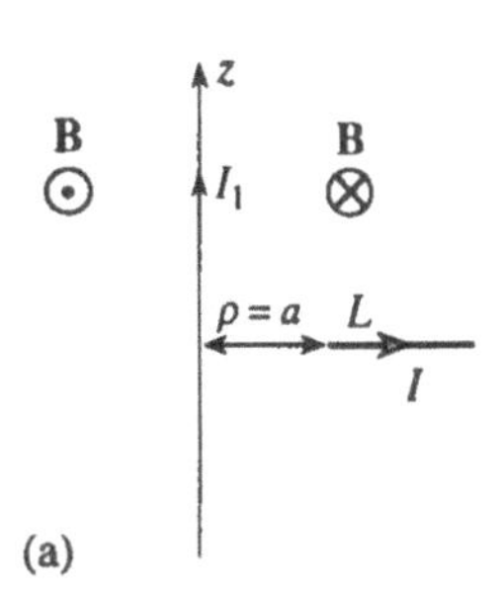

Fig 6.10a

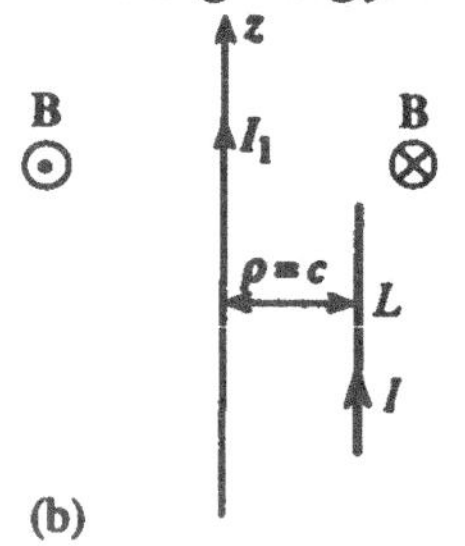

Fig 6.10b

by another straight wire on the z-axis carrying a current I_1 in the positive z-direction. Determine the magnetic force on the segment when it is aligned (a) radially from the z-axis with its ends at $\rho = a$ and $\rho = a + L$ and with the current I directed away from the z-axis (Fig 6.10a), and (b) parallel to the z-axis and at a distance c from it with I in the positive z-direction (Fig 6.10b).

6.3 LINE INTEGRALS ALONG PARAMETERISED CURVES

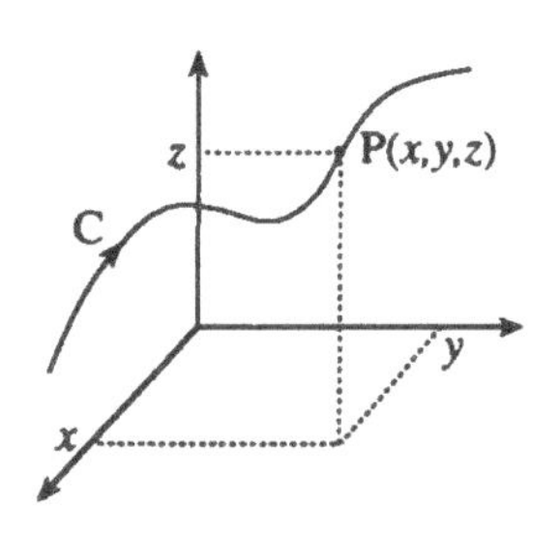

Fig 6.11
A fixed value of x is sufficient to determine the position of a point P on the curve C.

The three coordinates of a point on a curve cannot be chosen independently of one another. The position of the point P on the curve C shown in Fig 6.11 is fixed once its x-coordinate is specified. In other words, there is only one independent variable for points on the curve. This is the key idea in the evaluation of line integrals. It was used implicitly in Section 6.2 where the interdependence of the two coordinates of a point on a plane curve C was expressed by the equation of the curve ($y = 3x$ on path OB in Fig 6.5c and $y = x^2$ in Example 2.2). The equation was used to express the integrands in terms of the corresponding integration variables. We now introduce a systematic technique for evaluating line integrals based on the idea of expressing the three coordinates of a point on a curve in terms of a single coordinate, or parameter, which is then made the variable of integration.

6.3.1 Parameterisation of a curve

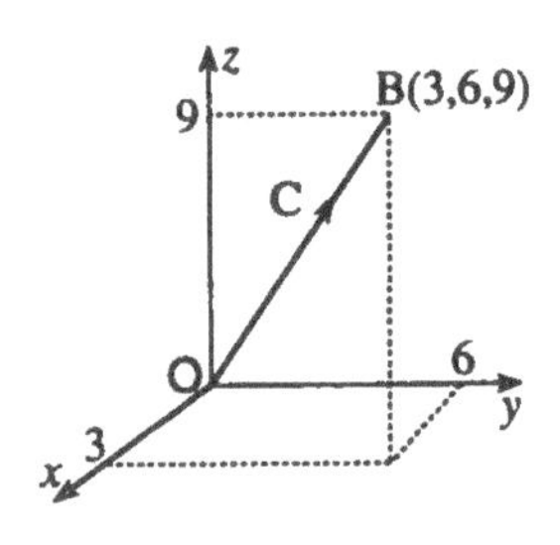

Fig 6.12
The straight-line path OB.

A plane curve is described by an equation relating x and y, for example, the parabola path in Fig 6.7 is described by $y = x^2$. More generally a path of integration may be a curve in three dimensions (such as the helix shown in Fig 6.13). Such curves are best described by introducing a single variable t called the **parameter** of the curve. The cartesian coordinates of a point on the curve can then be expressed in terms of t.

Consider the straight line path C from the origin O(0,0,0) to the point B(3,6,9) (Fig 6.12). We can describe this path by the three **parametric equations**

$$x = 3t, \quad y = 6t, \quad z = 9t \qquad (0 \le t \le 1;\ t \text{ increasing}) \tag{6.20}$$

Thus the coordinates (x,y,z) of any point on the curve can be expressed as $(3t,6t,9t)$. As the parameter t increases through its range from 0 to 1, the point $(3t,6t,9t)$ moves from the origin along the straight line to B. This parameterisation is not unique since we could obviously have chosen instead: $x = t$, $y = 2t$ and $z = 3t$ with t ranging from 0 to 3.

The statement "t increasing" in Eq (6.20) distinguishes the directed path with beginning point O and end point B from the reversed path from B to O. This statement is normally omitted with the understanding that the direction of the path is that of increasing t.

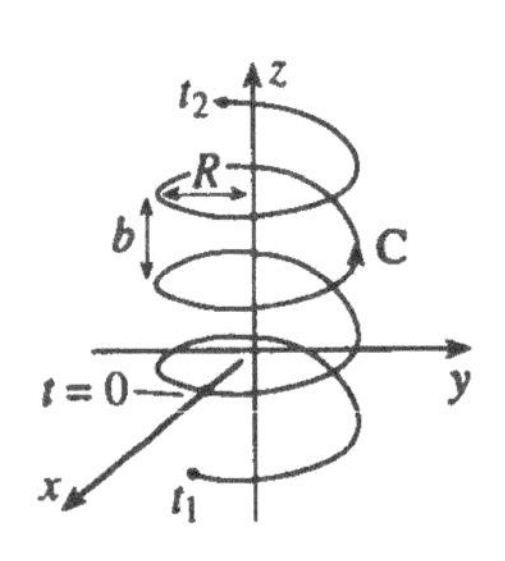

Fig 6.13
A helix.

A more complex path is the helix in Fig 6.13 of radius R and pitch b. A segment of this helix is described by the parametric equations

$$x = R\cos t,\quad y = R\sin t,\quad z = (b/2\pi)t \qquad (t_1 \le t \le t_2) \tag{6.21}$$

where t_1 and t_2 define the beginning and end points respectively of the segment.

6.3.2 A systematic technique for evaluating line integrals

Suppose we wish to evaluate the scalar line integral of the vector field

$$\mathbf{F}(x,y,z) = -y\mathbf{i} + x\mathbf{j} + z^2\mathbf{k}$$

along the straight path C described by parametric equations (6.20). We first write the components of **F** in terms of the parameter t,

$$F_x = -y = -6t,\quad F_y = x = 3t,\quad F_z = z^2 = (9t)^2$$

Substituting into Eq (6.15) gives

$$\int_C \mathbf{F}(\mathbf{r}).\mathrm{d}\mathbf{r} = \int_C (-6t)\,\mathrm{d}x + \int_C 3t\,\mathrm{d}y + \int_C (9t)^2\,\mathrm{d}z \tag{6.22}$$

These integrals cannot yet be evaluated since the integrands are expressed in terms of t while the variables of integration, indicated by the symbols dx, dy and dz, are still cartesian. We can proceed by making t the variable of integration in all three integrals. To do this we express dx, dy and dz in terms of dt using the following rule

$$\mathrm{d}x = \left(\frac{\mathrm{d}x}{\mathrm{d}t}\right)\mathrm{d}t,\quad \mathrm{d}y = \left(\frac{\mathrm{d}y}{\mathrm{d}t}\right)\mathrm{d}t,\quad \mathrm{d}z = \left(\frac{\mathrm{d}z}{\mathrm{d}t}\right)\mathrm{d}t \tag{6.23}$$

The derivatives, dx/dt etc., are found by differentiating the parametric equations. Thus differentiating Eqs (6.20) we obtain dx/dt = 3, dy/dt = 6 and dz/dt = 9. Eqs (6.23) now give

$$\mathrm{d}x = 3\mathrm{d}t,\quad \mathrm{d}y = 6\mathrm{d}t,\quad \mathrm{d}z = 9\mathrm{d}t$$

Finally, remembering that the beginning and end points of the path specified in Eq (6.20) correspond to $t = 0$ and $t = 1$, the line integral, Eq (6.22), becomes

$$\int_C \mathbf{F}(\mathbf{r}).\mathrm{d}\mathbf{r} = \int_0^1 (-6t)3\mathrm{d}t + \int_0^1 (3t)6\mathrm{d}t + \int_0^1 (9t)^2\,9\mathrm{d}t$$

$$= \int_0^1 (-18t + 18t + 729t^2)\mathrm{d}t = \frac{1}{3}\left[729t^3\right]_0^1 = 243$$

This example illustrates the systematic approach to line integrals based on the parametric equations of the path and Eq (6.15). Assuming the parametric equations of the curve are given, the procedure is:

(i) Use the parametric equations to express each integrand in Eq (6.15) in terms of the parameter t,

(ii) Differentiate the parametric equations and use Eqs (6.23) to express dx, dy and dz in terms of dt.

(iii) Express the line integral as a definite integral with respect to t and evaluate it.

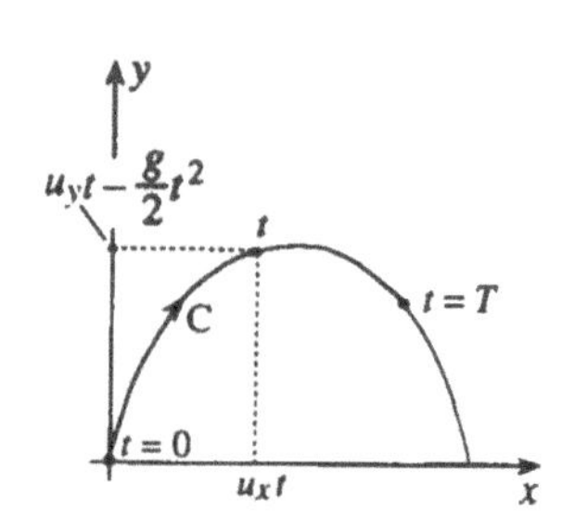

Fig 6.14
The path of a projectile.

The parameter of a curve sometimes has physical significance. In Eq (6.21) for example, t occurs as the argument of sine and cosine functions and is therefore an angle; in fact it is the angular displacement about the z-axis, increasing by 2π for each cycle of the helix. In problems involving moving particles, time is a natural parameter. The space curve traced out by a moving particle has the form $\mathbf{r}(t) = (x(t), y(t), z(t))$ where t here is the elapsed time. This is an example of a curve parameterised by time. Consider the case of a projectile moving under gravity. Let the vertical plane of the motion be the x-y plane ($z = 0$), and suppose $t = 0$ when the particle is at the origin (Fig 6.14). Then the coordinates of the particle up to time $t = T$ are given by the parametric equations

$$x = u_x t, \quad y = u_y t - \frac{1}{2}gt^2, \quad z = 0 \qquad (0 \le t \le T) \qquad (6.24)$$

where u_x and u_y are respectively the horizontal and vertical components of the velocity at $t = 0$.

Summary of section 6.3

- A curve in space can be parameterised by expressing the coordinates of points on the curve in terms of a single variable t called the **parameter**. The equations $x = x(t)$, $y = y(t)$, $z = z(t)$ are called **parametric equations**.

- A scalar line integral of a vector field $\mathbf{F}$ (Eq (6.15)) can be expressed as a definite integral with respect to the curve parameter t by writing the scalar components of $\mathbf{F}$ in terms of t and making t the variable of integration by using Eqs (6.23).

Example 3.1 (*Objective 2*) The parametric equations for a certain path C are

$$x = t, \quad y = t^2, \quad z = 0 \qquad (0 \le t \le 3)$$

(a) Determine the cartesian coordinates of the points for which t has the values 0, 1, 2 and 3. What shape is the path?

(b) Evaluate the line integral of $\mathbf{f}(x,y) = -y\mathbf{i} + x\mathbf{j}$ along C.

Solution 3.1

(a) The coordinates of any point on the curve are $(x,y,z) = (t,t^2,0)$. For $t = 0$, 1, 2, and 3 the coordinates are respectively: (0,0,0), (1,1,0), (2,4,0) and (3,9,0). By eliminating t from the parametric equations we obtain the cartesian equations $y = x^2$, $z = 0$ $(0 \le t \le 3)$. Thus the path C is a segment of a parabola in the x-y plane (Fig 6.7).

(b) The line integral of $\mathbf{f}(x,y) = -y\mathbf{i} + x\mathbf{j}$ along this parabolic path is found by expressing the components of $\mathbf{f}$ in terms of t and making t the variable of integration using Eqs (6.23). Thus we write

$$\int_C \mathbf{f}(\mathbf{r}) \cdot \mathrm{d}\mathbf{r} = \int_C (-y)\,\mathrm{d}x + \int_C x\,\mathrm{d}y$$

$$= \int_0^3 (-t^2)\left(\frac{\mathrm{d}x}{\mathrm{d}t}\right)\mathrm{d}t + \int_0^3 t\left(\frac{\mathrm{d}y}{\mathrm{d}t}\right)\mathrm{d}t$$

$$= \int_0^3 (-t^2)\,\mathrm{d}t + \int_0^3 t(2t)\,\mathrm{d}t = \int_0^3 t^2\mathrm{d}t = \left[\frac{t^3}{3}\right]_0^3 = 9$$

(This result is in agreement with the solution of Example 2.2.)

Example 3.2 (*Objective 2*)

(a) Determine the scalar line integral of the vector field $\mathbf{g}(\mathbf{r}) = \alpha\mathbf{i}$ (α is a constant) along the semicircular path S of radius 5 (Fig 6.15) described by the parametric equations

$$x = 5\cos t, \quad y = 5\sin t, \quad z = 0 \qquad (0 \le t \le \pi)$$

(b) Determine the scalar line integral of g along a path C following the x-axis from $x = 5$ to $x = -5$.

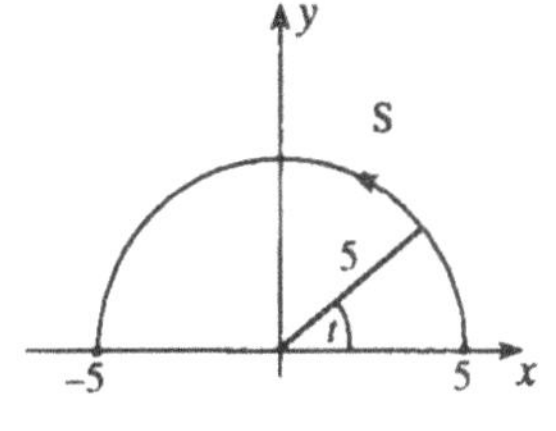

Fig 6.15
A semicircular path S of radius 5.

Solution 3.2

(a) For this constant vector field we have $g_x = \alpha$, $g_y = g_z = 0$, and so Eq (6.15) yields

$$\int_S \mathbf{g}(\mathbf{r}) \cdot \mathrm{d}\mathbf{r} = \int_S \alpha\,\mathrm{d}x = \alpha\int_S \mathrm{d}x \qquad (6.25)$$

Note that the curve parameter t here is actually the azimuthal angle ϕ in a polar coordinate system.

Using Eq (6.23) and the given parameterisation we can put $\mathrm{d}x = (\mathrm{d}x/\mathrm{d}t)\mathrm{d}t = (-5\sin t)\mathrm{d}t$. Thus we can express the line integral as a definite integral with respect to t with lower limit $t = 0$ and upper limit $t = \pi$,

$$\int_S \mathbf{g}(\mathbf{r}) \cdot d\mathbf{r} = \alpha \int_0^{\pi} (-5 \sin t) dt = 5\alpha[\cos t]_0^{\pi} = 5\alpha((-1) - 1) = -10\alpha$$

(b) A simple parameterisation of the path C along the x-axis is $x = -5t$, $y = 0$, $z = 0$ $(-1 \le t \le 1)$ from which $dx = (dx/dt)dt = -5dt$, and so we have

$$\int_C \mathbf{g}(\mathbf{r}) \cdot d\mathbf{r} = \alpha \int_C dx = \alpha \int_{-1}^{1} (-5dt) = -5\alpha[t]_{-1}^{1} = -10\alpha$$

Example 3.3 (*Objective 2*) Evaluate the scalar line integral of the vector field

$$\mathbf{G}(x, y, z) = \frac{x\mathbf{i} + y\mathbf{j}}{x^2 + y^2} + z^2\mathbf{k}$$

along the helix defined by

$$x = \cos t,\ y = \sin t, \quad z = t/2\pi \qquad (0 \le t \le 2\pi)$$

Solution 3.3 We use the parametric equations to express the components of **G** in terms of t. Thus, using the identity $\cos^2 t + \sin^2 t = 1$, we have

$$G_x = \frac{x}{x^2 + y^2} = \cos t,\ G_y = \frac{y}{x^2 + y^2} = \sin t,\ G_z = z^2 = \left(\frac{t}{2\pi}\right)^2$$

Then, using Eqs (6.23), we can express the line integral as

$$\int_C \mathbf{G}(\mathbf{r}) \cdot d\mathbf{r} = \int_0^{2\pi} \cos t \left(\frac{dx}{dt}\right) dt + \int_0^{2\pi} \sin t \left(\frac{dy}{dt}\right) dt + \frac{1}{(2\pi)^2} \int_0^{2\pi} t^2 \left(\frac{dz}{dt}\right) dt$$

where $dx/dt = d(\cos t)/dt = -\sin t$, $dy/dt = \cos t$ and $dz/dt = 1/2\pi$. We now find that the first two integrals above cancel, leaving

$$\int \mathbf{G}(\mathbf{r}) \cdot d\mathbf{r} = \frac{1}{(2\pi)^3} \int_0^{2\pi} t^2 dt = \frac{1}{(2\pi)^3} \left[\frac{t^3}{3}\right]_0^{2\pi} = \frac{1}{3}$$

Example 3.4 (*Objective 2*) Calculate the line integral of the two-dimensional vector field

$$\mathbf{G}(x,y) = -2y\mathbf{i} + (x - y)\mathbf{j}$$

along the straight line path C (Fig 6.16) from the origin to the point (3,6) using the parameterisation $x = 3t$, $y = 6t$ $(0 \le t \le 1)$.

Solution 3.4 We have $G_x = -2y = -12t$ and $G_y = (x - y) = (3t - 6t) = -3t$; also $dx/dt = 3$ and $dy/dt = 6$. Thus the line integral is

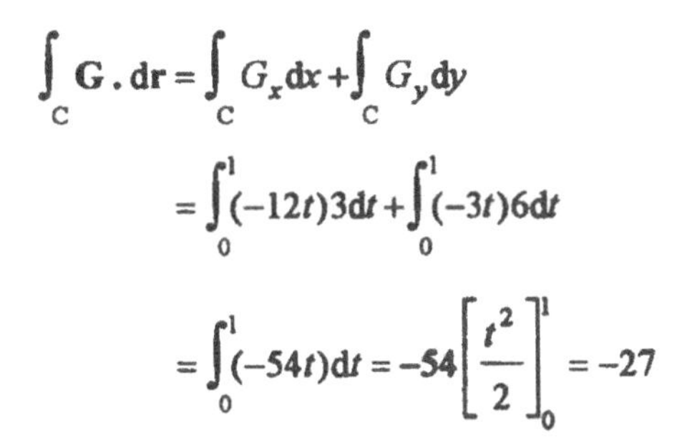

$$\int_C \mathbf{G}\,.\,d\mathbf{r} = \int_C G_x dx + \int_C G_y dy$$

$$= \int_0^1 (-12t)3dt + \int_0^1 (-3t)6dt$$

$$= \int_0^1 (-54t)dt = -54\left[\frac{t^2}{2}\right]_0^1 = -27$$

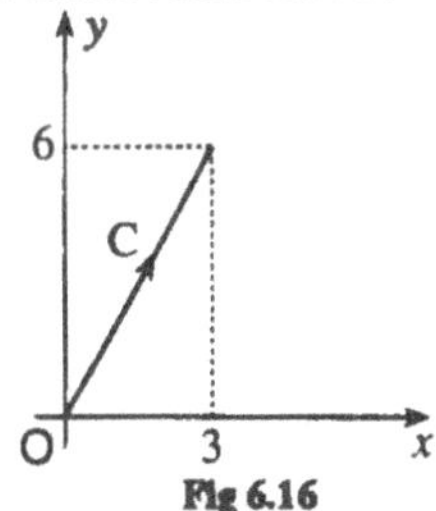

Fig 6.16
The path of integration for Example 3.4.

Problem 3.1 (*Objective 2*) If $\mathbf{F} = 2xy\mathbf{i} - yz\mathbf{j} + z^2\mathbf{k}$, evaluate $\int_C \mathbf{F}(\mathbf{r}).d\mathbf{r}$ where the path C is given by $x = 3t,\ y = 5t - 2,\ z = t^2\ (0 \le t \le 1)$.

Problem 3.2 (*Objective 2*)

(a) Make a rough sketch of the paths C_1 and C_2 described by the following parametric equations

$$C_1 : x = 3t,\ \ y = 3 - 3t \qquad (0 \le t \le 1)$$
$$C_2 : x = 3t,\ \ y = 3(1 - t^2) \qquad (0 \le t \le 1)$$

(b) Determine the scalar line integrals of the vector field

$$\mathbf{f}(x,y) = -2y\mathbf{i} + 5x\mathbf{j}$$

along each of the paths C_1 and C_2.

(c) Find the circulation of the vector field **f** around the closed loop comprising the paths C_1 and $-C_2$, where $-C_2$ denotes the path C_2 traversed in the negative sense, i.e. with t decreasing from $t = 1$ to $t = 0$.

Problem 3.3 (*Objective 2*) Consider the vector field $\mathbf{F}(\mathbf{r}) = x\mathbf{i} + y\mathbf{j}$ and the path Γ described by parametric equations

$$x = \cos t,\ \ y = \sin t,\ \ z = 0 \qquad (0 \le t \le \pi/4)$$

(a) Explain, without evaluating any integrals, why $\int_\Gamma \mathbf{F}.d\mathbf{r} = 0$.

(b) Determine the line integral of **F** along the path C described by

$$x = y = t \qquad (0 \le t \le 1)$$

Problem 3.4 (*Objective 2*) A projectile moves along a parabolic path in the x-y plane ($z = 0$). Its coordinates as a function of time are given by

$$x = 8t,\quad y = 8t - 5t^2,\quad z = 0$$

Determine the value of the line integral of the constant vector field $\mathbf{g} = -10\mathbf{j}$ along the segment of the parabola defined by $(0 \le t \le 1)$.

Problem 3.5 (*Objective 2*) Evaluate the line integral given by Eq (6.22) by the methods of Section 6.2.2, i.e. use the inverses of Eqs (6.20) ($t = x/3$, $t = y/6$ and $t = z/9$) to express the three integrands in terms of the corresponding integration variables.

6.4 CONSERVATIVE FIELDS

We have already met the "curl test" for identifying conservative vector fields (Eq 5.50). In this section we define conservative fields in terms of their line integrals. The equivalence of these two views of a conservative field involves some subtle features that we shall examine in Section 6.6.

We have seen several examples where the scalar line integral of a vector field has different values for different paths between two fixed points. Such fields are said to be non-conservative. A vector field **G** is conservative if the scalar line integral has the same value for all possible paths between any two fixed points (Fig 6.17). Hence the value of the line integral of a conservative field depends only on the field **G** and the positions of the beginning point A and the end point B of the path, and we can write

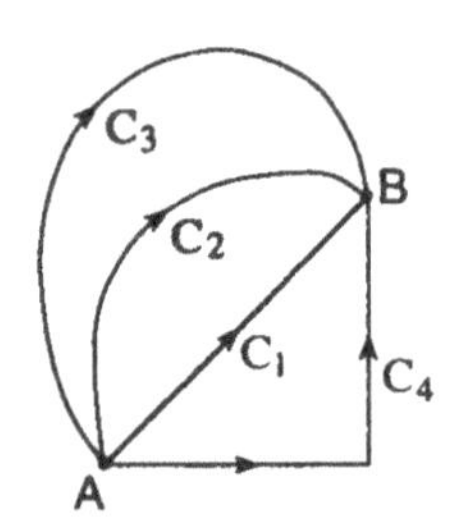

Fig 6.17
The line integral of a conservative field is the same for all paths C_1, C_2, etc., between two fixed points A and B.

$$\int_C \mathbf{G}(\mathbf{r})\,.\,d\mathbf{r} = \int_A^B \mathbf{G}(\mathbf{r})\,.\,d\mathbf{r} = W_{AB} \qquad \text{(conservative field } \mathbf{G}) \qquad (6.26)$$

where the path label C on the integral sign is replaced by writing the beginning point A as the lower limit and the end point B as the upper limit, all other details of the path being irrelevant.

Note that for the reverse path, beginning at B and ending at A, the upper and lower limits are interchanged and the line integral simply changes sign, i.e. $W_{BA} = -W_{AB}$. It follows that the circulation of a conservative field is zero around any closed loop,

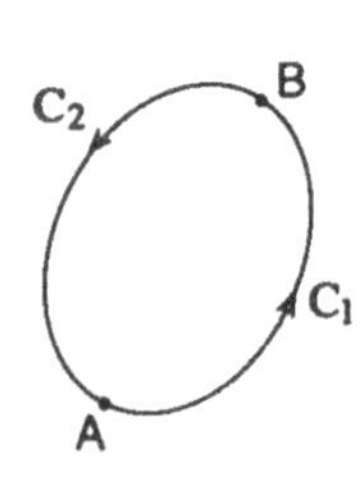

Fig 6.18
A and B are fixed points on the loop $C = C_1 + C_2$.

$$\oint \mathbf{G}(\mathbf{r})\,.\,d\mathbf{r} = 0 \qquad \text{(conservative field } \mathbf{G}) \qquad (6.27)$$

To see this, suppose A and B are any two fixed points on the loop (Fig 6.18). Then the circulation integral in Eq (6.27) is the sum of two parts: a contribution W_{AB} along the segment C_1 of the loop from A to B and a contribution $W_{BA} = -W_{AB}$ back along the other segment C_2, and so the sum of the two contributions is zero.

When the field **G** is a conservative force **F** acting on a particle at **r**, W_{AB} is the work done by the field when the particle moves from A to B. It follows from Eq (6.26) that if the particle returns from B to A, along any path whatever, the work done by the field along the return path is $W_{BA} = -W_{AB}$ where the minus sign indicates that the energy W_{AB} is returned to the field during the return journey. Thus we can associate a fixed energy value U_A with the point A and a fixed energy value U_B with the point B such that

$$U_A - U_B = W_{AB} = \int_A^B \mathbf{F}(\mathbf{r}) . d\mathbf{r} \qquad (6.28)$$

The energy associated with a point in a conservative force field is called **potential energy**. Equation (6.28) defines only the potential energy difference between two points. We can associate a definite potential energy value with each point by arbitrarily specifying the potential energy value at any one particular fixed point, point A say, and then using Eq (6.28) to define the potential energy at any other point B. Thus if we specify $U_A = 0$, the potential energy is defined at any other point B by

$$U_B = -W_{AB} = -\int_A^B \mathbf{F}(\mathbf{r}) . d\mathbf{r} \qquad (U_A = 0;\ \text{B is any point}) \qquad (6.29)$$

Since the potential energy defined by Eq (6.29) has a unique value at each point B in the domain of **F**, it is a scalar field, and we would normally write it as $U(\mathbf{r})$ where **r** is the position vector of B. The arbitrarily chosen point A is called the **point of zero potential energy**. When the force is zero at very large distances it is often convenient to take the point of zero potential energy to be "at infinity".

As an example of using Eq (6.29) we calculate the potential energy of a charged particle carrying a charge q in the electrostatic field produced by a fixed charge Q situated at the origin. By Coulomb's law, the electrostatic force acting on the charge q at **r**, in spherical polar coordinates, is

All electrostatic fields are conservative.

$$\mathbf{F}(r,\theta,\phi) = \frac{C\mathbf{e}_r}{r^2} \qquad (r > 0) \qquad (6.30)$$

where $C = Qq/4\pi\epsilon_0$. We take the point of zero potential energy to be at infinity, where the force vanishes, and use Eq (6.29) to calculate the potential energy at an arbitrary point B a distance r_1 from Q. Because the electrostatic force is conservative we are free to choose any path of integration from infinity to B. For convenience we choose a radial path on which the line element is simply $d\mathbf{r} = \mathbf{e}_r dr$ (Fig 6.19) and Eq (6.29) gives

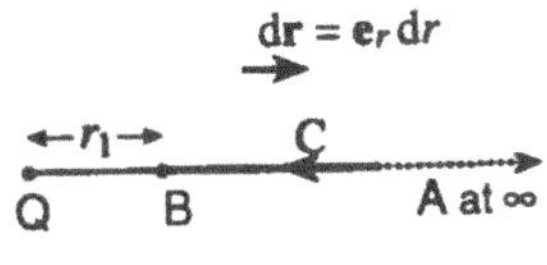

dr = e$_r$ dr
r_1
C
Q B A at ∞

Fig 6.19
A radial path of integration from r = "infinity" to $r = r_1$.

$$U_B = -C\int_\infty^{r_1} \frac{\mathbf{e}_r}{r^2} . \mathbf{e}_r dr = -C\int_\infty^{r_1} \frac{1}{r^2} dr = C\left[\frac{1}{r}\right]_\infty^{r_1} = \frac{C}{r_1}$$

where ∞ is the symbol for infinity and we have put $1/\infty = 0$. Using the symbol r in place of r_1 for the coordinate of point B, we can write the electrostatic potential energy associated with the force of Eq (6.30) in the familiar form

$$U(r,\theta,\phi) = \frac{C}{r} = \frac{Qq}{4\pi\epsilon_0 r} \qquad (r > 0)$$

The potential energy U associated with a conservative force **F** is a particular example of a scalar potential field. Given any conservative field **G**, not necessarily a force, we can define a **scalar potential** field ϕ by

$$\phi_A - \phi_B = \int_A^B \mathbf{G} \cdot d\mathbf{r} \tag{6.31}$$

together with a statement specifying the value of ϕ at some fixed point (e.g. $\phi_A = 0$).

If we know the scalar potential field ϕ we can evaluate the left-hand side of Eq (6.31) and hence obtain the value of the line integral on the right-hand side (Example 4.2). This is one of the main uses of the scalar potential.

Eq (6.31) defines the scalar potential ϕ as a line integral and is equivalent to the differential definition (Section 5.2.3),

$$\mathbf{G} = -\text{grad}\phi = -\left(\mathbf{i}\frac{\partial\phi}{\partial x} + \mathbf{j}\frac{\partial\phi}{\partial y} + \mathbf{k}\frac{\partial\phi}{\partial z}\right) \tag{6.32}$$

The scalar potential can be defined to be $\psi = -\phi$ so that $\mathbf{G} = \text{grad}\psi$. The left-hand side of Eq (6.31) is then $\psi_B - \psi_A$.

which also defines ϕ only to within a constant.

Finally we bring together various definitions of a **conservative field G**.

(a) $\text{curl}\mathbf{G} = 0$ everywhere in the domain of **G**

(b) $\int_A^B \mathbf{G} \cdot d\mathbf{r}$ is independent of the path between A and B

(c) $\oint \mathbf{G} \cdot d\mathbf{r} = 0$ for all loops.

(d) $\mathbf{G} = -\text{grad}\phi$

For many vector fields all four statements above are equivalent. There are however some important examples of fields that satisfy the "differential" statements (a) and (d) but fail the "integral" statements (b) and (c). We shall return to this point in Section 6.6.

Summary of section 6.4

- The line integral of a **conservative field** depends only on the field itself and the beginning point and end point of the path. It follows that the circulation of a conservative field is zero around any closed loop.

- If **G** is any conservative vector field we can define a **scalar potential** field ϕ by Eqs (6.31) or (6.32) with a statement specifying ϕ at some arbitrary point.

- When the conservative field is a force **F** acting on a particle, the line integral of **F** for any path from A to B is the work done by the field when the particle moves from A to B, and the scalar potential is the **potential energy** of the particle. The **point of zero potential energy** is chosen arbitrarily.

Example 4.1 (*Objective 3*) Determine the work done when the conservative force

$$\mathbf{G}(x,y) = (2x - 1)\mathbf{i}$$

acts on a particle which undergoes a displacement from A(0,2) to B(3,1).

Solution 4.1 The work done is the line integral found by applying Eq 6.16 with $\mathbf{G}_x = 2x - 1$ and $\mathbf{G}_y = 0$,

$$W_{AB} = \int_A^B \mathbf{G}(\mathbf{r}) \cdot \mathrm{d}\mathbf{r} = \int_A^B (2x-1)\mathrm{d}x \qquad (6.33)$$

Since the force is conservative, the line integral is independent of the particular path from A to B and so we are free to choose a path that will lead to simple definite integrals. We choose the path APB (Fig 6.20) made up of the two segments AP and PB. Then $W_{AB} = W_{AP} + W_{PB}$. But W_{AP} is zero (because $\mathbf{G}_y = 0$). Thus $W_{AB} = W_{PB}$. Now x varies on PB from $x = 0$ at P to $x = 3$ at B, and so we have

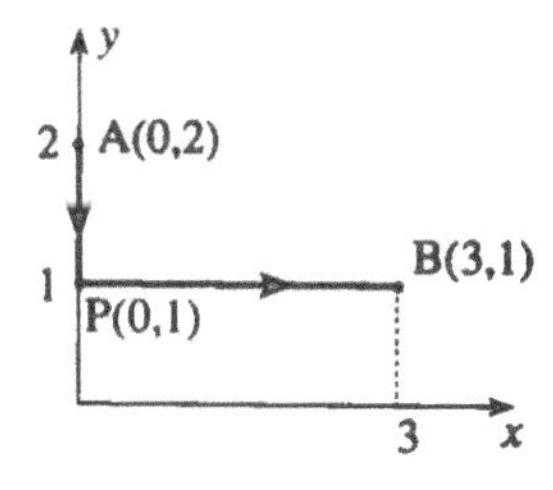

Fig 6.20
The path of integration APB.

$$W_{AB} = W_{PB} = \int_P^B (2x-1)\mathrm{d}x = \int_0^3 (2x-1)\mathrm{d}x = \left[x^2 - x\right]_0^3 = 6$$

Example 4.2 (*Objective 3*)

(a) Use Eq (6.32) to confirm that

$$\phi(x,y,z) = -\frac{1}{3}(x^3 + y^3 + z^3)$$

is a scalar potential of

$$\mathbf{G}(x,y,z) = x^2\mathbf{i} + y^2\mathbf{j} + z^2\mathbf{k}$$

(b) Hence find the line integral of **G** along a path from A(1,0,–1) to B(3,5,0).

Solution 4.2

(a) $\text{grad}\phi = \mathbf{i}\dfrac{\partial\phi}{\partial x} + \mathbf{j}\dfrac{\partial\phi}{\partial y} + \mathbf{k}\dfrac{\partial\phi}{\partial z}$ where

$$\frac{\partial\phi}{\partial x} = -\frac{1}{3}3x^2 = -x^2, \quad \frac{\partial\phi}{\partial y} = -y^2, \quad \frac{\partial\phi}{\partial z} = -z^2$$

Hence $\text{grad}\phi = -\mathbf{G}$.

(b) Knowing the potential ϕ, we can use Eq (6.31) to write the line integral as

$$\int_A^B \mathbf{G}\,.\,d\mathbf{r} = \phi_A - \phi_B = \phi(1,0,-1) - \phi(3,5,0)$$

$$= -\frac{1}{3}(1^3 + 0^3 + (-1)^3) - \left(-\frac{1}{3}\right)(3^3 + 5^3 + 0^3) = \frac{152}{3}$$

Example 4.3 (*Objective 3*) The conservative force $\mathbf{F} = -C\mathbf{r}$ (C is a constant) acts on a particle at **r**. Determine the potential energy U of the particle when it is at a point P a distance a from the reference point $\mathbf{r} = \mathbf{0}$ where the potential energy is zero. Explain why it would be inappropriate in this example to take the point of zero potential energy to be at infinity.

Solution 4.3 We use spherical polar coordinates in which the force is given by $\mathbf{F}(r,\theta,\phi) = -C\mathbf{e}_r r$. To calculate the potential energy we use Eq (6.29) with point A at the origin O. For convenience we choose a radial path from O to P on which $d\mathbf{r} = \mathbf{e}_r dr$. Thus

$$U_P = -\int_O^P \mathbf{F}(\mathbf{r})\,.\,d\mathbf{r} = \int_O^P (C\mathbf{e}_r r)\,.\,\mathbf{e}_r dr = C\int_0^a r\,dr = C\left[\frac{r^2}{2}\right]_0^a = \frac{C}{2}a^2$$

If we were to put $U_\infty = 0$, the potential at P would be $U_P = C\int_\infty^a r\,dr$ which cannot be evaluated to give a finite number.

Problem 4.1 (*Objectives 2,3*) The time-dependent coordinates of a particle moving under a force $\mathbf{T} = 10\mathbf{j}$ are $(x,y,z) = (2t, 3t^2, 7)$. Determine the work done by the force in the first 5 seconds.

Problem 4.2 (*Objective* 3) Confirm that $U(x,y) = x - x^2$ is a scalar potential of the vector field **G** of Example 4.1 and hence verify the answer to that Example.

Problem 4.3 (*Objective 3*) Find the line integral of the conservative vector field

$$\mathbf{T}(x,y,z) = \cos(x)\cos(y)\mathbf{i} - \sin(x)\sin(y)\mathbf{j} - \sin^2(z)\mathbf{k}$$

along a path from $A(\pi/2,0,0)$ to $B(0,\pi/2,\pi/2)$.

Problem 4.4 (*Objective 3*) The **electrostatic potential** at a point B is defined by

$$V_B = -\int_A^B \mathbf{E}(\mathbf{r}) \cdot d\mathbf{r} \qquad (V_A = 0)$$

where **E** is the electrostatic field.

The electrostatic field outside a long fibre of radius a lying with its axis on the z-axis and carrying a static electric charge distributed uniformly with line density λ (Cm^{-1}) is given by

$$\mathbf{E}(\rho,\phi,z) = \frac{\lambda \mathbf{e}_\rho}{2\pi\epsilon_0 \rho} \qquad (\rho > a)$$

Calculate the electrostatic potential at a distance l from the axis of the fibre where $b > l > a$ and $\rho = b$ is a cylinder on which the electrostatic potential is zero.

6.5 SURFACE INTEGRALS

Surface integrals involve the summation of field values over a surface, an important example being the flux of a vector field across a surface. Flux was introduced in Section 4.5.1 for the special case where the component of the field vector normal to the surface was constant everywhere on the surface. Section 6.5.1 defines flux more generally as a surface integral and introduces other kinds of surface integral. Section 6.5.2 shows how surface integrals are expressed as double integrals and evaluated.

6.5.1 Introducing surface integrals

We introduce surface integrals by considering the flux of a vector field across a surface. Let $\mathbf{F}(\mathbf{r})$ be a vector field and S a surface of area A. Then the component of $\mathbf{F}(\mathbf{r})$ normal to the surface at a point P is $\mathbf{n} \cdot \mathbf{F}(\mathbf{r}) = F_n(\mathbf{r})$, where **n** is the unit surface normal at P pointing from the negative side of the surface to the positive side. In the special case where $\mathbf{n}\,.\,\mathbf{F}(\mathbf{r})$ has the constant value F_n everywhere on the surface, the flux of $\mathbf{F}$ across S is simply the product of F_n and the area A of the surface, as described in Section 4.5.1. More generally $\mathbf{n}\,.\,\mathbf{F}(\mathbf{r})$ varies over the surface. The approach then is to divide the surface S into N surface elements such that $\mathbf{n}\,.\,\mathbf{F}(\mathbf{r})$ is approximately constant over any one element. The total flux Φ across the whole surface S is then given approximately by the sum of the fluxes across all N surface elements. Thus we write

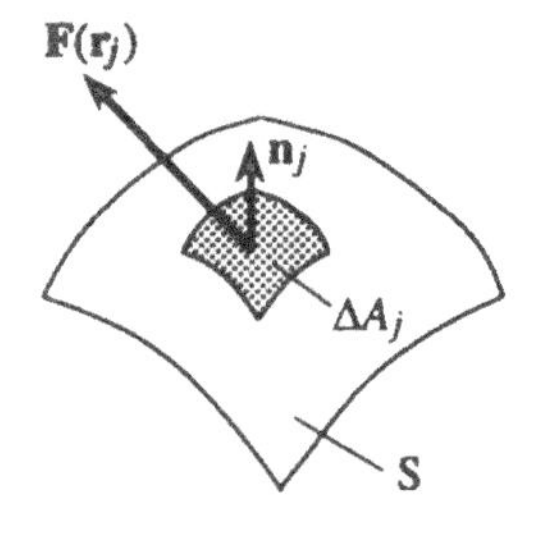

Fig 6.21
A surface element of area ΔA_j on a surface S.

$$\Phi \approx \sum_{j=1}^{N} \mathbf{n}_j . \mathbf{F}(\mathbf{r}_j) \Delta A_j \tag{6.34}$$

where ΔA_j is the area of the jth surface element and $\mathbf{n}_j$ is the unit surface normal at any point $\mathbf{r}_j$ on the element (Fig 6.21). As $N \to \infty$ and all $\Delta A_j \to 0$, this sum tends to a limit which is independent of how the surface is subdivided. This limit is called **the scalar surface integral** of $\mathbf{F}$ over S, and is denoted by $\int_S \mathbf{n}\,.\,\mathbf{F}(\mathbf{r})\, dA$ or $\int_S F_n(\mathbf{r})\, dA$. Thus

$$\Phi = \lim_{N \to \infty} \left[\sum_{j=1}^{N} \mathbf{n}_j . \mathbf{F}(\mathbf{r}_j) \Delta A_j \right] = \int_S \mathbf{n}\,.\,\mathbf{F}(\mathbf{r})\, dA \tag{6.35a}$$

$$= \int_S F_n(\mathbf{r})\, dA \tag{6.35b}$$

It is useful to think of the symbol dA as the area of a very small element of the surface S. We refer to dA as the **scalar surface element** or simply the surface element. We can also define a **vector surface element** $\mathbf{dA} = \mathbf{n}dA$, and write Eq (6.35a) as

$$\Phi = \int_S \mathbf{F}(\mathbf{r})\,.\,\mathbf{n}dA = \int_S \mathbf{F}(\mathbf{r}).\mathbf{dA} \tag{6.35c}$$

The three integrals in Eqs 6.35(a), (b) and (c) are just different ways of writing the scalar surface integral.

When the surface is closed we take the outside face to be the positive face so that $\mathbf{n}$ is directed from the inside to the outside and the flux of $\mathbf{F}$ across the closed surface is called the **outward flux** Φ_o. We often indicate the outward flux across a closed surface by a small circle on the integral sign, the same integral symbol that is used to indicate a line integral around a closed loop (Section 6.3). Thus we write $\Phi_o = \oint \mathbf{n}\,.\,\mathbf{F}(\mathbf{r})dA = \oint F_n dA$.

In the special case where the normal component of the field has a constant value $\mathbf{n}\,.\,\mathbf{F} = F_n$ everywhere on the surface, we can take the constant F_n outside the surface integral in Eq (6.35b) which then becomes

$$\Phi = F_n \int_S dA = F_n A \qquad (F_n \text{ is constant}) \qquad (6.36)$$

where $A = \sum_{j=1}^{N} \Delta A_j$, the total area of the surface S. Eq (6.36) was used in Section 4.5.1 where Examples and Problems can be found.

The flux of a vector field across a surface is a scalar surface integral of a vector field. We can define, in a similar way, the scalar surface integral of a scalar field $f(\mathbf{r})$ on a surface S. Thus

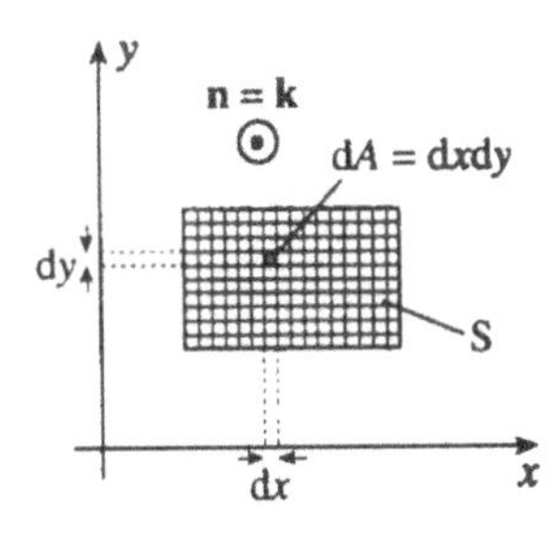

Fig 6.22
The surface S is divided into N surface elements by lines of constant x and lines of constant y. In the limit ($N \to \infty$) we write the area of a surface element as $dA = dxdy$.

$$\lim_{N\to\infty}\left[\sum_{j=1}^{N} f(\mathbf{r}_j)\Delta A_j\right] = \int_S f(\mathbf{r})dA \qquad (6.37)$$

An example is the total electric charge Q on the surface S of a charged body. If the surface charge density (Cm^{-2}) is $\sigma(\mathbf{r})$, then $Q = \int_S \sigma(\mathbf{r})dA$. Vector surface integrals also exist but we shall not consider any examples.

6.5.2 Expressing surface integrals as double integrals and evaluating them

We shall be concerned mainly with cases where the surface S is rectangular, circular, cylindrical or spherical.

For a plane rectangular surface S we introduce a cartesian coordinate system with the surface in the x-y plane ($z = 0$) (Fig 6.22). The surface element dA and the unit surface normal $\mathbf{n}$ are then

$$dA = dxdy \quad \text{and} \quad \mathbf{n} = \mathbf{k} \qquad (6.38)$$

We have made the arbitrary choice of $\mathbf{n} = \mathbf{k}$ rather than $\mathbf{n} = -\mathbf{k}$.

and the surface integral of a vector field $\mathbf{F}$ is

$$\Phi = \int_S \mathbf{n}\,.\,\mathbf{F}(\mathbf{r})dA = \int_S \mathbf{k}\,.\,\mathbf{F}(x,y,0)dxdy = \int_S F_z(x,y,0)dxdy \qquad (6.39)$$

where the integrand $F_z(x,y,0)$ is the normal component of $\mathbf{F}$ on the surface and is a function of x and y only. Equation (6.39) indicates integrations with respect to both x and y. You can think of the x integral as a sum of elements along a thin horizontal strip AB (Fig 6.23) and the y integral as the sum over all strips in S. Thus the limits on the x integral are the lines $x = a$ and $x = b$, and the limits on the y integral are the lines $y = c$ and $y = d$. We now express the surface integral over S as a **double integral** over x and y:

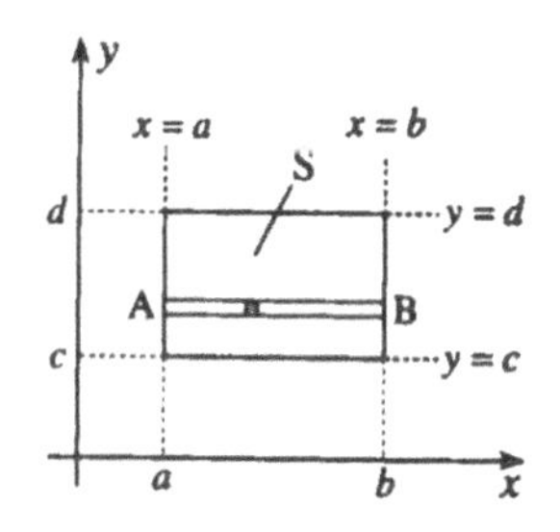

Fig 6.23
Integration with respect to x represents summing surface elements along the strip AB; integration with respect to y represents summing all strips in S.

$$\Phi = \int_{y=c}^{y=d}\left(\int_{x=a}^{x=b} F_z(x,y,0)dx\right)dy \qquad (6.40)$$

The integral over x (inside the brackets) can be evaluated first and then the y integral evaluated.

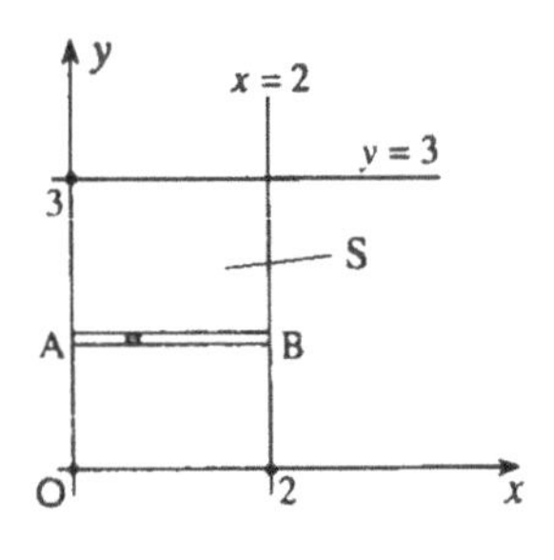

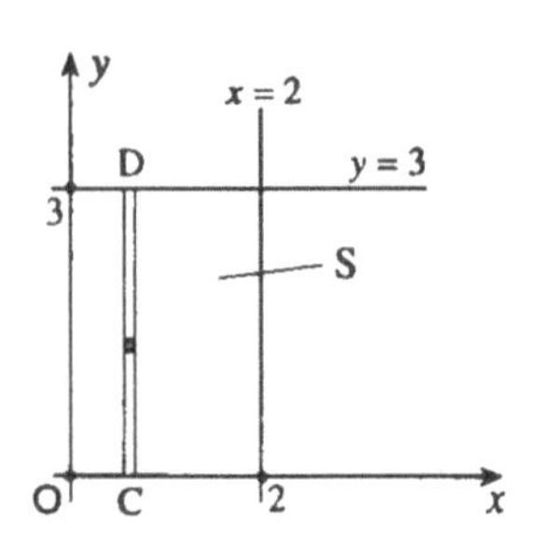

Fig 6.24
(a) Integrating over x first.
(b) Integrating over y first.

We now illustrate the evaluation of a double integral by calculating the flux of the vector field

$$\mathbf{F}(x,y,z) = (x^2 - z^2)\mathbf{i} - y^2\mathbf{j} + (xy - z^2)\mathbf{k}$$

across the rectangular surface S enclosed by the x- and y-axes and the lines $x = 2$ and $y = 3$ (Fig 6.24a). This surface is in the x-y plane ($z = 0$) and so the field values on it are given by $\mathbf{F}(x,y,0) = x^2\mathbf{i} - y^2\mathbf{j} + xy\mathbf{k}$, and the normal components are given by $\mathbf{k} \cdot \mathbf{F} = F_z(x,y,0) = xy$. Thus we have the double integral

$$\Phi = \int_{y=0}^{y=3} \left(\int_{x=0}^{x=2} xy \, \mathrm{d}x \right) \mathrm{d}y \tag{6.41}$$

We first evaluate the integral over x (representing a sum of elements along strip AB of Fig 6.24a) treating it as an ordinary definite integral and regarding y as a constant (the constant height of the strip AB). An indefinite integral of x is $x^2/2$ and so an integral of xy, treating y as a constant, is $(x^2/2)y$. Thus we obtain

$$\Phi = \int_{y=0}^{y=3} \left(\int_{x=0}^{x=2} xy \, \mathrm{d}x \right) \mathrm{d}y = \int_0^3 \left[\frac{x^2 y}{2} \right]_{x=0}^{x=2} \mathrm{d}y = \int_0^3 (2y) \mathrm{d}y$$

We now have a single definite integral with respect to y (representing the sum of all strips in S) which is evaluated to give

$$\Phi = \left[y^2 \right]_0^3 = 9$$

The order of integration can be reversed. Referring to Fig 6.24b, this corresponds to first summing elements along a vertical strip CD on which x is constant (integrating with respect to y) and then summing over all vertical strips in S (integrating with respect to x). The limits are the same as before and the same answer is obtained with equal ease (Example 5.1a). For non-rectangular surfaces the limits on the cartesian double integral will not both be constants and may depend on the order of integration (Example 5.2b).

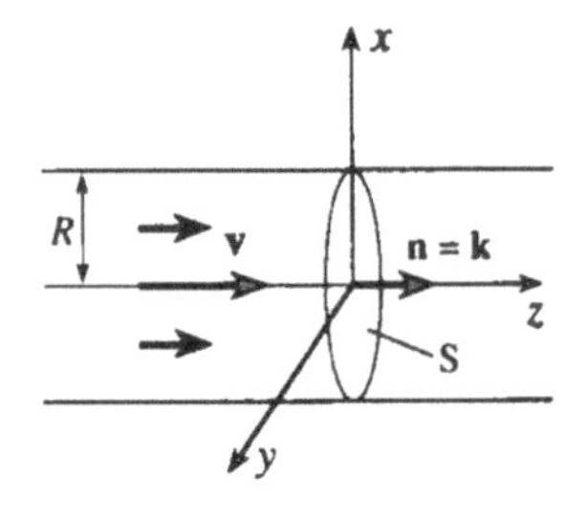

Fig 6.25
Flow through a cylindrical pipe.

For problems involving circular or cylindrical surfaces it is usually better to use a cylindrical polar coordinate system. Consider the steady flow of water through a cylindrical pipe of radius R (Fig 6.25). With the axis of the pipe lying on the z-axis, the velocity flow field is $\mathbf{v}(\rho,\phi,z) = \mathbf{k}D(R^2 - \rho^2)$ where D is a constant. You can see that this flow field has its maximum speed at the centre of the pipe ($\rho = 0$), zero speed at the pipe wall ($\rho = R$) and is independent of distance z along the pipe. The total volume flow rate through the pipe is the flux Φ of $\mathbf{v}$ across any surface bounded by the wall of the pipe such as the circular cross-section S of radius R in the plane $z = 0$. Taking the unit surface normal on S to be in the direction of flow, we have $\mathbf{n} = \mathbf{k}$ and the flux is

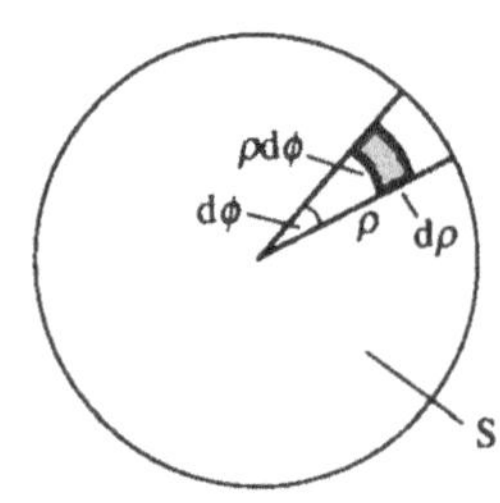

Fig 6.26
The surface is divided into N surface elements (only one shown) by lines of constant ρ (circles) and lines of constant ϕ (radial lines). In the limit ($N \to \infty$) we write the area of a surface element as $\mathrm{d}A = \rho \mathrm{d}\phi \mathrm{d}\rho$.

$$\Phi = \int_S \mathbf{n} \cdot \mathbf{v}(\mathbf{r}) \mathrm{d}A = \int_S v_z(\rho,\phi,0) \mathrm{d}A = \int_S D(R^2 - \rho^2) \, \mathrm{d}A \tag{6.42}$$

The surface element on the plane $z = 0$ in cylindrical polar coordinates is (Fig 6.26)

$$dA = \rho\, d\phi d\rho \tag{6.43}$$

We are effectively using plane polar coordinates here. Eq (6.43) is the surface element in plane polar coordinates.

Thus our surface integral (Eq (6.42)) can be expressed as a double integral over ρ and ϕ,

$$\Phi = D\int_{\rho=0}^{\rho=R}\left(\int_{\phi=0}^{\phi=2\pi}(R^2-\rho^2)\rho d\phi\right)d\rho$$

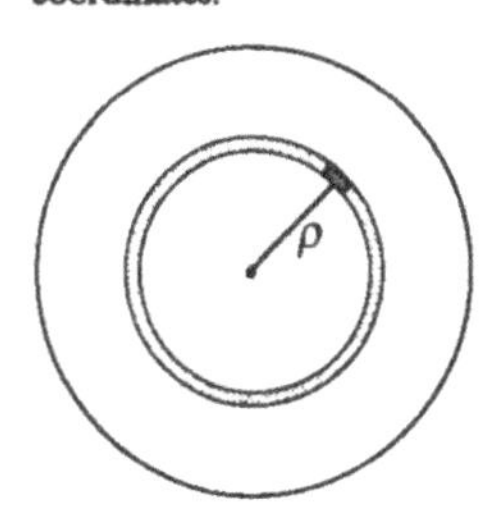

Fig 6.27
The ring is the sum of surface elements around a circle of radius ρ.

We have chosen to integrate over ϕ first, as indicated by the brackets. You can think of the ϕ-integration as a sum of elements around a ring of radius ρ with ϕ varying from $\phi = 0$ to $\phi = 2\pi$ (Fig 6.27). The integration over ρ then represents a sum of all rings with ρ varying from $\rho = 0$ to $\rho = R$, thereby covering the whole of S. We treat ρ (the radius of a ring) as a constant when integrating over ϕ and so we can take the factor $(R^2 - \rho^2)\rho$ outside the ϕ integral which is then simply $\int_0^{2\pi} d\phi = [\phi]_0^{2\pi} = 2\pi$. Thus

$$\Phi = 2\pi D\int_0^R (R^2\rho-\rho^3)d\rho \;=\; 2\pi D\left[\frac{R^2\rho^2}{2}-\frac{\rho^4}{4}\right]_0^R = \pi DR^4/2$$

The same result is obtained when the order of integration is reversed with the same limits.

It is sometimes required to find the flux of a vector field across the curved surface S of a cylinder (Example 5.4 and Problem 5.3). Fig 6.28 shows a cylinder of radius R and height h with its axis on the z-axis and its base on the x-y plane. The surface element is

$$dA = Rd\phi dz \tag{6.44}$$

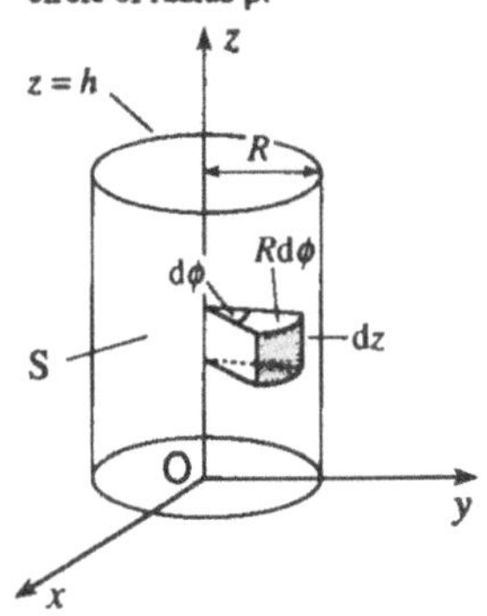

Fig 6.28
The surface is divided into N surface elements (only one shown) by lines of constant ϕ (parallel to the z-axis) and lines of constant z (circles). In the limit ($N \to \infty$) we write $dA = Rd\phi dz$.

The outward surface normal is $\mathbf{n} = \mathbf{e}_\rho$ and so the flux of a vector field $\mathbf{F}(\mathbf{r})$ across the curved surface of the cylinder is

$$\int_S \mathbf{n}\,.\,\mathbf{F}(\mathbf{r})dA = \int_{z=0}^{z=h}\left(\int_{\phi=0}^{\phi=2\pi}F_\rho(R,\phi,z)Rd\phi\right)dz \tag{6.45}$$

where again the limits on the ϕ-integral (summing around a ring) are $\phi = 0$ and $\phi = 2\pi$, while the lower and upper limits on the z integral (summing over all rings) are the values of z at the base and top of the cylinder respectively.

For surface integrals over spherical surfaces (Problem 5.2) we need the surface element and unit surface normal in spherical polar coordinates. These are (Fig 6.29)

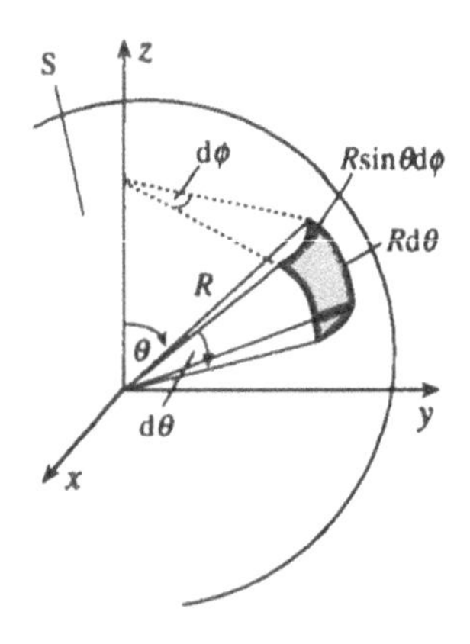

Fig 6.29
The surface element on a sphere.

$$dA = R^2 \sin\theta \, d\theta \, d\phi \quad \text{and} \quad \mathbf{n} = \mathbf{e}_r \qquad (6.46)$$

For integrals over the whole spherical surface the limits are $\phi = 0$ to $\phi = 2\pi$ and $\theta = 0$ to $\theta = \pi$.

Summary of section 6.5

- The flux Φ of a vector field $\mathbf{F}$ across a surface S is the **scalar surface integral** of the vector field $\mathbf{F}$

  $$\Phi = \int_S \mathbf{n} \cdot \mathbf{F}(\mathbf{r}) \, dA = \int_S \mathbf{F}(\mathbf{r}) \cdot d\mathbf{A} = \int_S F_n \, dA$$

  $\mathbf{n}$ is the unit surface normal, dA is the **scalar surface element** and $d\mathbf{A} = \mathbf{n} dA$ is the **vector surface element**. The **outward flux** across a closed surface is $\Phi_0 = \oint F_n dA$.

- When F_n is constant everywhere on the surface the flux is $F_n A$ where A is the surface area.

- Surface integrals are evaluated by choosing a suitable coordinate system, expressing the scalar surface element and the unit surface normal in coordinate form and then evaluating the **double integral**.

Example 5.1 (*Objective 5*)

(a) Evaluate the double integral of Eq (6.41) by integrating with respect to y first.

(b) Evaluate the following double integrals

(i) $\displaystyle\int_{y=0}^{y=1}\left(\int_{x=0}^{x=3} \cos x \, dx\right)dy,$

(ii) $\displaystyle\int_{y=1}^{y=2}\left(\int_{x=0}^{x=b} x \exp(3y) dx\right)dy$ (b is a constant).

Solution 5.1

(a) We write the double integral as

$$\Phi = \int_{x=0}^{x=2}\left(\int_{y=0}^{y=3} xy \, dy\right)dx$$

Evaluating the y integral, remembering to treat x as a constant, gives

$$\Phi = \int_{x=0}^{x=2} \left[\frac{xy^2}{2} \right]_{y=0}^{y=3} dx = \int_0^2 \left(\frac{9x}{2} \right) dx$$

We now evaluate the x integral to give $\Phi = 9$.

(b) (i) The x integral (in the brackets) is

$$\int_{x=0}^{x=3} \cos x \, dx = [\sin x]_0^3 = \sin 3 - \sin 0 = 0.141$$

In calculus all angles are routinely in radians.

The double integral is now reduced to the y integral

$$\int_0^1 0.141 \, dy = 0.141 \, [y]_0^1 = 0.141$$

(ii) The x integral is

$$\int_{x=0}^{x=b} x \exp(3y) dx = \exp(3y) \left[\frac{x^2}{2} \right]_0^b$$

$$= \exp(3y) \left(\frac{b^2}{2} \right)$$

where we have treated y as a constant. The double integral is now reduced to the single integral

$$\left(\frac{b^2}{2} \right) \int_1^2 \exp(3y) dy = \left(\frac{b^2}{2} \right) \left[\frac{1}{3} \exp(3y) \right]_1^2 = \left(\frac{b^2}{6} \right) (\exp(6) - \exp(3))$$

$$= 63.9 b^2$$

Example 5.2 (*Objectives 4,5*) Determine the flux of the vector field

$$\mathbf{F}(x,y,z) = y^2 \mathbf{i} + x^2 \mathbf{k}$$

across (a) the rectangle OABC in Fig 6.30a, and (b) the triangle OAC in Fig 6.30b. Take the unit surface normal to be in the positive z-direction.

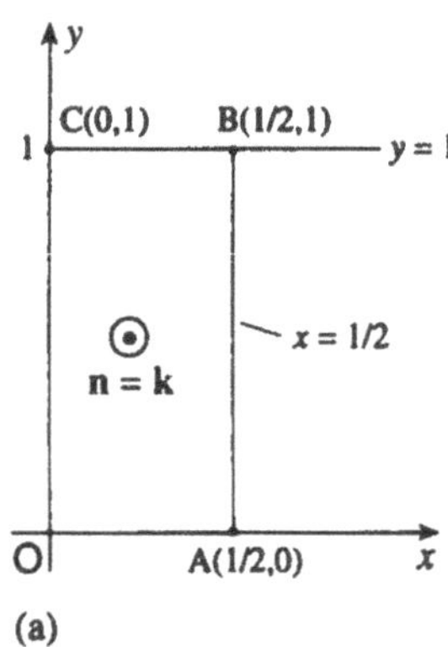

(a)

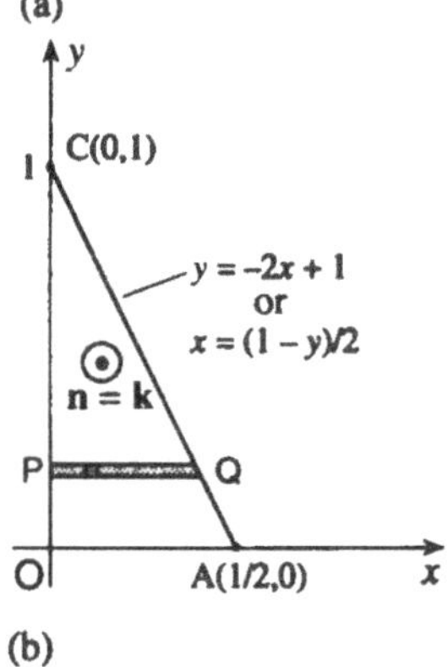

(b)

Fig 6.30
(a) The rectangle OABC.
(b) The triangle OAC.

Solution 5.2

(a) The flux of **F** across the rectangle will be the surface integral $\Phi = \int_{\mathrm{OABC}} \mathbf{k} \cdot \mathbf{F} dA = \int_{\mathrm{OABC}} x^2 dA$. In cartesian coordinates this is the double integral

$$\Phi = \int_{y=0}^{y=1}\left(\int_{x=0}^{x=1/2} x^2 dx\right)dy = \int_{y=0}^{y=1}\left[\frac{x^3}{3}\right]_{x=0}^{x=1/2} dy = \int_0^1\left(\frac{1}{24}\right)dy$$

$$= \frac{1}{24}[y]_0^1 = \frac{1}{24}$$

(b) For the triangular surface the limits depend on the order of integration. Choosing to integrate over x first corresponds to first summing elements along a horizontal strip PQ (Fig 6.30b) starting at P on the y-axis, where $x = 0$, and finishing at Q on the line $y = -2x + 1$ or $x = (1 - y)/2$. The x limits are therefore $x = 0$ and $x = (1 - y)/2$. The integral over y corresponds to summing strips from $y = 0$ to $y = 1$. Thus the double integral is

$$\Phi = \int_{y=0}^{y=1}\left(\int_{x=0}^{x=(1-y)/2} x^2 dx\right) dy = \int_{y=0}^{y=1}\left[\frac{x^3}{3}\right]_{x=0}^{x=(1-y)/2} dy$$

$$= \frac{1}{3}\int_0^1\left(\frac{1-y}{2}\right)^3 dy = \frac{1}{24}\int_0^1(1-3y+3y^2-y^3)\, dy$$

$$= \frac{1}{24}\left[y - \frac{3y^2}{2} + \frac{3y^3}{3} - \frac{y^4}{4}\right]_0^1 = \frac{1}{96}$$

Note that in this example, where one of the x limits depend on y, the x-integral yields a function of y which is then the integrand of the y-integral.

Example 5.3 (*Objectives 4,5*) The centre of a cube of edge length a is located at the point $\mathbf{r} = \mathbf{0}$. Find the outward flux of the vector field $\mathbf{F}(\mathbf{r}) = \mathbf{r}$ across one square face of the cube. Hence state the net outward flux of **F** across the cube.

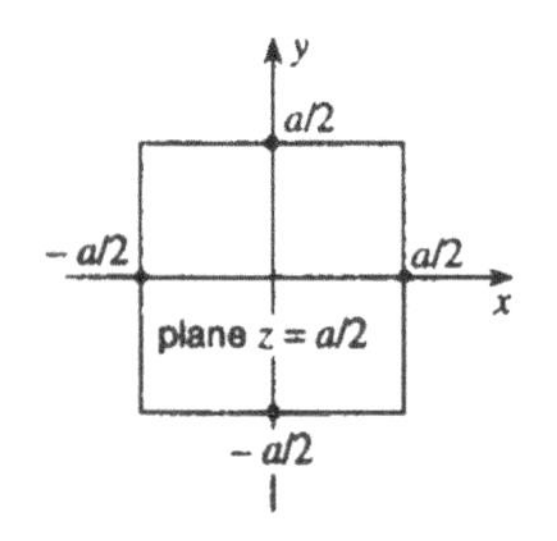

Fig 6.31
The square section of a cube.

Solution 5.3 Introduce a cartesian coordinate system with origin at $\mathbf{r} = \mathbf{0}$ and the positive z-axis passing through the centre of one face of the cube. This face then lies in the plane $z = a/2$ (Fig 6.31). In this cartesian system the given vector field is $\mathbf{F}(x,y,z) = x\mathbf{i} + y\mathbf{j} + z\mathbf{k}$ and the value of the field on the square face is $\mathbf{F}(x,y,a/2) = x\mathbf{i} + y\mathbf{j} + a\mathbf{k}/2$. The outward surface normal on this face is $\mathbf{n} = \mathbf{k}$ and so $\mathbf{n} \cdot \mathbf{F} = F_z = a/2$ and the flux across the square face is

$$\Phi = \int_{\text{square}} \mathbf{n} \cdot \mathbf{F}(\mathbf{r})\mathrm{d}A = \int_{y=-a/2}^{y=a/2} \left(\int_{x=-a/2}^{x=a/2} \frac{a}{2}\,\mathrm{d}x \right)\mathrm{d}y$$

$$= \int_{y=-a/2}^{y=a/2} \left[\frac{ax}{2} \right]_{x=-a/2}^{x=a/2} \mathrm{d}y = \int_{-a/2}^{a/2} \left(\frac{a^2}{2} \right)\mathrm{d}y = \frac{a^3}{2}$$

(We could have got the answer more directly by noting that F_z has the constant value of $a/2$ everywhere on the square face and using Eq (6.36).)

By symmetry the outward flux across each of the other five faces is the same and so the net outward flux across the cube is $3a^3$ or three times the volume of the cube.

Example 5.4 (*Objectives 4,5*) Find the outward flux of the vector field

$$\mathbf{f}(\rho,\phi,z) = \mathbf{e}_\rho \frac{a}{\rho} \cos\left(\frac{\pi z}{a} \right)$$

across the curved surface of a cylinder of radius a and height a located with its centre at the origin and its axis on the z-axis. What is the flux of **f** across the plane circular end faces of the cylinder?

Solution 5.4 The outward normal on the curved surface of the cylinder is $\mathbf{n} = \mathbf{e}_\rho$ and the surface element is $\mathrm{d}A = a\mathrm{d}\phi\mathrm{d}z$ (Eq (6.44)), and so the flux of **f** across the curved surface S is

$$\Phi = \int_S \mathbf{n} \cdot \mathbf{f}(\mathbf{r})\mathrm{d}A = \int_S f_\rho(a,\phi,z)\mathrm{d}A$$

$$= \int_{z=-a/2}^{z=a/2} \left(\int_{\phi=0}^{\phi=2\pi} \cos\left(\frac{\pi z}{a} \right) a\mathrm{d}\phi \right)\mathrm{d}z = \int_{-a/2}^{a/2} 2\pi a \cos\left(\frac{\pi z}{a} \right)\mathrm{d}z$$

where we have used $\int_0^{2\pi} \mathrm{d}\phi = 2\pi$, there being no ϕ-dependent factors in the integrand. Finally we evaluate the z integral to obtain

$$\Phi = 2\pi a \frac{a}{\pi} \left[\sin\left(\frac{\pi z}{a} \right) \right]_{-a/2}^{a/2} = 4a^2$$

The flux of **f** across the circular end faces is zero since $\mathbf{k} \cdot \mathbf{f} = 0$.

Example 5.5 (*Objectives 4,5*) The rate of flow of electromagnetic energy down a waveguide is the flux of a vector **P**, called the cycle-averaged Poynting vector, across a section of the waveguide.

A certain waveguide consists of a long copper tube of rectangular cross-section with sides of length a and b. Find the rate of flow of electromagnetic energy along the waveguide when one long edge lies on the z-axis and

$$\mathbf{P}(x,y,z) = I\sin^2(\pi y/b)\mathbf{k} \qquad (0 \le x \le a;\ 0 \le y \le b)$$

where I is a constant. (Make use of the trig. identity: $\sin^2 x = (1 - \cos 2x)/2$.)

Solution 5.5 Consider a plane rectangular section S of the waveguide (Fig 6.32) with unit normal $\mathbf{n} = \mathbf{k}$ and area ab. The flux of $\mathbf{P}$ across S is

$$\Phi = \int_S P_z \, dA = I\int_{y=0}^{y=b}\left(\int_{x=0}^{x=a}\sin^2\left(\frac{\pi y}{b}\right)dx\right)dy$$

There are no x-dependent factors in the integrand and so the x-integral yields $\int_0^a dx = [x]_0^a = a$ and we have

$$\Phi = Ia\int_{y=0}^{y=b}\sin^2\left(\frac{\pi y}{b}\right)dy = \frac{Ia}{2}\int_0^b\left(1-\cos\left(\frac{2\pi y}{b}\right)\right)dy$$

$$= \frac{Ia}{2}\left[y - \frac{b}{2\pi}\sin\left(\frac{2\pi y}{b}\right)\right]_0^b = \frac{Iab}{2}$$

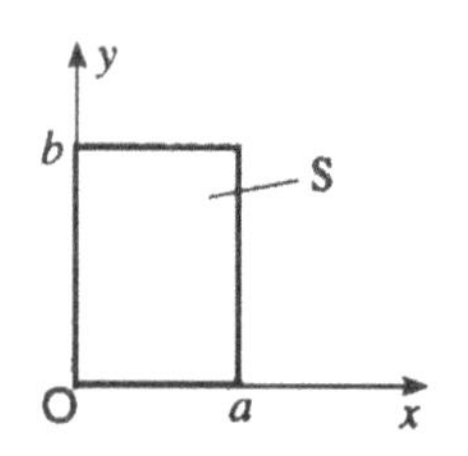

Fig 6.32
The surface S is a rectangular cross-section of a waveguide.

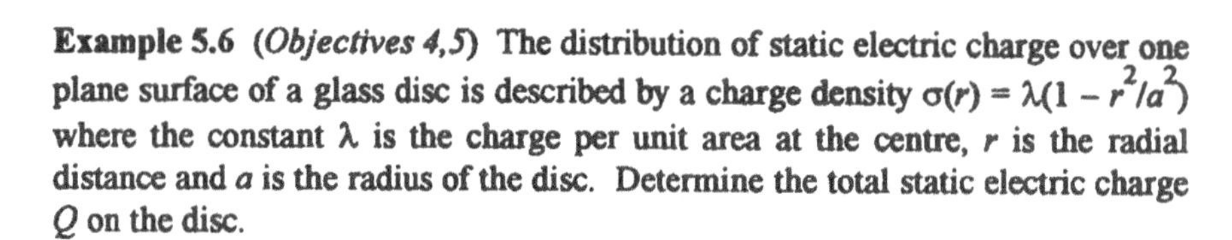

Example 5.6 (*Objectives 4,5*) The distribution of static electric charge over one plane surface of a glass disc is described by a charge density $\sigma(r) = \lambda(1 - r^2/a^2)$ where the constant λ is the charge per unit area at the centre, r is the radial distance and a is the radius of the disc. Determine the total static electric charge Q on the disc.

Solution 5.6 The total charge is the surface integral of the charge density over the surface of the disc, an example of a surface integral of a scalar field, Eq (6.37). Using plane polar coordinates we put the radial distance $r = \rho$, use the surface element of Eq (6.43) and obtain the double integral

$$Q = \lambda\int_{\rho=0}^{\rho=a}\left(\int_{\phi=0}^{\phi=2\pi}\left(1-\frac{\rho^2}{a^2}\right)\rho\, d\phi\right)d\rho$$

There are no ϕ-dependent factors in the integrand and so the ϕ integral gives a factor $\int_0^{2\pi} d\phi = 2\pi$, and we obtain

$$Q = 2\pi\lambda\int_0^a\left(1-\frac{\rho^2}{a^2}\right)\rho\, d\rho$$

$$= 2\pi\lambda\left[\frac{\rho^2}{2} - \frac{\rho^4}{4a^2}\right]_0^a = \frac{\pi\lambda a^2}{2}$$

Problem 5.1 (*Objective 5*) Evaluate the following double integrals:

(a) $\int_{y=0}^{y=1}\left(\int_{x=0}^{x=3}(x-y)\mathrm{d}x\right)\mathrm{d}y$

(b) $\int_{y=c}^{y=1}\left(\int_{x=a}^{x=1}y^3\mathrm{d}x\right)\mathrm{d}y$ (*c* and *a* are constants)

(c) $\int_{\phi=0}^{\phi=\pi}\left(\int_{\rho=0}^{\rho=3}\rho^2\sin\phi\mathrm{d}\rho\right)\mathrm{d}\phi$

(d) $\int_{y=1}^{y=2}\left(\int_{x=y}^{x=y^2}\mathrm{d}x\right)\mathrm{d}y$

Problem 5.2 (*Objectives 4,5*) Find, by evaluating a surface integral, the inward flux of the Earth's gravitational field **g** across the roof of St Paul's cathedral assumed to be a hemispherical dome of radius $R = 34$ m. Take $g = 10$ ms^{-2}.

Problem 5.3 (*Objectives 4,5*) The centre of a cylinder of radius a and height $2a$ is located at the point $\mathbf{r} = \mathbf{0}$. Determine the flux of the vector field $\mathbf{F}(\mathbf{r}) = C\mathbf{r}$ where C is a constant, across (a) one plane circular end face of the cylinder and (b) the curved surface of the cylinder. (c) Show that for $C = 1$ the net outward flux of **F** across the closed surface of the cylinder is three times the volume of the cylinder.

Problem 5.4 (*Objective 5*) Find the mass of a circular plate of radius a and mass per unit area given by $\sigma(\rho) = \sigma_0(1 - \rho/2a)$ where ρ is the radial distance from the centre and σ_0 is a constant.

Problem 5.5 (*Objective 5*) Evaluate the double integrals

(a) $\Phi = \int_{y=0}^{y=1/3}\left(\int_{x=1}^{x=3}\cos(2x+3y)\mathrm{d}x\right)\mathrm{d}y$

(b) $\Phi = \int_{y=0}^{y=1}\left(\int_{x=y^2}^{x=1}x^2y\,\mathrm{d}x\right)\mathrm{d}y$

Problem 5.6 (*Objective 5*) Repeat Example 5.5b by integrating with respect to y first.

6.6 STOKES'S THEOREM

Stokes's theorem is an identity between the scalar line integral of a vector field and certain surface integrals of the curl of the field. Curl was introduced in Section 5.4.1 by considering the circulation per unit enclosed area in a two-dimensional vector field. This led to the "differential" definition of curl for two- and three-dimensional vector fields in terms of partial derivatives (Eqs (5.40) and (5.43)). We now return to the idea of circulation per unit area which we apply to three-dimensional vector fields to obtain an alternative "integral" definition of curl. We then present an outline derivation of Stokes's theorem followed by applications.

6.6.1 An integral form of curl

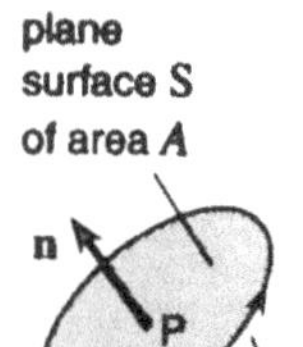

Fig 6.33
A plane loop C enclosing a plane surface S of area A. The orientation of the loop is specified by the unit surface normal **n**.

The limit (6.47) exists and is independent of the manner in which the limit is approached provided the partial derivatives of **F** exist everywhere on S.

Section 5.4.1 introduced the curl of a two-dimensional vector field $\mathbf{f}(x,y)$ as another vector field curl**f** pointing in the positive or negative z-direction. The magnitude of curl**f** is equal to the magnitude of the limit of the circulation per unit area, $\lim_{A\to 0}(W_0/A)$ where W_0 is the anticlockwise circulation of **f** around a loop enclosing an area A in the x-y plane. We now extend this idea to obtain an integral form of curl**F** for a three-dimensional vector field **F**.

Let C be a plane closed loop in the domain of **F**, and A the area of the plane surface S enclosed by the loop (Fig 6.33). The orientation of the loop in the field is specified by the direction of the unit surface normal **n** on S. We are free to choose the direction of **n** to be one of the two directions normal to S. This choice determines the positive sense of the circulation W_0 of **F** around C by the right-hand rule. Now let P be a fixed point on S and consider the limit of W_0/A as the loop shrinks to the point P, like a tightening noose. The value we obtain for this limit is the component of curl**F** in the direction of the unit surface normal **n**. If we write the circulation W_0 as a line integral of **F** around the loop we have

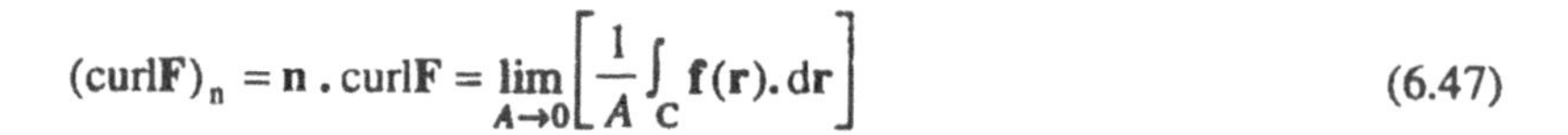

$$(\text{curl}\mathbf{F})_n = \mathbf{n}\,.\,\text{curl}\mathbf{F} = \lim_{A\to 0}\left[\frac{1}{A}\int_C \mathbf{f}(\mathbf{r}).\,d\mathbf{r}\right] \qquad (6.47)$$

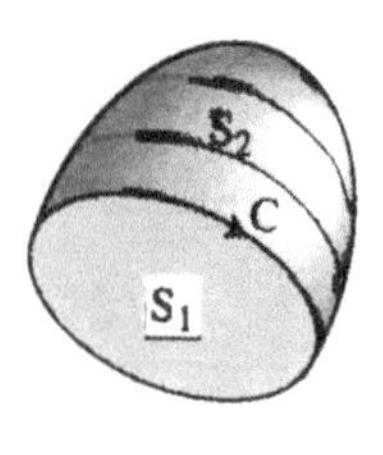

Fig 6.34
S_1 and S_2 are just two of an infinite number of surfaces that span a given closed curve C.

Eq (6.47) is an integral definition of the component of curl**F** in the direction specified by a unit vector **n**. We can define the three cartesian components of curl**F** in this way and so obtain an integral form of the definition of curl**F** which is equivalent to the differential form of curl**F** given in Eq (5.43).

6.6.2 Deriving Stokes's theorem

A formal derivation and statement of Stokes's theorem is beyond the scope of this book. We give a simplified version which brings out the main features.

Consider a closed loop C in the domain of a vector field **F**. There are many surfaces that span the loop (Fig 6.34). Let S be any one of them. We shall assume that **F** varies sufficiently smoothly that curl**F** is defined everywhere on the surface. We also assume that the loop C is the only boundary curve of S, i.e. the surface S has no holes (Fig 6.35). Now imagine S to be divided into a large number of nearly plane surface elements. Give each surface element a numerical label j and let the jth element have area ΔA_j and be bounded by a curve ΔC_j (Fig 6.36). It follows from Eq (6.47) that the component of curl**F** in the direction of the unit surface normal $\mathbf{n}_j$ at a point on the jth element can be approximated by

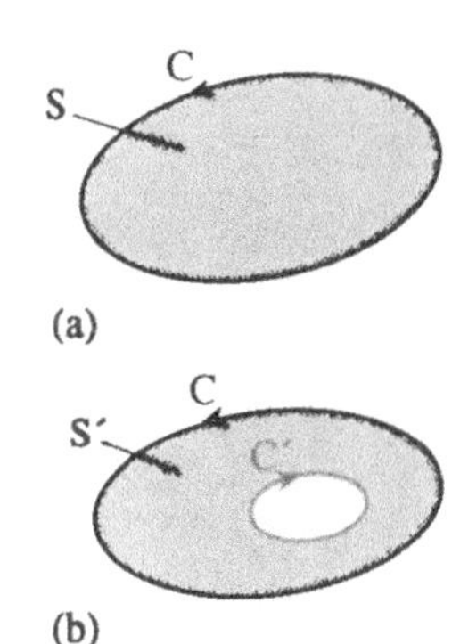

Fig 6.35
(a) The surface S has only one boundary curve C. (b) The surface S′ has two boundary curves; the boundary C′ encloses a hole.

$$\mathbf{n}_j.\text{curl}\mathbf{F} \approx \frac{1}{\Delta A_j}\int_{\Delta C_j} \mathbf{F}(\mathbf{r}).\,\mathbf{dr} \tag{6.48}$$

This approximation is good when the number N of subdivisions of S is very large and the ΔA_j very small. Now multiply both sides of Eq (6.48) by ΔA_j and sum over all surface elements. This gives

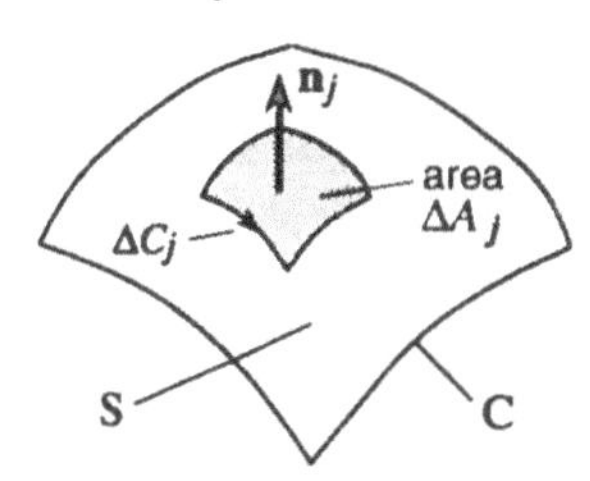

Fig 6.36
The surface element has a boundary curve ΔC_j and area ΔA_j.

$$\sum_{j=1}^{N}\Delta A_j \mathbf{n}_j.\text{curl}\mathbf{F} \approx \sum_{j=1}^{N}\int_{\Delta C_j} \mathbf{F}(\mathbf{r}).\,\mathbf{dr} \tag{6.49}$$

We now take the limit of both sides of (6.49) as $N \to \infty$ and $\Delta A_j \to 0$. You can see from Eq (6.34) that in this limit the left-hand side of (6.49) is the scalar surface integral (or flux) of the vector field curl**F** across the surface S. The right-hand side of (6.49) is a sum of circulations around all the surface elements comprising S. Referring to Fig 6.37 you can see that any boundary line between two surface elements contributes twice to this sum, with opposite signs in the two contributions. This is true of all boundary lines between surface elements and so the only lines making a net contribution to the sum are those lying on the single perimeter curve C. Thus the right-hand side of (6.49) is just the scalar line integral of **F** around the perimeter curve C of S. We can now write the limit of (6.49) as

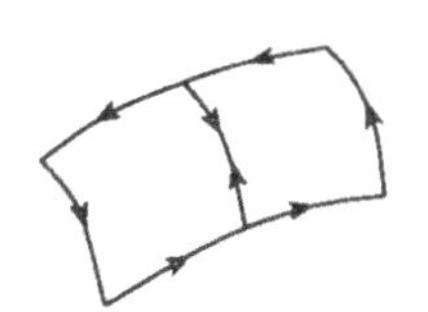

Fig 6.37
Two adjacent surface elements. The common boundary line is traversed in opposite directions.

$$\int_S (\text{curl}\mathbf{F})_n\, dA = \int_C \mathbf{F}(\mathbf{r}).\,\mathbf{dr} \tag{6.50}$$

This is **Stokes's theorem**; it states that the flux of curl**F** across a surface S is equal to the circulation of **F** around the surface boundary C, the sense of the circulation being related to the direction **n** by the right-hand rule. We have assumed that curl**F** is defined everywhere on S and that S has no holes. If there is a hole in S we can extend Stokes's theorem by taking account of the circulation around the additional boundary curve C′ surrounding the hole (Fig 6.35b), but we shall not go into the details of this.

We could write $\int_C \mathbf{F}(\mathbf{r}).\,\mathbf{dr}$ in Eq. 6.50 as $\oint \mathbf{F}(\mathbf{r}).\,\mathbf{dr}$ since the curve C is closed.

We have seen that there are many different surfaces that span any given perimeter curve (Fig 6.34) and you may find it surprising that the flux of curl**F** is the same across all of them, as required by Stokes's theorem. This can be understood however when we recall that any curl is divergence-free (solenoidal); this follows from the vector identity div(curl**F**) = 0 for any **F**. The field lines of curl**F** are therefore continuous lines without sources or sinks, and so the same number of lines (i.e. the same flux) of curl**F** must cross any two surfaces bounded by the same perimeter (Fig 6.38).

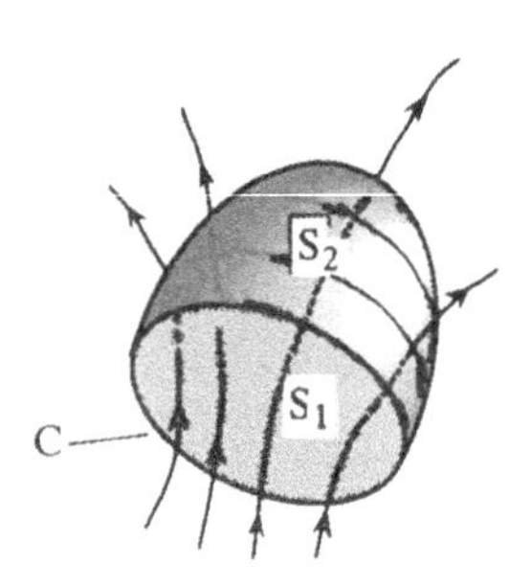

Fig 6.38
The flux (number of field lines) of a solenoidal field is the same across all surfaces spanning a given loop C.

6.6.3 Using Stokes's theorem

An obvious practical application of Stokes's theorem is the calculation of scalar line integrals by expressing them as surface integrals which may be easier to evaluate (or vice versa). For example, suppose you are asked to find the circulation of the vector field

$$\mathbf{G}(x,y,z) = yz\mathbf{i} - xz\mathbf{j} + \mathbf{k} \tag{6.51}$$

in the anticlockwise sense around a circular path specified by $x^2 + y^2 = 1$, $z = 3$; this path is the circle of unit radius centred on the z-axis in the plane $z = 3$. You could evaluate the line integral directly (Problem 6.3) but it is rather easier to evaluate the scalar surface integral (flux) of curl**G** across the plane circular surface S in the normal direction $\mathbf{n} = \mathbf{k}$. The curl is easily found (Eq 5.43) to be

$$\text{curl}\mathbf{G} = x\mathbf{i} + y\mathbf{j} - 2z\mathbf{k} \tag{6.52}$$

and the component of this curl in the direction of $\mathbf{n} = \mathbf{k}$ is $(\text{curl}\mathbf{G})_z = -2z$. The value of this component on the surface $z = 3$ is $-2 \times 3 = -6$. The flux of curl**G** across the circular surface S is therefore $-6 \times \pi(1^2) = -6\pi$, and so by Stokes's theorem the required anticlockwise circulation of **G** around the circle is -6π.

We now discuss a theoretical application of Stokes's theorem.

Conservative fields revisited

We can define a conservative field **G** as one that satisfies the curl test (Eq 5.50),

$$\text{curl}\mathbf{G} = 0 \qquad \text{everywhere in the domain of } \mathbf{G} \tag{6.53}$$

If curl**G** = 0 everywhere then the flux of curl**G** across any surface S must be zero and so by Stokes's theorem (6.50) the circulation of **G** around any closed loop C must be zero,

$$\int_C \mathbf{G} \, . \, d\mathbf{r} = 0 \qquad \text{for all closed loops C} \tag{6.54}$$

We could write Eq 6.54 as $\oint \mathbf{G} \, . \, d\mathbf{r} = 0$ since C here is closed.

Conversely, if Eq (6.54) is true for all possible loops, Stokes's theorem requires that Eq (6.53) is also true. Thus Eqs (6.53) and (6.54) provide us with equivalent definitions of a conservative force, provided Stokes's theorem can be applied.

However, there are some fields for which it may not be possible to find a hole-free "Stokes's" surface that spans a given loop. Stokes's theorem cannot then be applied in the form we have derived, and Eqs (6.53) and (6.54) may not be equivalent statements when applied to these fields. An example is the magnetic field $\mathbf{B}_1$ outside an infinitely long straight wire of radius a carrying a current I. With the axis of the wire lying on the z-axis and the current I in the positive z-direction (Fig 6.39), the field $\mathbf{B}_1$ is given by

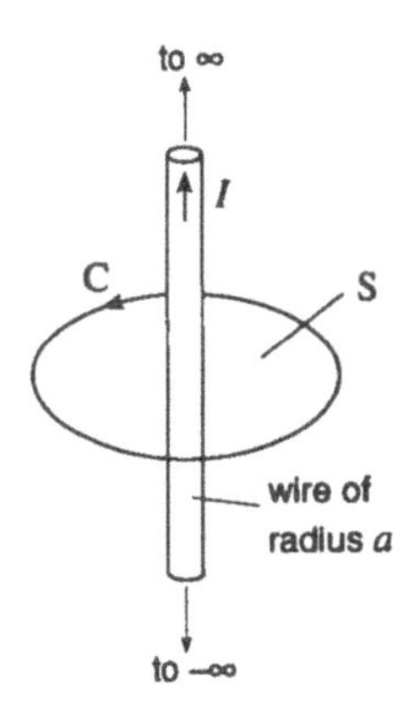

Fig 6.39
The infinitely long wire of radius a carries a current I in the z-direction.

$$\mathbf{B}_1(\rho,\phi,z) = \frac{\mu_0 I \mathbf{e}_\phi}{2\pi\rho} \qquad (\rho > a) \tag{6.55a}$$

Consider a closed loop C that encircles the wire (Fig 6.39). Any surface S that spans the loop is pierced by the wire which occupies the solid cylinder $\rho \le a$. This cylinder is not in the domain of the field (Eq 6.55a) and so the surface S has a hole. Stokes's theorem, in the form derived above, cannot be applied and Eqs (6.53) and (6.54) are not equivalent when applied to this field. In fact we can easily show (Example 6.3) that curl$\mathbf{B}_1 = \mathbf{0}$ everywhere in the domain of $\mathbf{B}_1$ but the line integral around a loop enclosing the wire has the non-zero value of $\mu_0 I$. Thus Eq (6.53) is satisfied but Eq (6.54) is not.

Is the field $\mathbf{B}_1$ conservative? A satisfactory answer becomes clear when we realise that the magnetic field produced by the current exists everywhere, inside the wire as well as outside, and so Eq (6.55a) is only part of the overall magnetic field produced by the current. When the current is distributed uniformly in the wire the magnetic field inside the wire is found to be

If you have studied electromagnetism you will know that the circulation of a static magnetic field is always equal to μ_0 times the enclosed current. This is Ampère's circuital law.

$$\mathbf{B}_2(\rho,\phi,z) = \frac{\mu_0 I \rho \mathbf{e}_\phi}{2\pi a^2} \qquad (\rho \le a) \tag{6.55b}$$

This field $\mathbf{B}_2$ has a non-zero curl (Example 5.5.4). The magnetic field $\mathbf{B}$ produced by the current is the sum of the fields in the two regions, i.e. $\mathbf{B} = \mathbf{B}_1 + \mathbf{B}_2$. The overall field $\mathbf{B}$ is a non-conservative field; it fails both (6.53) and (6.54). We now see that the difficulty arose because we artificially separated a physical field into mathematically-convenient parts and considered one part, $\mathbf{B}_1$, in isolation.

Although magnetic fields are not conservative it is often convenient when calculating magnetic fields produced by currents and magnets to introduce a scalar magnetostatic potential V_M defined by $\mathbf{B}_1 = -\text{grad}V_M$ in the curl-free regions.

Summary of section 6.6

Stokes's theorem states the flux of curl$\mathbf{F}$ across a surface S is equal to the circulation of $\mathbf{F}$ around the boundary curve C, the sense of the circulation being related to the direction $\mathbf{n}$ by the right-hand rule. Thus

$$\int_S (\text{curl}\mathbf{F})_n \, dA = \int_C \mathbf{F}(\mathbf{r}).d\mathbf{r} \qquad (6.50)$$

Example 6.1 (*Objective 6*) Verify Stokes's theorem for the vector field $\mathbf{F}(x,y,z) = -y\mathbf{i} + x\mathbf{j} + z\mathbf{k}$ and a circle described by $x^2 + y^2 = 9$.

Solution 6.1 We evaluate the line integral of $\mathbf{F}$ anticlockwise around the circle and the flux of curl$\mathbf{F}$ in the positive z-direction across the circular area, and show that they are equal.

We can evaluate the line integral in cylindrical polar coordinates. Recognising $-y\mathbf{i} + x\mathbf{j}$ as $\rho\mathbf{e}_\phi$ (Eqs 4.14 and 4.16) we write the given vector field as $\mathbf{F}(\rho,\phi,z) = \rho\mathbf{e}_\phi + z\mathbf{k}$. The unit tangent vector on the circle is $\mathbf{e}_\phi$ and so the tangential component of $\mathbf{F}$ on the circular path of radius 3 is $\mathbf{e}_\phi \,.\, \mathbf{F}(3,\phi,z) = 3$, and the circulation is simply $2\pi 3 \times 3 = 18\pi$.

Now the curl of $\mathbf{F}$ is easily found using Eq (5.43) to be curl$\mathbf{F} = 2\mathbf{k}$. The normal component of curl$\mathbf{F}$ on the surface is $(\text{curl}\mathbf{F})_z = \mathbf{k} \,.\, 2\mathbf{k} = 2$, and so the flux of curl$\mathbf{F}$ across the circle is simply $\pi(3^2) \times 2 = 18\pi$. Thus we have shown that both sides of Eq (6.50) are equal to 18π, verifying Stokes's theorem for the given field.

Example 6.2 (*Objective 6*) Determine the value of the line integral of

$$\mathbf{A}(x,y) = (e^x \sin y)\mathbf{i} - (e^x \cos y)\mathbf{j}$$

around the closed rectangular path OABC where the corner points are O(0,0), A(1,0), B(1,π/2), C(0,π/2).

Solution 6.2 The order of letters OABC indicates the anticlockwise sense around the path. We could evaluate the line integral directly (Problem 6.1) but it is quicker to use Stokes's theorem and evaluate the flux of the curl across the rectangle. We find curl$\mathbf{A} = (-2e^x \cos y)\mathbf{k}$, and so

$$\int_{\text{OABC}} (\text{curl}\mathbf{A})_z \, dA = \int_{y=0}^{y=\pi/2} \left(\int_{x=0}^{x=1} (-2e^x \cos y) dx \right) dy$$

$$= \int_0^{\pi/2} (-2(e-1)\cos y) dy = 2(1-e) = -3.44$$

This is the value of the required line integral.

Problem 6.1 (*Objective 6*) Evaluate directly the line integral of $\mathbf{A}(x,y) = (e^x \sin y)\mathbf{i} - (e^x \cos y)\mathbf{j}$ around the closed rectangular path OABC where the corner points are O(0,0), A(1,0), B(1,π/2), C(0,π/2), and hence verify the answer to Example 6.2.

Problem 6.2 (*Objective 6*) Use Stokes's theorem to find the anticlockwise circulation of the vector field $\mathbf{f}(x,y) = x^3\mathbf{j}$ around the circle $x^2 + y^2 = 4$.

Problem 6.3 (*Objective 6*) Consider the vector field **G** given by Eq (6.51). By evaluating the line integral directly, verify that the anticlockwise circulation of **G** around the circle $x^2 + y^2 = 1$, $z = 3$ is -6π.

Problem 6.4 (*Objective 6*) The integral form of Ampère's circuital law for steady currents (in the absence of magnetic materials) states that the circulation of a magnetic field **B** around any closed loop C is equal to $\mu_0 I$ where I is the total current enclosed by the loop, the sense of the circulation and the direction of I being related by the right-hand rule,

$$\oint \mathbf{B} \cdot d\mathbf{r} = \mu_0 I \qquad \text{(integral form of Ampère's law)} \qquad (6.56a)$$

(a) Given that the enclosed current is the flux of the current density vector **J** across any surface S spanning C, use Stokes' theorem to obtain the differential form of Ampère's law

$$\text{curl}\mathbf{B} = \mu_0 \mathbf{J} \qquad \text{(differential form of Ampère's law)} \qquad (6.56b)$$

$\mu_0 = 4\pi \times 10^{-7}$ Tm A^{-1}.

(b) Use the differential form of Ampère's law to determine the value of curl**B** at (i) a point inside a wire where the current density is uniform and of magnitude 50 kA m^{-2}, and (ii) at a point in space just outside the wire where there are no currents.

Problem 6.5 (*Objective 6*) Use Stokes's theorem to show that the circulation of any gradient vector field $\mathbf{V} = \text{grad}U$ is always zero when U exists everywhere in space.

Problem 6.6 (*Objective 6*) Consider the vector field function

$$\mathbf{B}(\rho, \phi, z) = \frac{C\mathbf{e}_\phi}{\rho} \qquad (\rho > 0;\ C \text{ is a constant})$$

(a) Evaluate directly the circulation of **B** around (i) the circle of radius 1 in the x-y plane centred on the origin, and (ii) the closed curve in the x-y plane bounded by radial lines $\phi = \phi_1$, $\phi = \phi_1 + \Delta\phi$ and the circular arcs $\rho = \rho_1$, $\rho = \rho_1 + \Delta\rho$ (Fig 6.40).

(b) Find curl**B** (see Eq (5.44) for the cylindrical polar form of curl).

(c) Explain why Eqs (6.53) and (6.54) are not equivalent when applied to this field.

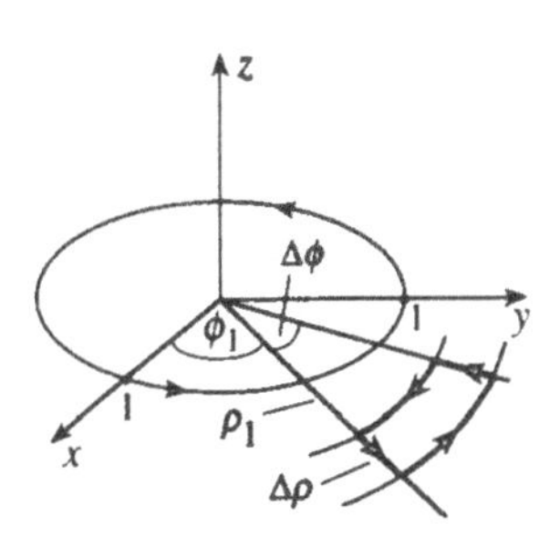

Fig 6.40
A circular path of radius 1, and a closed path bounded by two radial lines and two circular arcs.

6.7 VOLUME INTEGRALS

Volume integrals are defined as limits of sums in a similar way as line integrals and surface integrals. Let $U(\mathbf{r})$ be a three-dimensional scalar field and V the volume of a region R in the domain of U. We divide the region R into N volume elements ΔV_j (Fig 6.41) such that all $\Delta V_j \to 0$ as $N \to \infty$. Then if $U(\mathbf{r}_j)$ is the value of U at a point $\mathbf{r}_j$ in the element ΔV_j, the **volume integral** of the scalar field U over the region R is defined to be the limit

$$\lim_{N\to\infty}\left[\sum_{j=1}^{N} U(\mathbf{r}_j)\Delta V_j\right] = \int_R U(\mathbf{r})dV \tag{6.57}$$

If $U(\mathbf{r})$ is the density of a solid body (mass per unit volume) then $U(\mathbf{r})dV$ can be interpreted as the mass of a volume element dV, and the volume integral of $U(\mathbf{r})$ over the solid gives the total mass.

When $U(\mathbf{r})$ has the constant value U everywhere, then U can be taken outside the summation symbol in Eq (6.57), and the volume integral is simply UV. More generally a volume integral is evaluated by introducing a coordinate system and expressing the volume integral as a **triple integral** which can then be evaluated by methods that are simple extensions of those described in Section 6.5.2 for double integrals. For this purpose it is necessary to express the **volume element** dV in coordinate form. Referring to Fig 6.42 you can see that

$$dV = dx\, dy\, dz \qquad \text{(cartesian coordinates)} \tag{6.58}$$

$$dV = \rho\, d\phi\, d\rho\, dz \qquad \text{(cylindrical polar coordinates)} \tag{6.59}$$

$$dV = r^2 \sin\theta\, d\phi\, d\theta\, dr \qquad \text{(spherical polar coordinates)} \tag{6.60}$$

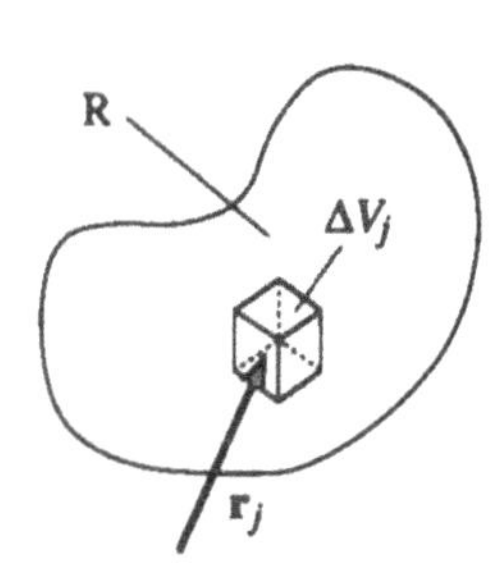

Fig 6.41
A volume element ΔV_j. $\mathbf{r}_j$ is the position vector of any point in the volume element.

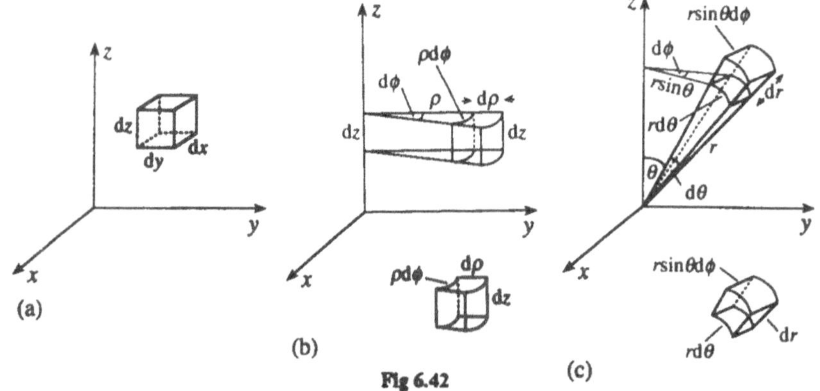

Fig 6.42
(a) The cartesian volume element is $dV = dxdydz$. (b) The volume element in cylindrical polar coordinates is $dV = (\rho d\phi)d\rho dz$. (c) The volume element in spherical polar coordinates is $dV = (r\sin\theta d\phi)(rd\theta)dr$.

We shall consider only rectangular, cylindrical and spherical regions and use appropriately chosen coordinate systems in which the boundary surfaces of the region lie on coordinate surfaces (Fig 6.43). For such cases the limits of the integrations are constants and independent of the order of integration. This is illustrated by the cartesian triple integral representing the volume integral of $U(\mathbf{r})$ over a rectangular block (Fig 6.44),

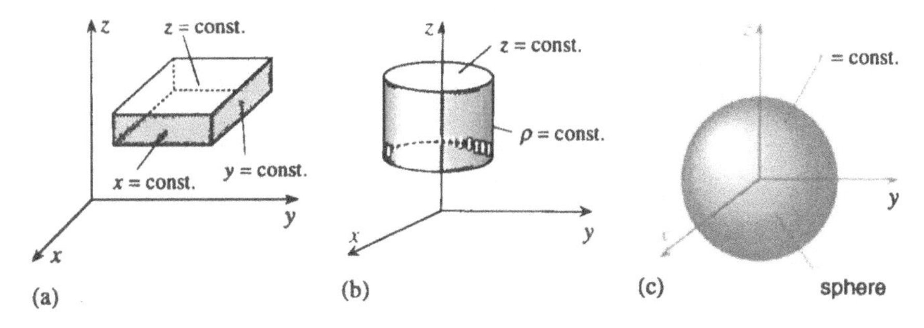

Fig 6.43
The surfaces of the three-dimensional regions lie on coordinate surfaces. (a) a rectangular block in a cartesian system. Three of the six coordinate planes are indicated. (b) The curved surface of the cylinder lies on a coordinate surface ρ = constant. The plane end faces lie on planes z = constant. (c) The surface of the sphere lies on a surface r = constant.

$$\int_{\text{block}} U(\mathbf{r})\mathrm{d}V = \int_{z=e}^{z=f}\left(\int_{y=c}^{y=d}\left(\int_{x=a}^{x=b} U(x,y,z)\mathrm{d}x\right)\mathrm{d}y\right)\mathrm{d}z \qquad (6.61)$$

where the brackets indicate a chosen order of integration, i.e. we work from the middle outwards, integrating first over x, then y and then z. The limits of the three integrals correspond to the coordinate planes $x = a$, $x = b$, $y = c$, etc., in which the six faces of the block lie. With these limits the volume integral covers the entire volume of the block.

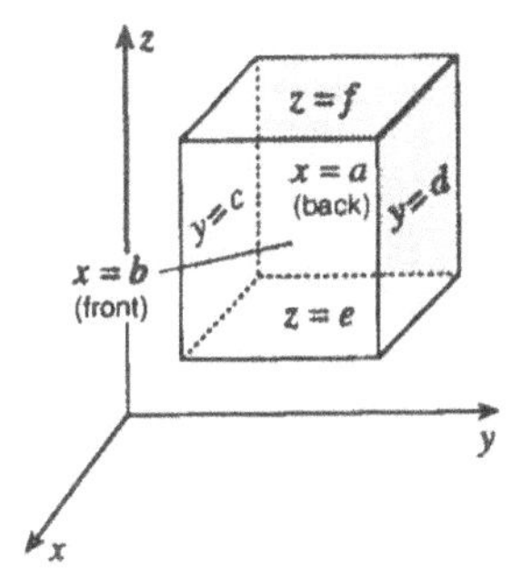

Fig 6.44
The six faces of the block lie on the coordinate planes $x = a$, $x = b$, etc. These planes specify the limits of integration.

In the special case where the scalar field is a constant, $U(\mathbf{r}) = C$, the three integrals are easily evaluated: the x integral yields the factor $(b - a)$, the y integral yields $(d - c)$ and the z integral yields $(f - e)$. Thus the value of the volume integral is $C(f - e)(d - c)(b - a)$ which is just C times the volume of the block. If C is the uniform density of a solid block then the volume integral is the mass of the block, and we have obtained the familiar result: mass = density × volume. In more interesting examples the density of a solid body may be non-uniform.

Consider a solid alloy sphere of radius 10 cm whose density D varies uniformly from 7.20 g cm^{-3} at the centre to 7.60 g cm^{-3} at the surface, i.e. D is given in spherical polar coordinates by

$$D(r) = 7.20 + 0.04r \qquad (0 \le r \le 10)$$

We often denote a spherically symmetric field such as $D(r,\theta,\phi)$ more simply as $D(r)$.

where r is the radial distance in cm. The total mass M of the sphere is the volume integral of the density over the sphere and is evaluated as a triple integral in spherical polar coordinates with the volume element of Eq (6.60). Thus

$$M = \int_{\text{sphere}} D(r)\mathrm{d}V = \int_{r=0}^{r=10}\left(\int_{\theta=0}^{\theta=\pi}\left(\int_{\phi=0}^{\phi=2\pi}(7.20+0.04r)r^2\sin\theta\mathrm{d}\phi\right)\mathrm{d}\theta\right)\mathrm{d}r$$

The limits on the integrals are such that the integration covers the whole volume of the sphere (Fig 6.45). We evaluate the ϕ integral first as indicated by the brackets. The factors inside the ϕ integral depend on r and θ but not on ϕ. These factors can therefore be taken outside the ϕ integral which is then easily evaluated to give $\int_0^{2\pi} d\phi = 2\pi$. Thus the triple integral is reduced to a double integral

$$M = \int_{\text{sphere}} D(r)dV = 2\pi \int_{r=0}^{r=10} \left(\int_{\theta=0}^{\theta=\pi} (7.20 + 0.04r)r^2 \sin\theta \, d\theta \right) dr$$

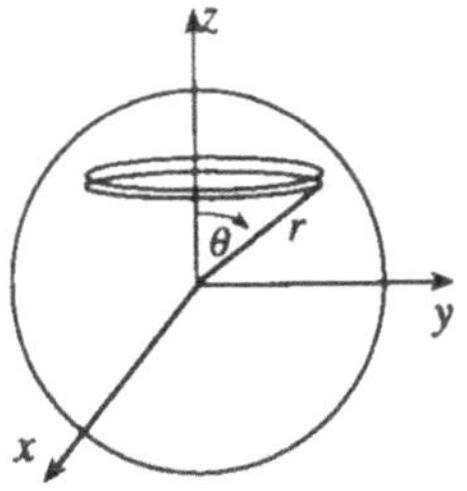

We now take the r dependent factors outside the θ integral which then gives $\int_0^{\pi} \sin\theta \, d\theta = [-\cos\theta]_0^{\pi} = 2$. Thus the two integrations over the angle variables amount to a factor of 4π, and we are left with the r integral

$$M = 4\pi \int_0^{10} (7.20 + 0.04r)r^2 dr = 4\pi \left[\frac{7.20r^3}{3} + \frac{0.04r^4}{4} \right]_0^{10}$$

$$= 4\pi \times 2500 = 3.14 \times 10^4$$

Fig 6.45
The region of integration is a sphere. The ϕ integral can be thought of as a sum of volume elements around a ring of radius $r\sin\theta$. The θ integral can be thought of as a sum of rings on a spherical shell of radius r. The r integral represents the sum of spherical shells in a solid sphere of radius 10.

i.e. the mass of the sphere is 31.4 kg.

Note that the integrals over θ and ϕ above involve essentially only the factor $\sin\theta$ that comes from the volume element (Eq (6.60)) since the spherically symmetric scalar field itself does not depend on the angular coordinates. In fact the volume integral of any spherically symmetric scalar field over a complete spherical region will contain the factor

$$\int_{\theta=0}^{\theta=\pi} \left(\int_{\phi=0}^{\phi=2\pi} \sin\theta d\phi \right) d\theta = 4\pi$$

The volume integral of any spherically symmetric scalar field $U(\mathbf{r}) = U(r)$ over the whole region R between radial surfaces $r = a$ and $r = b$ ($b > a$) is therefore

$$\int_R U(\mathbf{r})dV = 4\pi \int_a^b U(r)r^2 dr \qquad \text{(spherically symmetric field)} \qquad (6.62)$$

A common error in using Eq (6.62) is to forget the r^2 factor which comes from the spherical polar volume element (Eq (6.60)) – so beware!

This is a very useful result.

Summary of section 6.7

- **Volume integrals** are defined as limits of sums (Eq (6.57)) and are evaluated by expressing them as **triple integrals** with respect to coordinate variables; the **volume elements** are given by Eqs (6.58) to (6.60).

- The volume integral of any spherically symmetric scalar field $U(\mathbf{r})$ over the region R between radial coordinates $r = a$ and $r = b$ is

$$\int_{\mathbf{R}} U(\mathbf{r})\mathrm{d}V = 4\pi\int_a^b U(r)r^2\mathrm{d}r \qquad \text{(spherically symmetric field)} \qquad (6.62)$$

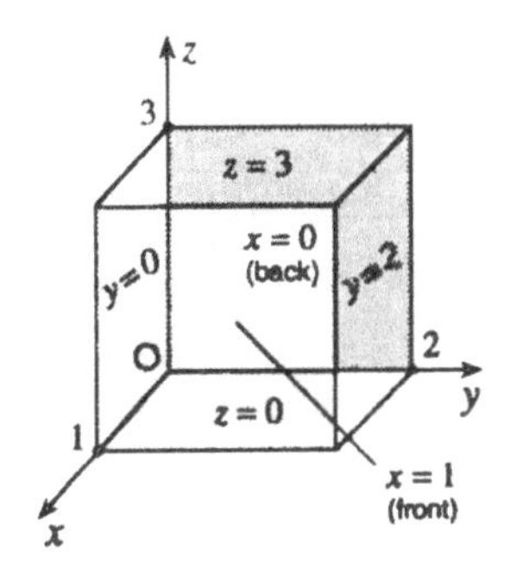

Fig 6.46
The region R of integration is a rectangular block.

Example 7.1 (*Objective 7*) Evaluate the volume integral of the scalar field $U(x,y,z) = x^2 + y^2 - 2z^2$ over a block whose faces are the planes $x = 0$, $x = 1$, $y = 0$, $y = 2$, $z = 0$, $z = 3$ (Fig 6.46).

Solution 7.1 This volume integral can be evaluated as a triple integral over the three cartesian variables. Thus

$$\int_{\text{block}} U(\mathbf{r})\mathrm{d}V = \int_{z=0}^{z=3}\left(\int_{y=0}^{y=2}\left(\int_{x=0}^{x=1}(x^2+y^2-2z^2)\,\mathrm{d}x\right)\mathrm{d}y\right)\mathrm{d}z$$

The x-integral (which is the integral in the innermost brackets) yields $\left[x^3/3+y^2x-2z^2x\right]_0^1 = 1/3+y^2-2z^2$. We are now left with a double integral:

$$\int_{\text{block}} U(\mathbf{r})\mathrm{d}V = \int_{z=0}^{z=3}\left(\int_{y=0}^{y=2}\left(\frac{1}{3}+y^2-2z^2\right)\mathrm{d}y\right)\mathrm{d}z$$

$$= \int_{z=0}^{z=3}\left[\frac{y}{3}+\frac{y^3}{3}-2z^2y\right]_{y=0}^{y=2}\mathrm{d}z$$

$$= \int_0^3\left(\frac{10}{3}-4z^2\right)\mathrm{d}z = \left[\frac{10z}{3}-\frac{4z^3}{3}\right]_0^3 = -26$$

Example 7.2 (*Objective 7*) Evaluate the volume integral of the scalar field $U(\mathbf{r})$ of Example 7.1 over the volume of a cylinder of unit height and radius 2 standing on the x-y plane with its axis on the positive z-axis.

Solution 7.2 We use cylindrical polar coordinates for integration over the cylindrical volume, with the volume element given by Eq (6.59). The polar form of the field is $U(\rho,\phi,z) = \rho^2 - 2z^2$ and so the volume integral is given by the triple integral

$$\int_{z=0}^{z=1}\left(\int_{\rho=0}^{\rho=2}\left(\int_{\phi=0}^{\phi=2\pi}(\rho^2-2z^2)\rho\,\mathrm{d}\phi\right)\mathrm{d}\rho\right)\mathrm{d}z$$

There are no ϕ-dependent factors and so the ϕ-integral yields $\int_0^{2\pi}\mathrm{d}\phi = 2\pi$, and we are left with the double integral

$$2\pi\int_{z=0}^{z=1}\left(\int_{\rho=0}^{\rho=2}(\rho^2-2z^2)\rho\mathrm{d}\rho\right)\mathrm{d}z = 2\pi\int_{z=0}^{z=1}\left[\frac{\rho^4}{4}-z^2\rho^2\right]_{\rho=0}^{\rho=2}\mathrm{d}z$$

$$= 2\pi\int_0^1(4-4z^2)\mathrm{d}z = 2\pi\left[4z-\frac{4z^3}{3}\right]_0^1 = 2\pi\times\frac{8}{3} = \frac{16\pi}{3}$$

Example 7.3 (*Objective 7*) The constant scalar field $U(\mathbf{r}) = C$ is a simple example of a spherically symmetric field. Use Eq (6.62) to find the volume integral of U over a sphere of radius R centred at the origin.

Solution 7.3 In spherical polar coordinates the field is $U(r,\theta,\phi) = U(r) = C$, and so we use Eq 6.62 to obtain

$$\int_{\text{sphere}} U(\mathbf{r})\mathrm{d}V = 4\pi\int_0^R Cr^2\mathrm{d}r = 4\pi C\left[\frac{r^3}{3}\right]_0^R = \frac{4\pi CR^3}{3}$$

This is just C times the volume of the sphere, as expected.

Example 7.4 (*Objective 7*) Electric charge is distributed in space with charge density (Cm^{-3}) described by the scalar field

$$\sigma(\mathbf{r}) = \frac{q_0}{R^3} \qquad (|\mathbf{r}| < R)$$

$$= \frac{q_0 R}{|\mathbf{r}|^4} \qquad (|\mathbf{r}| \geq R)$$

where q_0 and R are constants. Determine the total electric charge in (a) the spherical region $|\mathbf{r}| < R$ and (b) all space.

Solution 7.4

(a) The charge density has the constant value q_0/R^3 everywhere in the spherical region $|\mathbf{r}| < R$ and so the total charge in this region is simply $(q_0/R^3) \times 4\pi R^3/3 = 4\pi q_0/3$. (This result can also be obtained by evaluating a triple integral as in Example 7.3 with C replaced by q_0/R^3.)

(b) We have to find the volume integral of the charge density over "all space". We use spherical polar coordinates and interpret all space as being a spherical region of infinite radius. We already have the charge in the region $|\mathbf{r}| < R$ from part (a). To this we must add the total charge Q in the region $|\mathbf{r}| \geq R$. This additional charge is the volume integral of the spherically symmetric field $\sigma(r) = q_0R/r^4$ evaluated using Eq (6.62) with lower limit $r = R$ and upper limit $r = \infty$. Thus

$$Q = 4\pi q_0 R\int_R^\infty \left(\frac{1}{r^4}\right) r^2 dr = 4\pi q_0 R\left[-\frac{1}{r}\right]_R^\infty = 4\pi q_0 R\left[-\frac{1}{\infty}+\frac{1}{R}\right] = 4\pi q_0$$

(We have manipulated the infinity symbol as if it were a number and have interpreted $1/\infty$ as zero.) The total charge in all space is therefore $4\pi q_0/3 + 4\pi q_0 = 16\pi q_0/3$.

Problem 7.1 (*Objective 7*) Find the volume integral of the scalar field $U(x,y,z) = x + y + z$ over a cube of unit edge length situated with its centre at (1/2,1/2,1/2) and three faces in the planes $x = 0$, $y = 0$ and $z = 0$.

Problem 7.2 (*Objective 7*) Find the mass of a sphere of radius R whose density varies with radial distance r according to $\rho(r) = \alpha(R^2 - r^2/5)$ where α is a constant.

Problem 7.3 (*Objective 7*) A rod of length 1m and square cross-section of side length 1 cm has a density that increases uniformly along its length from 5.0 g cm^{-3} at one end to 5.3 g cm^{-3} at the other. Express the mass of the rod as a volume integral and evaluate it.

Problem 7.4 (*Objective 7*) Electric charge resides in the space between two concentric cylinders of height h and radii a and b ($b > a$). When the axes of the cylinders lie on the z-axis with the midpoints of the cylinders at the origin, the charge density (Cm^{-3}) is given by $\sigma(\rho,\phi,z) = \sigma_0\exp(-\alpha\rho^2)\cos(\pi z/h)$ where σ_0 and α are constants. Determine the total electric charge residing between the cylinders.

Problem 7.5 (*Objective 7*) Find the volume integral of div**F** over the sphere of unit radius centred on the origin, where $\mathbf{F}(\mathbf{r}) = C\mathbf{r}$ and C is a constant.

Problem 7.6 (*Objective 7*) Find, where possible, the volume integral of $f(\mathbf{r}) = 1/|\mathbf{r}|^n$ ($\mathbf{r} \neq \mathbf{0}$ and n is a positive integer) over the whole of the space outside a sphere of radius a centred at $\mathbf{r} = \mathbf{0}$.

6.8 GAUSS'S THEOREM (THE DIVERGENCE THEOREM)

Gauss's theorem, also called the divergence theorem, is an identity between a surface integral and a volume integral, and plays a role in vector calculus in some ways similar to that of Stokes's theorem. We give a simplified derivation and Gauss's theorem and show some of its applications. The first task is to obtain an integral expression for divergence.

An integral expression for divergence

In Chapter 5 the divergence of a vector field **F** at a point P was introduced as the limit of the outward flux Φ_o of **F** across a closed surface S enclosing P divided by the enclosed volume V, as the surface collapsed onto P, i.e. $\text{div}\mathbf{F} = \lim_{V\to 0}\left(\frac{\Phi_o}{V}\right)$. This led to the differential definition of divergence in terms of partial derivatives (Eq (5.26)). If we now write the flux as a surface integral, we have

$$\text{div}\mathbf{F} = \lim_{V\to 0}\left[\frac{1}{V}\int_S \mathbf{n}\,.\,\mathbf{F}(\mathbf{r})\mathrm{d}A\right] \tag{6.63}$$

where **n** is the outward normal on S. Eq (6.63) is an integral expression for divergence at a point, and is equivalent to Eq (5.26).

Derivation of Gauss's theorem

To derive Gauss's theorem we consider a region R of volume V bounded by a single closed surface S. We assume that the region is in the domain of a vector field **F** which varies sufficiently smoothly that div**F** is defined everywhere. We divide the region into N subregions of volume ΔV_j enclosed by surfaces $\Delta \mathrm{S}_j$ of surface area ΔA_j. If the subregions are small enough we can use Eq (6.63) to give a good approximation for the divergence at a point $\mathbf{r}_j$ in the jth subregion,

Note that the left-hand side of Eq (6.64) is the value of the divergence of **F** at a point $\mathbf{r}_j$ in the jth subregion. For convenience we write it simply as div**F** rather than $(\text{div}\mathbf{F})(\mathbf{r}_j)$ or $\text{div}\mathbf{F}(\mathbf{r}_j)$.

$$\text{div}\mathbf{F} \approx \frac{1}{\Delta V_j}\int_{\Delta \mathrm{S}_j} \mathbf{n}\,.\,\mathbf{F}(\mathbf{r})\mathrm{d}A \tag{6.64}$$

Multiplying both sides of (6.64) by ΔV_j and summing over all j gives

$$\sum_{j=1}^{N} \text{div}\mathbf{F}\,\Delta V_j \approx \sum_{j=1}^{N}\int_{\Delta \mathrm{S}_j} \mathbf{n}\,.\,\mathbf{F}(\mathbf{r})\mathrm{d}A \tag{6.65}$$

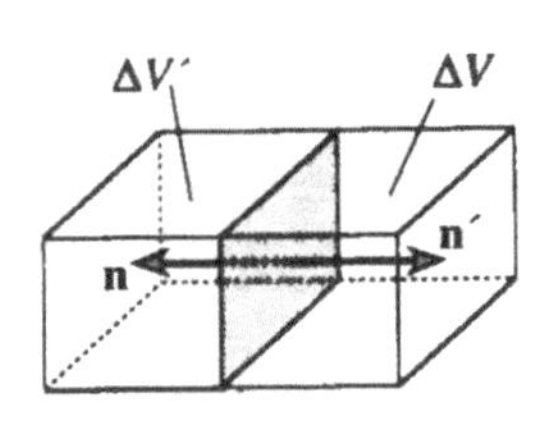

Fig 6.47
Adjacent volume elements ΔV and $\Delta V'$ share a common boundary wall. The outward normals on the common wall are **n** and $\mathbf{n}' = -\mathbf{n}$.

We now take the limit of both sides of (6.65) as $N \to \infty$ and $\Delta V_j \to 0$. The left-hand side of (6.65) then becomes the volume integral of div**F** over the region R (see Eq (6.57)). The right-hand side of (6.65) is a sum of the outward fluxes across the surfaces of all subregions. You can see from Fig 6.47 that the outward normals on the two faces of any boundary wall separating two volume elements are in opposite directions and so the two outward fluxes across the wall cancel, i.e. the outward flux from one volume element is an inward flux for the adjacent one. This is true of all boundary walls separating volume elements and so the only surfaces making a contribution to the outward flux are those lying on single closed surface S. Thus we can write the limit of (6.65) as

$$\int_R \text{div}\mathbf{F}\,\mathrm{d}V = \int_S \mathbf{n}\,.\,\mathbf{F}(\mathbf{r})\mathrm{d}A \tag{6.66}$$

This identity is known as **Gauss's theorem** or the **divergence theorem**. Note that the surface integral over the closed surface S could be written as $\oint \mathbf{n}.\mathbf{F}(\mathbf{r})\mathrm{d}A$. Gauss's theorem as stated in Eq (6.66) applies when the region is enclosed by a single closed surface. If the region has a hole, such as the region enclosed between two spherical surfaces S and S′ shown in Fig 6.48, Gauss's theorem must be modified by replacing the single surface integral in Eq (6.66) by the sum of two surface integrals, one over S and the other over S′ with surface normals **n** and **n′** directed away from the enclosed region (Problem 8.2).

Gauss's theorem, like Stokes's theorem, has practical and theoretical applications, some of which are explored in the Examples and Problems.

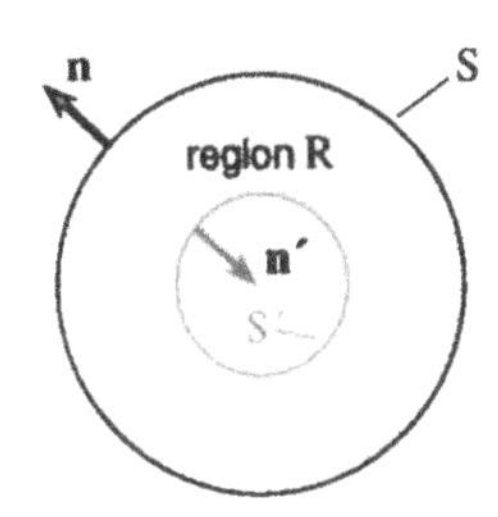

Fig 6.48
The region enclosed by two concentric spheres. The surface S′ encloses a hole.

Summary of section 6.8

Gauss's theorem (or the **divergence theorem**) is an identity between a volume integral and a surface integral

$$\int_{\mathrm{R}} \mathrm{div}\mathbf{F}\,\mathrm{d}V = \int_{\mathrm{S}} \mathbf{n}.\mathbf{F}(\mathbf{r})\mathrm{d}A \qquad (6.66)$$

Example 8.1 (*Objective 8*) Verify Gauss's theorem for the vector field $\mathbf{F}(\mathbf{r}) = C\mathbf{r}$ (C is a constant) and a spherical region of radius a centred on the origin.

Solution 8.1 We evaluate both sides of Eq (6.66) and show that they are equal.

div**F** is easily worked out in cartesian or spherical polar coordinates (Eqs (5.26) or (5.30)). Using cartesians we write $\mathbf{F} = C\mathbf{r} = C(x\mathbf{i} + y\mathbf{j} + z\mathbf{k})$ and obtain $\mathrm{div}(C\mathbf{r}) = 3C$ which is a constant scalar field. The volume integral of the divergence over the sphere is therefore simply $3C \times 4\pi a^3/3 = 4\pi Ca^3$.

To evaluate the surface integral (outward flux) of **F** it is best to use spherical polar coordinates in which the field is $\mathbf{F}(r) = Cr\mathbf{e}_r$ and the unit outward normal on the sphere is $\mathbf{n} = \mathbf{e}_r$. Thus on the surface ($r = a$) we have $\mathbf{n}\,.\,\mathbf{F}(a) = \mathbf{e}_r\,.\,Ca\mathbf{e}_r = Ca$ and the surface integral is simply $Ca \times 4\pi a^2 = 4\pi Ca^3$ which is equal to the volume integral of the divergence.

Example 8.2 (*Objective 8*) Evaluate the outward flux of the vector field $\mathbf{F}(x,y,z) = 3xy\mathbf{i} - 2zx\mathbf{k}$ across the cube bounded by the coordinate planes $x = 0$, $x = 1$, $y = 0$, $y = 1$, $z = 0$, $z = 1$ (Fig 6.49).

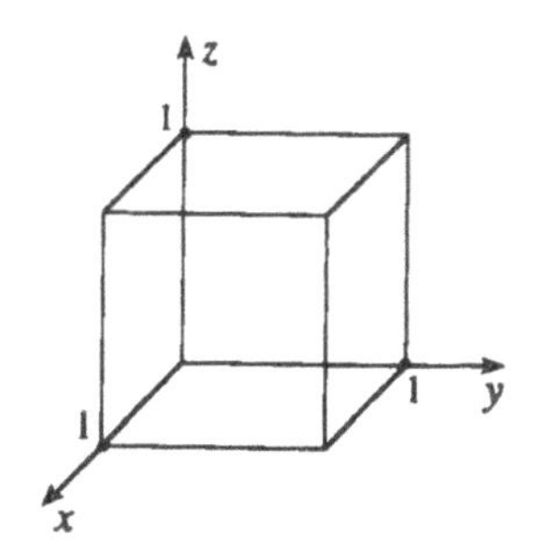

Fig 6.49
A cubic region.

Solution 8.2 We could evaluate the outward flux of **F** directly as a scalar surface integral but it is easier to evaluate the volume integral of div**F** over the volume of the cube and use Gauss's theorem. Using Eq (5.26) we find $\mathrm{div}\mathbf{F} = 3y - 2x$ and the volume integral is

$$\int_{z=0}^{z=1}\left(\int_{y=0}^{y=1}\left(\int_{x=0}^{x=1}(3y-2x)\,dx\right)dy\right)dz$$

$$=\int_{z=0}^{z=1}\left(\int_{y=0}^{y=1}\left[3yx-x^2\right]_{x=0}^{x=1}dy\right)dz = \int_{z=0}^{z=1}\left(\int_{y=0}^{y=1}(3y-1)\,dy\right)dz$$

$$=\int_{z=0}^{z=1}\left[\frac{3y^2}{2}-y\right]_{y=0}^{y=1}dz = \int_0^1\left(\frac{1}{2}\right)dz = \frac{1}{2}$$

Hence, by Gauss's theorem, the outward flux of **F** across the cube is 1/2.

Example 8.3 (*Objective* 8) Gauss's theorem is an identity and should be distinguished from physical laws that also bear Gauss's name. For example, the integral form of *Gauss's law of electrostatics* (in the absence of dielectric material) states that the net outward flux of the electrostatic field **E** across *any* closed surface S is equal to the total enclosed electric charge Q divided by ϵ_0,

$$\int_S \mathbf{n}\,.\,\mathbf{E}(\mathbf{r})\,dA = \frac{Q}{\epsilon_0} \qquad \text{(integral form of Gauss's law)}$$

Use Gauss's theorem to obtain a differential form of Gauss's law.

Solution 8.3 Gauss's theorem allows us to write the left-hand side of Gauss's law as the volume integral of div**E** over the region R enclosed by S. We can also express the total enclosed charge Q as the volume integral of the charge density $\sigma(\mathbf{r})$ (Cm^{-3}) over the region. Thus the integral form of Gauss's law and Gauss's theorem together yield

$$\int_R \text{div}\mathbf{E}\,dV = \frac{1}{\epsilon_0}\int_R \sigma(\mathbf{r})\,dV$$

For the two volume integrals in the above equation to be equal for an *arbitrary* region R it is necessary for the integrands to be equal. Thus

$$\text{div}\mathbf{E} = \frac{\sigma}{\epsilon_0}$$

This is the differential form of Gauss's law of electrostatics and is one of Maxwell's equations.

Problem 8.1 (*Objective 8*) Verify Gauss's theorem for the vector field $\mathbf{F}(\mathbf{r}) = C|\mathbf{r}|\mathbf{r}$ in the region of a sphere of radius a with its centre at the point $\mathbf{r} = 0$.

Problem 8.2 (*Objective 8*) Consider the region enclosed by two concentric spheres of radii a and b ($b > a$) with their centres at the origin.

(a) Evaluate directly the net outward flux of the vector field $\mathbf{F}(r,\theta,\phi) = C\mathbf{e}_r/r^2$ ($r > 0$ and C is a constant) across the two surfaces enclosing the region.

(b) Determine div**F** and hence the volume integral of div**F** over the region. Are your answers consistent with Gauss's theorem?

Problem 8.3 (*Objective 8*) Any magnetic field **B** has the property that the net outward flux across any closed surface is always zero. Show that this property is equivalent to div**B** = 0 everywhere.

Problem 8.4 (*Objective 8*) Consider the vector field $\mathbf{F} = 2xz\mathbf{i} - 3xy\mathbf{j} - z^2\mathbf{k}$ and the closed cylinder bounded by the surface $x^2 + y^2 = 4$ and the planes $z = 0$ and $z = 5$. Use Gauss's theorem to show that the flux of **F** across the closed cylinder is zero.

Appendix A SI units and physical constants

The **International System of Units (SI)** is nowadays used by most scientists and engineers. There are seven **SI base units** of which the five listed in Table A1 are used in this book.

Table A1 SI base units

Physical quantity	Name of unit	Symbol for unit
length	metre	m
mass	kilogram	kg
time	second	s
electric current	ampere	A
temperature	kelvin	K

Other SI units called **derived units** are obtained by combining the base units as products and quotients according to the algebraic formulas linking the corresponding physical quantities. Some derived units are given special names. See Table A2.

Table A2 Some SI derived units

Physical quantity	Name of unit	Symbol for unit
velocity		$m\ s^{-1}$
acceleration		$m\ s^{-2}$
momentum		$kg\ m\ s^{-1}$
force	newton	$N\ (1\ N = 1\ kg\ m\ s^{-2})$
pressure	pascal	$Pa\ (1\ Pa = 1\ N\ m^{-2})$
energy	joule	$J\ (1\ J = 1\ N\ m)$
power	watt	$W\ (1\ W = 1\ J\ s^{-1})$
angle	radian	rad
angular velocity		$rad\ s^{-1}$
angular momentum		$kg\ m^2\ s^{-1}$
torque		N m
electric charge	coulomb	$C\ (1\ C = 1\ A\ s)$
electric potential	volt	$V\ (1\ V = 1\ J\ C^{-1})$
electric field		$N\ C^{-1}$ or $V\ m^{-1}$
magnetic field	tesla	$T\ (1\ T = 1\ N\ s\ m^{-1}\ C^{-1})$
frequency	hertz	$Hz\ (1\ Hz = 1\ s^{-1})$

Prefixes are used in SI to represent powers of 10. See Table A3.

Table A3 Some SI prefixes

multiple	prefix (symbol)	multiple	prefix
10^9	giga (G)	10^{-2}	centi (c)
10^6	mega (M)	10^{-3}	milli (m)
10^3	kilo (k)	10^{-6}	micro (μ)

(The symbol m stands for milli as well as for metre, as in 1 mm = 1 millimetre = 10^{-3} metre.)

Some useful **conversions** involving SI and non SI units are:

1 radian = $(180/\pi)$ degrees
2π rad s^{-1} = 1 Hz = 1 revolution per second
1 hour = 3600 s
1 mile per hour = 0.447 m s^{-1}
0 kelvin = −273.15 degree Celsius (°C) = absolute zero of temperature. (A temperature change of 1 K is equal to a temperature change of 1 °C.)

Table A4 lists some common **physical constants**.

Table A4 Some physical constants

electric charge on a proton	$e = 1.602 \times 10^{-19}$ C
electric charge on an electron	$-e = -1.602 \times 10^{-19}$ C
mass of an electron	$m_e = 9.109 \times 10^{-31}$ kg
mass of a proton	$m_p = 1.673 \times 10^{-27}$ kg
speed of light in a vacuum	$c = 2.998 \times 10^8$ m s^{-1}
permittivity of free space	$\epsilon_0 = 8.854 \times 10^{-12}$ $C^2 N^{-1} m^{-2}$
	$(1/4\pi\epsilon_0 = 8.988 \times 10^9$ $Nm^2 C^{-2})$
permeability of free space	$\mu_0 = 4\pi \times 10^{-7}$ kg m C^{-2}
magnitude of acceleration of gravity	$g = 9.81$ m s^{-2}
radius of the Earth	$R_E = 6.37 \times 10^6$ m

Physical quantities are often denoted by symbols taken from the **Greek alphabet**.

Table A.5 Greek alphabet

alpha	α	A			
beta	β	B	nu	ν	N
gamma	γ	Γ	omicron	ο	O
delta	δ	Δ	pi	π	Π
epsilon	ϵ	E	rho	ρ	P
zeta	ζ	Z	sigma	σ	Σ
eta	η	H	tau	τ	T
theta	θ	Θ	upsilon	υ	Υ
iota	ι	I	phi	ϕ	Φ
kappa	κ	K	chi	χ	X
lambda	λ	Λ	psi	ψ	Ψ
mu	μ	M	omega	ω	Ω

Appendix B Mathematical conventions and useful results

Some common **mathematical signs** are:

$=$	equal to
$\neq$	not equal to
$\approx$	approximately equal to
$\propto$	proportional to
$>$	greater than
$<$	less than
$\geq$	greater than or equal to
$\leq$	less than or equal to
$\sum_{i=1}^{N} x_i$	the sum $x_1 + x_2 + \ldots + x_i + \ldots x_N$
$\lvert x \rvert$	the modulus or magnitude of x; $\lvert x \rvert$ is always ≥ 0.

Derivatives of functions can be found by referring to tables of derivatives (Table B1) and using **rules of differentiation** (Table B2).

Table B1 Derivatives of some elementary functions

function $f(x)$	derivative $f'(x)$ or df/dx
C (a constant)	0
x^n	nx^{n-1}
$\sin x$ (x in radians)	$\cos x$
$\cos x$ (x in radians)	$-\sin x$
$\tan x$ (x in radians)	$\sec^2 x = 1/\cos^2 x$
e^x (or $\exp(x)$	e^x (or $\exp(x)$)
$\log_e x$ (or $\ln x$)	$1/x$

Table B2 Rules of differentiation

Name of rule	function	derivative
Constant multiple rule (C is a constant)	$Cf(x)$	$Cf'(x)$
Sum rule	$f(x) + g(x)$	$f'(x) + g'(x)$
Product rule	$f(x)g(x)$	$f'(x)g(x) + f(x)g'(x)$
Reciprocal rule	$1/f(x)$	$-\dfrac{f'(x)}{f(x))^2}$
quotient rule	$f(x)/g(x)$	$\dfrac{g(x)f'(x) - f(x)g'(x)}{(g(x))^2}$
Composite function rule	$f(g(x))$	$f'(g(x))g'(x)$

An important special case of the composite function rule is when $g(x) = kx$ where k is any constant. Then the derivative of $f(kx)$ is $kf'(kx)$. For example, the derivative of $\sin kx$ is $k\cos kx$, and the derivative of $\exp(kx)$ is $k\exp(kx)$.

A **primitive** or **indefinite integral** of a function $f(x)$ is any function $F(x)$ that has a derivative equal to $f(x)$, i.e. $\mathrm{d}F/\mathrm{d}x = f(x)$. See Table B3.

Table B3 Primitives (indefinite integrals) of elementary functions

$f(x)$	$F(x)$ = primitive of $f(x)$
x^n $(n \neq -1)$	$\dfrac{x^{n+1}}{n+1}$
x^{-1}	$\log_e x$ (or $\ln x$)
$\sin x$ (x in radians)	$-\cos x$
$\cos x$ (x in radians)	$\sin x$
e^x (or $\exp(x)$)	e^x (or $\exp(x)$)
$\sin kx$ (k a constant)	$-(1/k)\cos kx$
$\cos kx$	$(1/k)\sin kx$
e^{kx} (or $\exp(kx)$)	$(1/k)\,e^{kx}$ (or $(1/k)\exp(kx)$)

Trigonometric identities and triangle formulas

For any angle x:

$$\sin^2 x + \cos^2 x = 1$$
$$\sin(2x) = 2\sin(x)\cos(x)$$
$$\cos^2 x = \frac{1}{2}(1 + \cos(2x))$$
$$\sin^2 x = \frac{1}{2}(1 - \cos(2x))$$

For any triangle

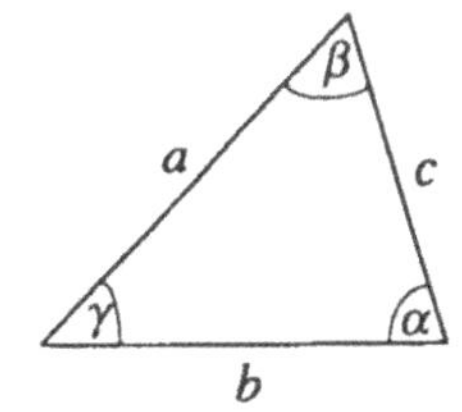

$$a^2 = b^2 + c^2 - 2bc\cos\alpha$$ (the cosine rule)

$$\frac{\sin\alpha}{a} = \frac{\sin\beta}{b} = \frac{\sin\gamma}{c}$$ (the sine rule)

For any right-angled triangle

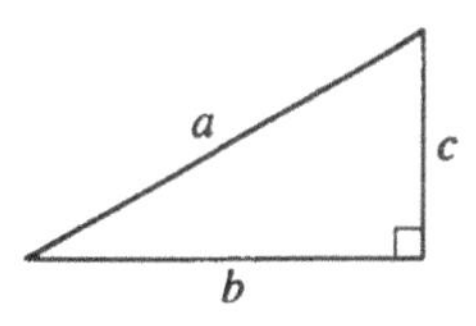

$$a^2 = b^2 + c^2$$ (Pythagoras's theorem)

Answers to selected Problems

CHAPTER 1

1.1 B, F and H are true.

2.1 (a) 5**l**. (b) –2**l**. (c) No.

2.2 E and F are incorrect.

3.1 (a) $|\mathbf{a}+\mathbf{c}| = 2^{1/2}k$, $|\mathbf{a}-\mathbf{c}| = 2^{1/2}k$.

(b) $|\mathbf{b}| = |\mathbf{a}+\mathbf{c}| = 2^{1/2}k$ and **b** is directed towards the south-east.

4.2 (a) **OP** = **OS** + **SQ** + **QP** = **b** + **c** + **a**.

(b) **QR** = **QP** + **PR** = **a** – **b**. **OP** + **QR** = 2**a** + **c**. (c) $|\mathbf{OP}| = 6^{1/2}|\mathbf{a}|$.

4.4 (a) **AG** = **a** + **b** + **c**, **HB** = **a** – **b** – **c**.

(b) **QP** = (–1/2)(**c** + **a**).

4.6 **AF** = (1/3)(**v** + 2**u**).

4.8 **a** = 2**b** – **c**/5, **b** = (5**a** + **c**)/10.

4.9 **OQ** = (m**OA** + l**OB**)/(l + m).

5.2 **i** = (1,0,0), **j** = (0,1,0), **k** = (0,0,1), **0** = (0,0,0).

5.4 $\alpha = -1/2$, $\beta = -3/2$, $\gamma = -7/2$.

6.1 $|\mathbf{p}| = 7.071$. $\theta_x = 135°$, $\theta_y = 64.9°$ and $\theta_z = 124.5°$.

6.3 D and E are incorrect, G is meaningless.

6.5 (a) **OS** = (–1,0,0), **OT** = (0,–7,–1), **ST** = (1,–7,–1).

(b) $(\mathbf{ST})_x = 1$, $(\mathbf{ST})_y = -7$, $(\mathbf{ST})_z = -1$, $|\mathbf{ST}| = 7.14$.

7.1 (a) **u** = 2**i** – **j**, **v** = –3**i** – **j** + 3**k**, **w** = 2**i** + 2**j** – **k**.

(b) **q** = –3**j**.

(c) $|\mathbf{q}| = |-3\mathbf{j}| = 3$. $\hat{\mathbf{q}} = \mathbf{q}/|\mathbf{q}| = -\mathbf{j}$.

(d) 90°.

7.3 **OP** = (6,–9,9). $\theta_z = 50.24°$.

8.1 **f** + **s** + **t** = (–1,–2,4)N. **u** = 0.2182(–1,–2,4).

8.3 T = 1.3 N and $\alpha = 67.38°$.

8.5 (a) 2.828 N. $\theta_x = 135°$ and $\theta_y = 45°$.

(b) 2.828 N. $\theta_x = 45°$ and $\theta_y = 135°$.

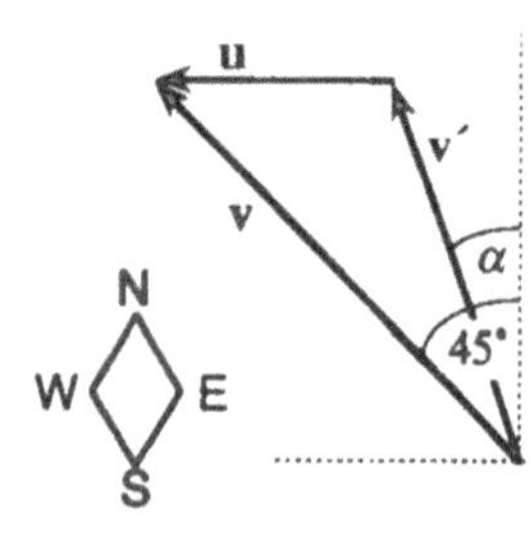

8.7 (a) See figure.

(i) $v_y = (240 \text{ km s}^{-1})\cos\alpha$.

(ii) $v_x = (-240 \text{ km s}^{-1})\sin\alpha - 30 \text{ kms}^{-1}$.

(b) $\alpha = 39.93°$ west of north.

8.9 42.29 km hr^{-1} (northerly component of **u**) and –15.4 km hr^{-1} (easterly component). $|\mathbf{v}'| = 55.14$ km hr^{-1}. $\theta_y = 140.1°$, $\theta_x = 50.07°$. (50.07° south of east.)

8.11 Resultant of zero. With one force reversed the magnitude of the resultant **R** would be $2F$.

8.13 $|\mathbf{F}| = 5$ N and **u** = 4**j**/5 + 3**k**/5.

8.15 1.269×10^4 N.

CHAPTER 2

1.2 The projection of **PS** onto **PQ** is 3.08. The projection of **PR** onto **PQ** is 6.08. The projection of **RS** onto **QR** is –1.85.

1.4 C, D, E, G and H are meaningless.

2.2 **A.B** = 14, **i.B** = –1, (**i** + **j**).(**A** – **B**) = 4, **A.**$(A_x + B_y)$**B** = 14.

3.1 A and E.

3.3 109.5° (or 70.5°).

3.4 (a) 42.1°. (b) 52.0°.

3.7 α = 39.23°. Angle of reflection is also 39.23°.

4.1 (a) **b** × **c** = 3 × 2sin55°**n** = 4.91**n** (**n** is into the plane of the figure).
Area of triangle = 2.46.
(b) **b** × (**b** × **c**) = 14.7**u** (**u** is is the plane of the figure normal to **b**).
(c) **b** – **c** is in the direction of **a**; **b** – **c** and **a** are therefore collinear vectors and so they have a vector product of zero.

4.3 A, D, E and G are meaningless.

4.5 **i** × **i** = **j** × **j** = **k** × **k** = 0; **i** × **j** = **k**, **j** × **k** = **i**, **k** × **i** = **j**; **j** × **i** = –**k**, **k** × **j** = –**i** and **i** × **k** = –**j**.

5.1 (a) **r** × **p** = (–3,7,–3).
(b) **i** × **A** = **k** + (–**j**).
(c) (**i** + **j**) × (**i** – **j**) = –2**k**.

5.3 (a) In the x-y plane normal to **q**.
(b) 3**j** + **i** (or any scalar multiple).

5.5 No, because **a**.(λ**a** × **b**) = 0 for any scalar λ.

6.1 (a) –49.
(b) (–55,23,3).
(c) (4,20,15).

6.3 (b) 4206.

7.2 (a) (3**i** + **j**) N, –2**k** Nm, –5**k** Nm.
(b) (3**i** + **j**) N, y = (5/3) m.

7.3 F = 45.5 N.

7.6 (a) **R** = (2**i** + **j**) N. **G** = –3**i** Nm.
(b) (**R** . **G**)**R**/|**R**|^2 = (–6/5)(2**i** + **j**) Nm.
G – (**R** . **G**)**R**/|**R**|^2 = (3/5)(2**j** – **i**) Nm. The vector projections of **G** in directions parallel to and orthogonal to **R**.

7.7 W = 9000 J. ΔU_g = –9000 J. v = 92.2 ms^{-1}.

7.9 **B** = (0.0118 T)(–**i**).

7.11 $F = w = 1.2 \times 10^4$ N and $f = P$ = 1000 N. Work done by **w**, **P**, **F** and **f** (respectively): 0 J, 10^6 J, 0 J, -10^6 J.

7.13 Torque = (1.7, –2.1, 0.0) × 10^{-3} Nm. Potential energy = –22J.

CHAPTER 3

1.1 $\mathbf{f}(0) = \mathbf{i}$, $\mathbf{f}(0.25) = (\mathbf{i} - \mathbf{j})/2^{1/2}$, $\mathbf{f}(0.5) = -\mathbf{j}$, $\mathbf{f}(0.75) = -(\mathbf{i} + \mathbf{j})/2^{1/2}$, $\mathbf{f}(1) = -\mathbf{i}$. $\mathbf{f}(2)$ does not exist.

1.3 (a) 5 cm and 2 cm.
(b) $s_x(t) = t$ and $s_y(t) = \sin(\pi t)$. The path is $y = \sin(\pi x)$ from $x = 0$ to $x = 5$.

2.1 1.19 ms^{-1} in a direction 19.0° east of south.

3.1 $\mathbf{v}(t) = 10\mathbf{j}\exp(-2t)$. $v(4) = |\mathbf{v}(4)| = 3.35 \times 10^{-3}$.

3.2 (a) $v_x(t) = u\cos\theta$, $v_y(t) = u\sin\theta - gt$. $v_x(0.2) = 4.015$, $v_y(0.2) = 3.734$. $|\mathbf{v}(0.2)| = (4.015^2 + (3.734)^2)^{1/2} = 5.48$.
(b) $\mathbf{v}(2).\mathbf{v}(2) = 30.06$. $v(2) = 5.48$.
(c) $\mathbf{a}(t) = d\mathbf{v}/dt = d(\mathbf{u} + \mathbf{g}t)/dt = \mathbf{g}$.

3.5 A circle of radius ωr.

3.7 $d\mathbf{q}/dt = (d\mathbf{s}/dt) \times \mathbf{v} + \mathbf{s} \times (d\mathbf{v}/dt) = \mathbf{v} \times \mathbf{v} + \mathbf{s} \times \mathbf{a} = 0 + \mathbf{s} \times \mathbf{a}$, where $\mathbf{a} = d^2\mathbf{s}/dt^2 = -\omega^2\mathbf{s}$. Hence $\mathbf{s} \times \mathbf{a} = 0$ and $d\mathbf{q}/dt = 0$.

4.1 $\omega = 1.745 \times 10^{-3}$ rad s^{-1}. $v = 1.745 \times 10^{-3}$ cm s^{-1}.

4.4 188 ms^{-1}.

5.2 ω and $\mathbf{b}$ must be orthogonal.

5.3 (a) $\omega = \pi$. $\mathbf{r}(0) = (0.1,0,-1)$, $\mathbf{r}(T/4) = (0,-0.1,-1)$, $\mathbf{r}(T/2) = (-0.1,0,-1)$.
(b) $\alpha = 5.71° = 6°$ to the nearest degree.
(c) $\mathbf{v} = -0.1\pi(\mathbf{i}\sin(\pi t) + \mathbf{j}\cos(\pi t))$, $\mathbf{a} = d\mathbf{v}/dt = -\omega^2\mathbf{r}$ and $|\mathbf{a}| = 1.0$ to two significant figures.

6.1 240 N (equivalent to 24 kg).

6.3 (a) $\mathbf{F}_{cor} = \mathbf{0}$ and $\mathbf{F}_{cen}$ is vertically upwards.
(b) $\mathbf{F}_{cor}$ and $\mathbf{F}_{cen}$ are vertically upwards.
(c) $\mathbf{F}_{cor}$ is horizontal and directed to the driver's left. $\mathbf{F}_{cen} = \mathbf{0}$.
(d) $\mathbf{F}_{cor} = \mathbf{0}$ and $\mathbf{F}_{cen}$ is directed 51.3° from the upward vertical towards the south.

CHAPTER 4

1.2 (a) Clockwise-directed circles centred on O.
(b) Straight lines parallel to the bank directed from left to right, more closely spaced near midstream.

1.3 (a) Equally spaced concentric circles.
(b) Concentric circles of separation increasing with radii.
(c) Equally spaced east-west horizontal lines.

2.1 The domain is an infintely long cylindrical region of radius 3 standing on the x-y plane with its axis on the positive z-axis. $f(1,2,3) = 0$, $f(-1,-2,0) = -3$, $f(-1,-2,-3)$ is not defined, $f(0,0,0) = 0$ and $f(0,0,3) = 3$.

2.3 (a) (i) $\phi(\mathbf{0}) = \delta$. (ii) $\phi((3\pi\hat{\boldsymbol{\eta}}/\eta) = 3\pi + \delta$. (iii) $\phi(\mathbf{r}) = \delta$.
(b) $\phi(x,y,z) = \eta z + \delta$. The contours are plane surfaces of constant z separated by $\pi/2\eta$.

2.5 (a) (i) 4×10^{-4} Wm^{-2}. (ii) 25 Wm^{-2}.
(b) Concentric spherical surfaces centred on the lamp of radii: 3.02 cm, 3.33 cm, 3.78 cm, 4.47 cm, 5.77 cm and 10.0 cm. The contours are concentric circles on a plane passing through the centre of the lamp.

3.1 $\mathbf{A}(-1,2,3) = -2\mathbf{i} + 3\mathbf{j} + 3\mathbf{k}$, $\mathbf{i.A}(3,1,0) = 3$. $A_y(x,y,z) = y^2 - x^2$.

3.3 **P** (Option C), **Q** (Option A), **S** (Option D), **T** (Option B), **U** (Option F), **V** (Option E).

3.5 $xy = 1$ (a hyperbola).

4.1

cartesian	cylindrical polar	spherical polar
(0,1,0)	$(1,\pi/2,0)$	$(1,\pi/2,\pi/2)$
(1,0,0)	(1,0,0)	$(1,\pi/2,0)$
(0,0,1)	(0,0,1)	(1,0,0)
(0,1,1)	$(1,\pi/2,1)$	$(2^{1/2},\pi/4,\pi/2)$
(1,0,−1)	(1,0,−1)	$(2^{1/2},3\pi/4,0)$
(0,0,−1)	(0,0,−1)	$(1,\pi,0)$

4.3 (i) $\rho(0) = 7 \times 10^3$ kg m^{-3}. (ii) $\rho(a) = 5 \times 10^3$ kg m^{-3}.
$\rho(a/3,\pi/6,\pi) = 6.333 \times 10^3$ kg m^{-3}.

4.5 (a) $M/(2\pi\epsilon_0 d^3)$.
(b) $-M/(4\pi\epsilon_0 d^3)$.
(c) $M/(8\pi\epsilon_0 d^3)$.
(d) 0.
(e) $-M/(8 \times 2^{1/2}\pi\epsilon_0 s^3)$.
(f) 0.

5.1 −3 (with $\mathbf{n} = \mathbf{k}$).

5.3 $\mathbf{E}(\rho) = \lambda\mathbf{e}_\rho/(2\pi\epsilon_0\rho)$.

5.5 (a) −2.4 m^2 s^{-1}.
(b) 0. The field is not conservative.

5.7 (a) $W = [f(x_1 + a) - f(x_1)]b - [g(y_1 + b) - g(y_1)]a$.
(b) $W/ab = [f(x_1 + a) - f(x_1)]/a - [g(y_1 + b) - g(y_1)]/b$.
(c) $W/ab \to \dfrac{df}{dx}(x_1) - \dfrac{dg}{dy}(y_1)$

CHAPTER 5

1.1 $\partial p/\partial x = 2x + 3y$, $\partial p/\partial y = 2y + 3x$, $\partial p/\partial z = 2z$.
$\partial q/\partial x = -5\sin(5x)\sin(3y)$, $\partial q/\partial y = 3\cos(5x)\cos(3y)$, $\partial q/\partial z = 0$.
$\partial t/\partial x = 2x + y$, $\partial t/\partial y = x + 2y$, $\partial t/\partial z = 0$.
$\partial c/\partial x = \partial c/\partial y = 0$, $\partial c/\partial z = -6$.

1.3 $n = 1/2$.

1.5 $\partial\psi/\partial\rho = -2\rho\psi$, $\partial\psi/\partial\phi = 2\exp(-\rho^2)\cos(2\phi)$, $\partial\psi\partial z = 0$.

2.1 (a) grad $f = 2xy\mathbf{i} + (x^2 - 2y)\mathbf{j}$.
(b) grad $\phi = \mathbf{i}(\cos y + \sin z) - x(\mathbf{j}\sin y - \mathbf{k}\cos z)$.
(c) grad $h = 2\mathbf{e}_r$.
(d) grad $q = [\mathbf{e}_\rho(-1/a)\cos(m\phi) - \mathbf{e}_\phi(m/\rho)\sin(m\phi)]\exp(-\rho/a)$.

2.2 grad $f = 2x\mathbf{i} + z\mathbf{j} + (y + 2z)\mathbf{k}$. grad $f(1,2,3) = 2\mathbf{i} + 3\mathbf{j} + 8\mathbf{k}$. The required directional derivative is 7.51. $|\text{grad} f| = 77^{1/2} = 8.77 > 7.51$.

2.5 $(\text{grad}\,\phi)_x$ at $(0,y) = 2\cos y$.

2.7 Force is 6×10^{-4} N directed radially inwards towards the origin.

3.1 (a) div$\mathbf{F} = 0$.
(b) div$\mathbf{G} = -\sin x + \cos x$.
(c) div$\mathbf{H} = 0$.
(d) div$\mathbf{C} = 0$.
(e) div$\mathbf{V} = 0$.
(f) div$\mathbf{W} = (3 - r)\exp(-r)$.

3.3 div$\mathbf{E}(x,y,-3) = \exp(-3^2) = 1.23 \times 10^{-4}$. $(x,y,-3)$ is a plane parallel to and 3 units below the x-y plane.

3.6 $n = 1$.

4.1 curl$\mathbf{F} = -z\mathbf{j} + x\mathbf{k} = (0,-z,x)$. $(\text{curl}\mathbf{F})_z = x$ which has the value 1 at the point $(1,-1,3)$. curl$\mathbf{F} = 0$ where $(0,-z,x) = (0,0,0)$, i.e. on the y-axis.

4.3 curl$\mathbf{q} = \mathbf{e}_\phi(p\sin\phi)/r^3$.

4.4 (a) All n.
(b) $n = 1$.
(c) $n = 0$.

4.6 curl$\mathbf{v} = 2(-xz\mathbf{i} - yz\mathbf{j} + z^2\mathbf{k})$. On the z-axis curl$\mathbf{v} = 2z^2\mathbf{k}$.

5.1 (a) $\mathbf{i}(3x^2y)$, a vector function.
(c) $(x^3y - 2y)\mathbf{i}(3x^2y)$, a vector function.
(e) $-\mathbf{k}x$, a vector function.
(g) $\mathbf{k}3x^2$, a vectof function.
(i) $\partial^2/\partial x^2$, a scalar differential operator.

5.2 $\nabla(x^2 + y^2) = \text{grad}(x^2 + y^2) = 2(x\mathbf{i} + y\mathbf{j})$. $\nabla\mathbf{h}$ is not defined. $\nabla.\boldsymbol{\phi}$ is not defined. $\nabla.\mathbf{h} = \text{div}\mathbf{h} = 2(x - y)$. $\nabla \times \mathbf{h} = \text{curl}\mathbf{h} = -x\mathbf{i} + y\mathbf{j}$. $\nabla^2\phi = 4$. $\nabla^2\mathbf{h} = 2(\mathbf{i} - \mathbf{j})$.

5.5 $(\mathbf{F}.\nabla)\phi = 2(x^2y - xy^2) + 10z^3$. $\mathbf{F}.(\nabla\phi) = (\mathbf{F}.\nabla)\phi$.

CHAPTER 6

1.1 (a) 0.
(b) 2/5.
(c) 0.
(d) −3.352.

2.1 160/3.

2.3 −1/4.

2.5 (a) $\mathbf{e}_z(\mu_0 I I_1/2\pi)\log_e[(a+L)/a]$.

(b) $-\mathbf{e}_\rho(\mu_0 I I_1 L/(2\pi c)$.

3.1 9.42.

3.3 (a) The force has no tangential component on the path. (b) 1.

3.5 243.

4.1 Work done = 750.

4.3 $-1-\pi/4=-1.785$.

5.1 (a) 3.

(b) $(1-a)(1-c^4)/4$.

(c) 18.

(d) 5/6.

5.3 (a) $C\pi a^3$.

(b) $4C\pi a^3$.

(c) Φ_0 is $6C\pi a^3$, $V=\pi a^2\times 2a=2\pi a^3$. Thus for $C=1$, $\Phi_0=3V$.

5.5 (a) -0.4898.

(b) 0.125.

6.2 12π.

6.6 (a) (i) $2\pi C$. (ii) $W_0=0$.

(b) curl$\mathbf{B}=\mathbf{0}$.

(c) We can apply Stoke's theorem in part (a) (ii) but not in part (a) (i) because there is a hole in the domain at $\rho=0$.

7.1 3/2.

7.3 0.515 kg.

7.5 $4\pi C$.

8.2 (a) $4\pi C$.

(b) div$\mathbf{F}=0$. There is a hole at $\mathbf{r}=\mathbf{0}$.

Index

For Product Safety Concerns and Information please contact our EU representative GPSR@taylorandfrancis.com Taylor & Francis Verlag GmbH, Kaufingerstraße 24, 80331 München, Germany

T - #0138 - 160425 - C0 - 241/191/16 - PB - 9780412627101 - Gloss Lamination